11 $\frac{25}{}$

The powder method
in x-ray crystallography

Leonid V. Azaroff

Professor of Metallurgical Engineering
Illinois Institute of Technology

Martin J. Buerger

Professor of Mineralogy and Crystallography
Massachusetts Institute of Technology

McGRAW-HILL BOOK COMPANY

1958   New York   Toronto   London

THE POWDER METHOD IN X-RAY CRYSTALLOGRAPHY

6 7 8 9 – M P – 0 9 8 7

02670

Dedicated to

Maria Y. Azároff

and

Julie W. Buerger

Preface

Instruction in the several laboratory methods of investigating crystals has long been offered to students of mineralogy and geology in the Massachusetts Institute of Technology. These include both optical and x-ray methods. While there are a number of available texts on optical methods, there has been no suitable one on the x-ray powder method. To supplement the lectures and to prepare the students for their laboratory work, we wrote a set of notes on the theory and practice of this method. Eventually we felt that these notes should be expanded into a book. This would not only enable us to discuss a number of matters at greater length, but would permit us to give our students some advanced material for their future use which could not be covered in a short course. This book is the result.

The practice of the powder method is inherently a simple matter, and the results it provides are so valuable that it is used as a routine tool not only in many branches of science but in many industries. It commonly happens that the person in charge of making and interpreting powder photographs for this routine purpose has had no training in it. We have written this book with such users in mind. We have tried to keep the discussion on a fundamental plane and have assumed no extensive background on the part of the reader. On the other hand, we have not discussed at length the elements of crystallography, the production of x-rays, or the details of x-ray apparatus (other than the powder cameras themselves) since this information is available in a number of books.

Methods of detecting x-rays are also not discussed. Since photographic film constitutes the simplest kind of detector, it is assumed in this book that this mode of recording diffraction spectra is used. In recent years, the employment of ionization-type detectors has become the vogue. Although somewhat different sample-preparation procedures are necessary in this case, we feel that little would be gained by complicating the discussion of the powder method by taking different detectors into account. The principles discussed in this book apply equally well regardless of experimental arrangement used. Photographic methods have the advantage of permanently recording diffraction spectra on a conveniently small record. We have emphasized the use of film methods because we believe that film methods will continue to be used in routine applications for years to come.

Many books have been written on the subject of x-ray diffraction, most of which include a chapter or two on the powder method. Naturally, the emphasis given to the powder method in such works is limited because it is only one of many subjects treated. During the preparation of our manuscript, two books devoted to the powder method have appeared: Klug and Alexander's *X-ray diffraction procedures* (1954), and Peiser, Rooksby, and Wilson's *X-ray diffraction by polycrystalline materials* (1955). Both books deal with the powder method, and are excellent general references to this field. Our objective in writing our own book has been to give a general introduction to the way the powder method can be made to yield crystallographic information. We have directed our treatment primarily to this end. Thus our work is naturally complementary not only to the above two large works, but also to other treatments of the powder method which have appeared.

We are indebted to several sources for permission to reproduce in this book certain material which has appeared before. Chapter 4 is an abridged version of a paper which appeared in the Journal of Applied Physics **16** (1945) 501–510. We are grateful to the editor for permission to use most of the original text and illustrations again in this book. Drs. J. B. Nelson and D. P. Riley kindly allowed us to use the numerical values of their extrapolation function in our Appendix 3. We are also glad to acknowledge the following sources of specific illustrations, mostly charts: Dr. W. L. Bond for Figs. 10, 14, and 17, Chapter 7; Dr. C. W. Bunn and the Clarendon Press for Fig. 9, Chapter 7; Dr. L. A. Carapella and the Journal of Applied Physics for Figs. 5, 6, 7, 8, and 9, Chapter 15; Dr. Wheeler P. Davey and the McGraw-Hill Book Company for Figs. 6, 12, and 16, Chapter 7; Drs. R. A. Harrington, J. C. Bell, and A. E. Austin and the Battelle Memorial Institute for Figs. 11 and 15, Chapter 7; Dr. G. W. Brindley and the Joint Committee on Chemical Analysis by Powder Diffraction Methods for Fig. 4, Chapter 13; Dr. M. E. Straumanis for Fig. 10, Chapter 15.

The other illustrations are new with this book. We are glad to acknowledge the help received from several people in their preparation. Mr. Tibor Zoltai made the final pencil sketches of five of the illustrations. We are also indebted to the Armour Research Foundation for a grant which defrayed the cost of putting most of the preliminary pencil sketches into final form. The powder photographs of Chapter 16 were prepared by Miss Irene Corvin and Miss Bernadine Zajda.

The preparation of the difficult typescript was the handiwork of Mrs. Doris Ahles and Mrs. Sylvia Garvin. Prof. I. Fankuchen and Dr. Benjamin Post kindly read the manuscript and corrected several errors. One of us (L. V. A.) also wishes to express his gratitude to the Armour Research Foundation, and particularly to Dr. James J. Brophy, Assistant Manager of Physics Research of that organization, for encouraging and providing facilities for this work.

Leonid V. Azároff
Martin J. Buerger

Contents

Contents

Contents

1

Introduction

Historical background

In 1912, Laue and his assistants discovered that crystals can act as diffraction gratings to x-rays. This made a new tool available to those who wished to investigate crystals. Immediately following this discovery it was natural that the investigators of crystals should treat them as individuals since each one could be regarded as an individual diffraction grating. This point of view was a very fruitful one indeed, and research on single crystals has been responsible for revealing the arrangements of atoms in a large fraction of the crystals of known inorganic compounds as well as the simpler organic compounds. The knowledge of the way atoms are geometrically arranged in crystals has created a substantial revolution in modern science, particularly in chemistry and solid-state physics.

Shortly after the discovery of the diffraction of x-rays by crystals, World War I broke out, and this succeeded in isolating major sections of the world. During this period scientists in two different parts of the world independently discovered that there existed a characteristic x-ray diffraction effect from a fine-grained crystalline aggregate. This discovery was made by Debye and Scherrer[1] in Germany and almost simultaneously by Hull[2,3] in the United States.

Although much information is lost or degraded by using an aggregate in place of a single crystal, this method of investigating crystals has proved to be exceedingly useful in those cases where one wishes to examine a crystalline material which is not in the form of discrete single crystals, for example, a metal. The method usually requires the individual crystals of the aggregate to be about the size of grains of a fine dust. If the material to be examined is not in this form already, it is usually ground to a powder. For this reason this way of examining crystals is commonly called the *powder method*. It is also known as the *Debye-Scherrer* method after its first discoverers.

1

Uses of the powder method

There are many applications of the powder method, but two of these are of primary importance. Fundamentally, the powder method provides a way of investigating, within limits, the crystallography of the crystal in the powder. Secondarily, since the powder *diffraction diagram* produced by a crystalline substance is a characteristic of that substance, the powder method can be used as a means of identification of crystals.

It will become evident in subsequent chapters that the powder method is a comparatively weak one in investigating the crystallographic characteristics of a crystal. Whenever possible therefore, a crystal should be studied by single-crystal methods.† There are many instances in which single crystals are not available or are difficult to obtain, or in which it is desirable to study a material without altering its form. In such cases the only way to study the crystallographic characteristics of the substance is by the powder method. It is important to realize the limitation of this method and yet to take advantage of its possibilities to the full. In this book an attempt is made to provide some of the important background for the intelligent use of the powder method in x-ray crystallography.

One of the most important uses of the powder method is in the identification of an unknown material. There are a number of ramifications of this. One usually has an unknown material in fine-grained form. If it is known to be a single crystalline substance, one may seek to establish its crystallographic characteristics as far as possible by the powder method. In the event that such a study leads to indefinite results, the powder diffraction diagram can still be regarded as a kind of fingerprint which it is unimportant to try to understand. If a set of standard diagrams of known substances, or tabular representations of them, are available, then it is usually possible to identify a pure substance with the aid of a set of rules for finding an unknown diagram among the standard diagrams.

A somewhat related problem arises when one knows that an aggregate contains crystalline compounds A and B and one wishes to know their relative proportions. In such instances one can make use of the fact that a powder diagram of a mixture of A and B is a mixture of the diagram of A and the diagram of B.

There are many secondary uses of the powder method. A most important one is the statistical study of the relative orientations of the individual crystals of an aggregate. Detailed discussion of such applications of the powder method are regarded as beyond the scope of this small book.

† Single-crystal methods of investigation are discussed in another work: M. J. Buerger, *X-ray crystallography* (John Wiley & Sons, Inc., New York, 1942).

Instrumentation

By far the simplest and most inexpensive way of practicing the powder method is to record the x-ray diffraction on photographic film, using a *powder camera*. A more elaborate way is to detect the diffracted radiation by means of a quantum counter, such as a Geiger counter. Counter methods require expensive bulky equipment whose maintenance and repair are beyond the capabilities of most routine users of the powder method. The use of counter, diffractometer, and recorder equipment is justified chiefly when one wishes to examine many different samples rapidly, as is the case, for instance, in an industrial laboratory. Such methods also have a real advantage whenever accurately measured intensities are necessary, for example, to ascertain quantitatively the relative amounts of several crystalline substances in a mixture.

Background for users of the powder method

It commonly falls to the lot of a chemist, physicist, metallurgist, or even someone without much scientific or technical background, to take powder photographs and interpret them. This book was written as a guide to such persons. It does not, however, provide all the background which the user ought to have. The more the user knows about crystallography, the more intelligently can he tackle the problems which confront him in interpreting powder photographs.

A knowledge of x-ray diffraction theory is also important. Only a brief discussion of this subject, especially as it concerns the powder method, is given in Chapter 2. A further study of x-ray diffraction methods, in general, will afford the user of the powder method a deeper insight into his own problems.

Literature

[1] P. Debye and P. Scherrer, *Interferenzen an regellos orientierten Teilchen im Röntgenlicht.*
    I. Nachr. kgl. Ges. Wiss. Göttingen, Math.-physik. Kl. 1916, 1–15. [Repeated in Physik. Z. **27** (1917) 277–283.]
    II. Nachr. kgl. Ges. Wiss. Göttingen, Math.-physik. Kl. 1916, 16–26.
    III. Physik. Z. **28** (1917) 290–301.
[2] A. W. Hull, *The crystal structure of iron*, Phys. Rev. (2) **9** (1917) 84–87.
[3] A. W. Hull, *A new method of x-ray analysis*, Phys. Rev. (2) **10** (1917) 661–696.

2

Elementary x-ray diffraction theory

The general nature of crystals

In the crystalline state the atoms are arranged in patterns which are characterized by periodic repetition in three dimensions. A good two-dimensional analogue of a crystal is a wallpaper pattern. In a wallpaper pattern the motif is an arbitrary design, whereas in a crystal the motif is not arbitrary, but rather a comparatively small collection of atoms representing the chemical composition of the crystal.

Figure 1 shows a two-dimensional analogue of a crystal pattern. The motif consists of several kinds of circles, which can be taken to represent as many kinds of atoms in an actual crystal. From a geometrical viewpoint, the pattern can be thought of as the repetition of the motif at intervals t_1 in one direction and t_2 in another direction. The geometrical motion of repetition is a pure translation; accordingly t_1 and t_2 are called *conjugate translations*. If the motif is a single geometrical point, its periodic repetition by the translations t_1 and t_2 generates an infinite collection of points, a small region of which is shown in Fig. 2. The set of such geometrical points is called a *lattice*. If a motif more complicated than a geometrical point is repeated periodically by translations t_1 and t_2, as in Fig. 1, the entire collection is a two-dimensional pattern (*not* a lattice).

These notions also apply to repetition in three directions. Repetition of a geometrical point generates a *space lattice* (or simply a *lattice*) and repetition of a more complex motif generates a space pattern. If the motif is a group of atoms, as in a crystal, the material body generated by the repetition defined by the three conjugate translations is called a *crystal structure*.

The region determined by the three conjugate translations t_1, t_2, and t_3 is a parallelepiped known as a *primitive cell*. The shape of this cell does not always have a symmetry as high as that of the pattern of the crystal. When it does not, a larger and more convenient unit parallelepiped is chosen to represent the pattern for most purposes. This unit

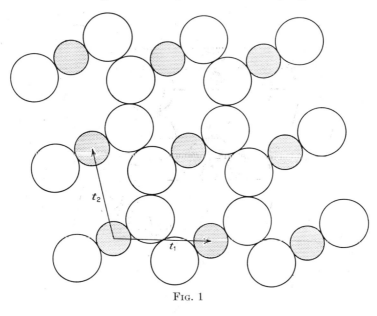

Fig. 1

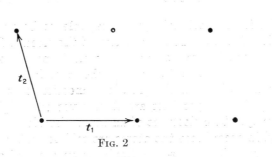

Fig. 2

parallelepiped, which may have a volume of one, two, three, or four primitive cells, is known as the *unit cell*.

For the purposes of discussing x-ray diffraction by crystals it is convenient to consider first a simple motif consisting of one atom. The periodic repetition of this motif by the three translations produces a pattern consisting of an identical atom at each point of a lattice. Let this simple pattern be termed a *lattice array* of atoms. Then, for a more complicated crystal, the entire crystal can be regarded as several lattice arrays of atoms each somewhat displaced from one another (Fig. 3). In

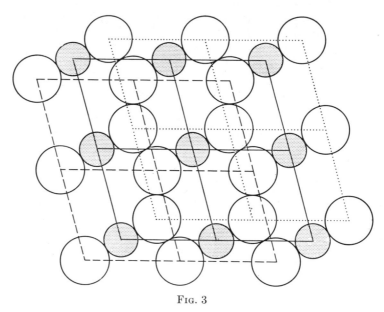

FIG. 3

this way any crystal structure can be decomposed into several parallel lattice arrays.

The diffraction of x-rays by a lattice array of atoms

An atom is an electrical system capable of being disturbed by an external electric field. The fluctuation of the electric field of an impinging electromagnetic wave displaces the electrons of an atom. For this reason they undergo vibration having the same frequency as the electromagnetic wave which, in the present connection, is x-radiation. These accelerating charged particles are themselves the origin of radiation of this frequency. The electrons of an atom, therefore, absorb and reemit x-rays, and in accordance the atoms are said to *scatter* x-radiation.

When a wave front of x-rays impinges on a set of atoms, each atom scatters the x-rays. If the atoms are centered on points in a plane, for example, a plane in a lattice array corresponding to a crystallographic plane (*hkl*), two directions of scattering have special properties, as shown in Fig. 4. In both these directions the distance from the original wave front, to an atom, and on to a new wave front is the same for all atom locations in the plane.   These directions correspond, respectively, to a continuation of the beam in the original direction, and to a reflection of the beam by the plane on which the atoms lie.   The scattering by atoms in a plane is, therefore, tantamount to reflection by the plane.

A lattice array of atoms can be regarded as an infinite stack of parallel, equally spaced planes (Fig. 5).   Any rational plane (hkl) of the lattice

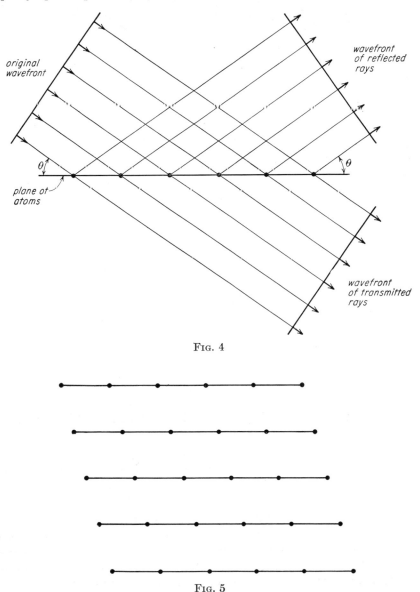

Fig. 4

Fig. 5

array can be chosen as the plane in question, and then the whole array can be thought of as a stack of planes parallel to this one. How does such a stack reflect x-rays? The condition for scattering-in-phase by one plane of the stack was established above. If two (or more) are considered (Fig. 6), it is evident that the path length from incoming wave

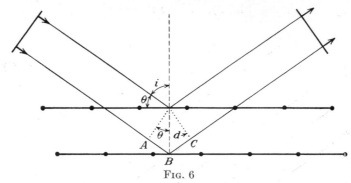

Fig. 6

front, to plane, to scattered wave front is longer in the case of the lower plane. The greater path difference is

$$\Delta = ABC$$
$$= 2AB. \tag{1}$$

Since
$$AB = d_{hkl} \sin \theta, \tag{2}$$

the total path difference is

$$\Delta = 2AB$$
$$= 2d_{hkl} \sin \theta. \tag{3}$$

If both these planes are to scatter in phase, the path difference Δ must be an integral number of wavelengths, that is, $n\lambda$ where n is an integer. Therefore, the condition for scattering-in-phase is

$$n\lambda = \Delta$$
$$= 2d_{hkl} \sin \theta. \tag{4}$$

This condition for scattering-in-phase is known as *Bragg's law*. Note that it is defined in terms of an interplanar spacing d_{hkl} of the lattice, and a *glancing angle* θ. This "glancing angle" is the complement of the angle of incidence i (Fig. 6) of a reflection in geometrical optics.

When a crystal diffracts x-rays in accordance with Bragg's law, the scattered x-rays are said to constitute a *reflection*. Since the reflection is attributed to the plane (hkl) the reflection itself is designated hkl, written without parentheses.

If (4) is solved for θ there results

$$\theta = \sin^{-1}\left(\frac{\lambda}{2} \frac{n}{d_{hkl}}\right). \tag{5}$$

The term $\lambda/2$ is constant for the experiment. The term n can only have the discrete values of integers, and the term d_{hkl} can only have the discrete values of the spacings of the planes (hkl). Therefore, θ can only have certain discrete values. The possible discrete values are further limited by the fact that the term in parentheses cannot exceed unity.

In most cases it is convenient to avoid the explicit use of n by incorporating it in the indices of the plane. This can be done as follows: Equation (5) can be rearranged to

$$\theta = \sin^{-1}\left(\frac{\lambda}{2\dfrac{d_{hkl}}{n}}\right). \tag{6}$$

The term d_{hkl}/n has a specific meaning. It signifies a spacing $1/n$th that of the spacing of the plane (hkl). This is the spacing of the plane $(nh\ nk\ nl)$. That is,

$$d_{nh\ nk\ nl} = \frac{d_{hkl}}{n}. \tag{7}$$

This can be substituted in the denominator of (6) to give

$$\theta = \sin^{-1}\left(\frac{\lambda}{2d_{nh\ nk\ nl}}\right). \tag{8}$$

One notes that the indices in (8) contain the common factor n. In classical crystallography such indices were not permitted. In x-ray crystallography, however, it is convenient to refer a reflection to a plane whether it has a common factor or not. If it does contain a common factor, this factor is the n of Bragg's law, (4) and (5).

The diffraction by the whole crystal structure

Any crystal structure can be regarded as several mutually displaced lattice arrays (Fig. 3). Each lattice array can diffract x-rays as if reflecting them from a plane (hkl), provided the glancing angle θ is adjusted so that it is one of the discrete solutions of (5). Now consider how the diffraction from several lattice arrays of the crystal structure interacts for a particular reflection (Fig. 7). Let the crystal structure be composed of only two lattice arrays, labeled 1 and 2 in Fig. 7. When (5) is satisfied, all atoms of lattice array 1 scatter in phase with each other, and all atoms of lattice array 2 scatter in phase with each other. But the path from the incoming wave front to array 2 is longer than to array 1. This means that array 2 contributes to the resultant wave scattered by the whole crystal a wave whose phase is behind that scattered by array 1. The resultant scattered wave is not destroyed unless this phase difference is π, and only then if the two amplitudes are equal.

From this several conclusions can be drawn: 1. The full crystal structure scatters at the same glancing angles θ as any of its component lattice
arrays.   2. The displacements between the component lattice arrays
cause phase differences in their contributions to the net scattered wave.

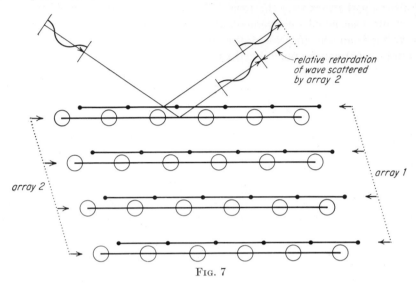

Fig. 7

3. These phase differences tend to reduce their contributions to the intensity from a value which would be obtained if all atoms of the structure
scattered in phase.

Positions and intensities of x-ray reflections

The conclusions of the last section can be reformulated in a way which
brings out some important characteristics of x-ray reflection.   It has
been shown that the glancing angles θ at which a crystal may reflect
x-rays depend fundamentally on the interplanar spacings d_{hkl} of the lattice.   These d's in turn depend only on the dimensions of the lattice.
(This relation will be developed in detail in Chapter 6.)   They are in no
way concerned with the arrangement of atoms in the repeated motif.
In the next chapter it will be shown that the position where an x-ray
reflection is recorded (or detected) is dependent on θ only.   This means
that the *set of positions of all the x-ray reflections from a crystal depends
only on the dimensional characteristics of its lattice* and does not depend on
the arrangement of its atoms.   As a consequence, two crystals having the
same type and dimensions of unit cell give the same x-ray diagram with
regard to location of reflections, even if they have completely unrelated
chemical compositions.   An example of such a pair is tin tetraiodide,

SnI_4, and rubidium aluminum alum, $RbAl(SO_4)_2 \cdot 12H_2O$. Both these compounds have primitive isometric unit cells of edge 12.2 Å.

On the other hand, the relative intensities of the various reflections *hkl* of a crystal depend upon the way the contributions from its several lattice arrays interfere with each other for the several reflections *hkl*. Therefore, the set of intensities of the reflections *hkl* depends entirely on the arrangement of atoms in the motif.

These conclusions can be brought together as follows: *The locations of the reflections of a crystal depend on the shape and type of its unit cell; the relative intensities of these reflections depend on the arrangement of the atoms within this cell.* The combination of the unit cell and the arrangement of atoms in it comprises the crystal structure itself. Therefore, the locations and relative intensities of the reflections of a crystal are characteristics of the crystal structure. Whether or not the powder diagram of an unknown crystal can be interpreted, at least this diagram is characteristic of the crystal and can be used like a fingerprint to distinguish it from other crystals, and hence to identify it. This is the philosophic basis for using the powder diagram of x-ray reflections in crystal identification.

3

Principles of powder photography

Collimating system

In all common x-ray diffraction methods, it is necessary to limit the x-radiation to a small pencil. This is accomplished with a "collimator," which may take the form of a *pinhole system*, or a *slit system*. The principle is illustrated in Fig. 1. The radiation emanating from the focal spot on the target of the x-ray tube is limited by a pair of holes, 1 and 2,

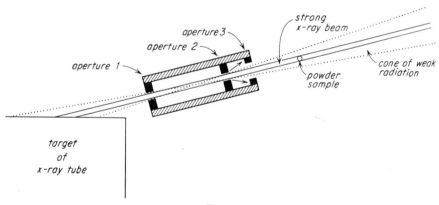

FIG. 1

of the collimator. The holes are small apertures, usually in lead. Since lead has its own x-ray powder diagram, diffraction by the aperture 2 would confuse the diffraction to be studied unless it is removed. Its removal is accomplished by adding another diaphragm, 3, to the sequence. This hole is designed to be large enough to pass the desired x-ray beam defined by apertures 1 and 2, and without being touched by it, but small enough so that the diaphragm intercepts the smallest cone of diffracted radiation arising from aperture 2.

12

Production of the powder diagram

The principles involved in the production of a powder diagram can be appreciated by considering the simplified experimental arrangement shown in Fig. 2. An x-ray beam is defined by the pinhole system, just described. A photographic film is then placed normal to the x-ray beam. The powder sample is introduced into the path of the x-ray beam. As the beam travels through the powder sample, it meets thousands of powder grains, each a tiny crystal in a different orientation. Among these

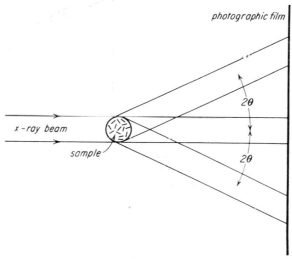

photographic film

x-ray beam

sample

2θ

2θ

Fig. 2

grains many are so oriented that a particular set of planes (*hkl*) makes the appropriate glancing angle θ (for that plane) with the x-ray beam. Such grains are in position to reflect x-rays. The reflection occurs in a direction making an angle 2θ with the direct x-ray beam. The locus of directions making an angle 2θ with a given direction is a cone of half-opening angle 2θ (Fig. 2). For each solution of the Bragg equation

$$\theta = \sin^{-1}\left(\frac{\lambda}{2}\frac{n}{d_{hkl}}\right) \tag{1}$$

there exists such a cone.

Considering a particular cone (Fig. 3), the separate reflections from all crystals which satisfy (1) for a particular n/d_{hkl} lie along the directrices of a certain cone of half angle 2θ. If the experimental arrangements are appropriate, these diffracted rays are sufficiently numerous so that the cone is densely outlined by rays. These rays cut the photographic plate in a circle which is continuous if the rays along the cone are sufficiently

dense.   From a measurement of the radius of the circle, and the known
crystal-to-film distance, it is an easy matter to compute the cone angle,
and eventually determine θ.

The use of a flat film severely limits the maximum angle 2θ which can
be recorded.   A much greater range of 2θ can be recorded if the film is
wrapped on a cylindrical form coaxial with the specimen, with the axis
of the cylinder at right angles to the x-ray beam, as shown in Fig. 4A.
Only a narrow strip of film is required.   With this arrangement, the cone

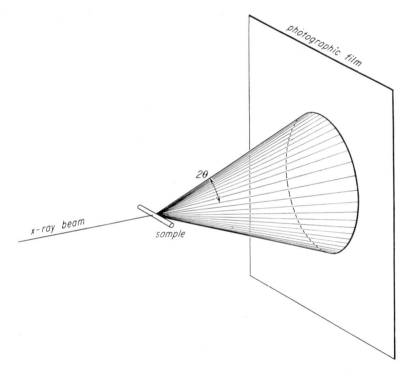

Fɪɢ. 3

intersects the cylinder in a curve; that part of the curve which is caught
by the narrow strip of photographic film is a nearly circular arc.   The
distance S between similar arcs (Fig. 4B) corresponds to 4θ, and if R is
the radius of the film, this distance is

$$S = R \cdot 4\theta, \tag{2}$$

(θ expressed in radians)

$$\therefore \ \theta = \frac{S}{4R}. \tag{3}$$

It is evidently an easy matter to determine θ for each cone by measuring
S after the film is opened out, as in Fig. 4B.

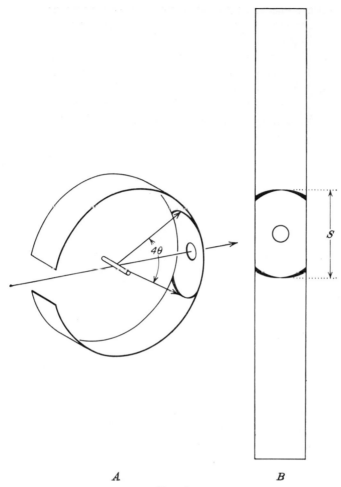

A B

Fig. 4

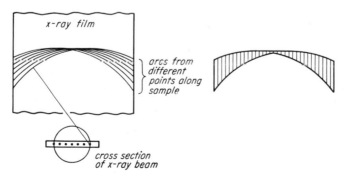

A B

Fig. 5

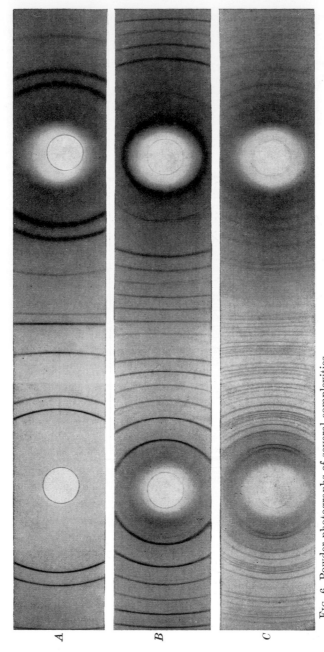

Fig. 6. Powder photographs of several complexities.
A. α-Brass, isometric, $a = 2.942$ Å.
B. PbTe, isometric, $a = 6.439$ Å.
C. PbCl, orthorhombic, $a = 4.535$ Å, $b = 7.62$ Å, $c = 9.05$ Å.

It is customary to use a specimen whose shape is that of a tiny cylinder coaxial with the film cylinder (Fig. 4A). This tends to give a "shape" to the arc recorded on the film. The reason for this is shown in Fig. 5A. Each element of length of the sample produces an arc, and these individual arcs are displaced depending on the location of the element of length producing them. The result is a "shaded arc" (Fig. 5B).

Some actual powder photographs are shown in Fig. 6.

4

The design and alignment
of powder cameras

Desirable features of a camera

Cameras for making powder photographs are so simple that their construction is often attempted without regard to efficiency and convenience. This was particularly true of the early forms, which are still copied.

Among the features required of a powder camera are the following:

1. It should be capable of being used repeatedly in such a way that conditions are duplicated each time. This implies, for example, that the specimen axis is fixed with respect to the film axis, a condition not fulfilled in early models and in some modern copies. It also implies that, when the camera is removed for the purpose of developing the film or changing the sample, exact alignment to the x-ray beam is recovered on returning it.

2. The film location should not only be fixed by being against the metal camera cylinder, but the film should be inserted and removed easily, and without fussing with frail paper envelopes or paper covers.

3. The direct-beam path should be shielded to prevent x-ray scattering by the air to the film.

4. The x-ray beam should be stopped in such a way as to protect the investigator, and yet not permit back scatter by the stop to the film.

5. The radius of the camera should be so selected that the location of the lines on the film can be measured with a millimeter scale, and then converted to θ or 2θ by using a simple factor (such as 1 or 10). This avoids computation or, at the very least, looking up a computed result recorded in a table. Some cameras forgo this convenience for the sake of preserving a tradition based upon an arbitrary decision made by an early camera designer.

Because of limitation of space, no attempt is made to describe a range of powder-camera designs. Instead, a specific powder camera designed by one of the authors[1] is described in some detail, partly because it was designed with the purpose of fulfilling the requirements already stated, and partly because this design has been so closely copied[2] by several commercial camera manufacturers as to be a standard powder camera in the United States.

Camera mounting and adjustment

It was a common fault of powder-camera mountings that they were difficult to adjust and that, after the powder camera was removed from

 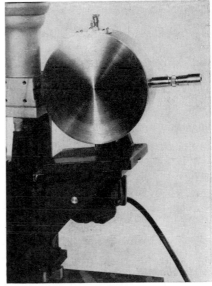

Fig. 1   Fig. 2

its place at the x-ray tube, it did not automatically recover duplicate alignment with the x-ray beam when returned to position. These difficulties are overcome in the design shown in Figs. 1 and 2.

An important feature of the camera mounting is that it is attached directly to the same support which holds the x-ray tube (Figs. 1 and 2). This arrangement prevents play between the x-ray tube and camera so that, once the pinhole system of the camera has been aligned to the x-ray beam, the alignment is permanent.

The cylindrical camera is attached to a base with respect to which it has certain adjustments. These are provided by a shaftlike extension of the camera which fits into a horizontal bearing in the base. The bearing is split, and is normally kept tightly clamped to the shaft by means of a screw. When the screw is loosened, the camera shaft may be rotated or translated in the bearing. When the screw is tightened, the camera-base assembly is a rigid unit.

This assembly is attached to the x-ray tube mounting through a bracket, which offers certain adjustments. The camera rests on the approximately horizontal surface of the bracket, and is slipped onto it with the aid of a longitudinal dovetail which serves two functions. It fixes the orientation of the camera so that, once adjusted, it maintains alignment with the x-ray beam in the vertical plane, and it also prevents the camera from being accidentally knocked off the bracket. The dove-tail can be loosened and its orientation on the horizontal surface of the bracket varied within small limits.

The vertical surface of the bracket rests against a corresponding surface on the x-ray tube mounting which is approximately normal to the x-ray beam. The two surfaces are held together by means of a single central screw. A large vertical movement of the bracket on the base is permitted when this screw is loosened, and rotation about an axis normal to the surface is prevented by a vertical keyway between the surfaces.

The mounting just described permits the following motions of the pin-hole axis with respect to the x-ray beam, during adjustment:

Translation in three orthogonal directions:

1. *Left and right* (when the screw in the split bearing of the camera base is loosened)

2. *Up and down* (when the screw in the x-ray tube mounting is loosened)

3. *Parallel with the beam* (as the camera is pushed back and forth on the dovetail; the forward motion is limited by an adjustable stop)

Rotations about two orthogonal axes perpendicular to the x-ray beam:

1. *About a horizontal axis* (when the screw in the split bearing of the camera base is loosened)

2. *About a vertical axis* (when the two screws holding the dovetail to the base are loosened)

It will be noted that to loosen the entire assembly for adjustment calls for loosening screws only at three points.

The initial adjustment of the camera is made easily, and requires a few minutes' time at most. A rough approximation can be first made by removing the pinhole system of the camera and aligning the coarse beam which comes through the resulting hole. Final adjustments are made after the pinhole system is returned to the camera.

In practice, all screws may be just barely loosened so that all adjustments can be made by the use of a little force. The camera may then be manipulated as if free in space, and its position and orientation can be varied by hand until the spot on a fluorescent screen shows the pinhole system to be pointing at the focal spot of the x-ray tube. This gives approximate adjustment. For final adjustment, each motion is varied independently a little in order to find the position giving maximum intensity of the x-ray beam as judged by the brightness of the spot on the fluorescent screen. When these adjustments have been completed, the vertical-translation coordinate of the bracket is fixed by a setscrew. This screws into a lug in the bracket and rests on a shelf on the x-ray tube mounting.

Once adjusted, the camera-base-bracket-tube is a rigid unit whose only degree of freedom is the camera translation parallel to the x-ray beam; this permits removing and replacing the camera base with respect to the bracket. The entire attachment scheme has important advantages in the event that the window of the x-ray tube is required for other diffraction apparatus. Should this contingency arise, the entire adjusted assembly can be instantly removed as a unit by removing one screw, leaving the table top clear for other apparatus, and it can also be instantly replaced in correct adjustment whenever required. This is because the only degree of freedom between bracket and x-ray tube mounting, the vertical translation along the keyway, is not operative because of the setscrew in the bracket. This was set after adjustment so that the vertical coordinate of the bracket is correct to align the pinhole system to the x-ray beam. When the bracket is returned, the alignment is recovered.

Fig. 3

The camera

General features. The photographic film is expanded against the internal wall of a hollow cylinder (Figs. 3, 4, and 5). Since the film is exposed directly to the diffracted radiation, the camera must be inherently light-tight. This is partly effected by a screw-on cover, partly by light-tight entrance and exit ports for the x-ray beam. In some commercial copies of this camera the cover is attached to the camera cylinder by a friction fit rather than a threaded fit. The friction is supplied by a rubber or plastic gasket permanently inserted on the inside lip of the cover.

It is common practice to make cameras of arbitrary diameter. Arbitrary diameters are a serious inconvenience to the user. There are good reasons why effective diameters of 57.3 and 114.6 mm should be used. For these diameters, the conversion factors from millimeters distance on the film to angle of arc are 2 and 1° per mm, respectively. The film

Fig. 4

record can thus be measured with a standard millimeter scale, and no computation is required to obtain angles. The angular significance of the millimeter measurements on each of these camera diameters is expressed in Table 1. When these camera diameters are used, the spacing d may be found immediately from each film reading either by direct computation or by consulting tables for converting θ to d for various values of wavelength.

Another advantage of these camera diameters is that commercial photographic films can be easily cut, without waste, to provide film strips for them. Thus the circumference of the 57.3-mm-diameter cam-

Table 1

Camera diameter	1 mm on film corresponds with	
	Arc, also deviation angle, 2θ	Bragg angle, θ
57.3 mm	2°	1°
114.6	1°	$\frac{1}{2}$°

era is 180 mm. Seven inches is 177.8 mm. The commercial "seven" inches of the 5- by 7-inch x-ray film is a trifle less than 7 inches (actual dimensions $4\frac{60 \pm 1}{64} \times 6\frac{60 \pm 2}{64}$ inches). Thus a 5- by 7-inch film can be cut into five approximately 1-inch strips, each of which comes just short of covering the circumference. For the 114.6-mm-diameter camera, either a 14-inch film strip may be used to cover the full circumference, or a 7-inch film may be used to cover any 180° range. The desirable 180° ranges are:

1. One complete half of the double-reflection record, i.e., the 180° range from exit port to entrance port, corresponding with a θ range from 0 to 90°.
2. The forward-reflection field only, i.e., 90° on each side of the exit port, corresponding with a θ range from 0 to 45°. This range is especially valuable for use with large-cell organic crystals, such as soaps, which give few, if any, reflections beyond this range.
3. The back-reflection field only, i.e., 90° on each side of the entrance port, corresponding with a θ range of 45 to 90°. This range is necessary when carrying out precision-spacing investigations.

The 57.3-mm-diameter camera (Figs. 1 and 3) is desirable for routine identification and, in general, for comparison work with inorganic crystals. The larger 114.6-mm-diameter camera (Figs. 2 and 4), of course, requires much longer exposures. It is a preferable size, however, when more complicated crystals are to be examined, or when the powder photograph, if complicated, is to be indexed.

Film arrangement. The film is expanded directly against the inside of the metal cylinder by means of fingers, one at each end of the film (Figs. 3, 4, and 5). One of these fingers is fixed to the camera; the other is attached to a slide which can be moved tangentially from the outside of the camera. It can be clamped in any position by turning a milled head which acts against a brake. When put into tangential compression by the movable finger, the film expands radially against the inside of the camera.

The two free ends of the film can be located in several positions. For most purposes, the free ends should be halfway between entrance and exit ports, as shown in Fig. 6. This film placement was first suggested by Straumanis and his coworkers,[3,4] who pointed out that, when this arrangement is used, the effective camera diameter, including shrinkage, can be determined individually for each film. This is because the diffraction rings about entrance and exit ports center at points differing by exactly half the circumference. The Straumanis film placement is desirable for two additional reasons:

1. The diffraction record is complete, without small missing regions near either 0 or 90° θ.

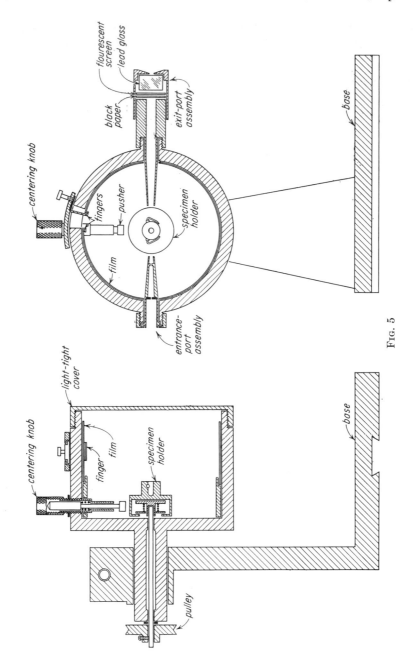

Fig. 5

2. A doubled back-reflection record is always available for precision determinations either by the method of Cohen or by a modification of the method of Bradley and Jay (see Chapter 15).

A modification of the Straumanis film arrangement has been suggested by A. J. C. Wilson,[5] for use with substances which do not give powder lines in the back-reflection region. This modification consists of placing the gap in the film cylinder in about the middle of the front-reflection

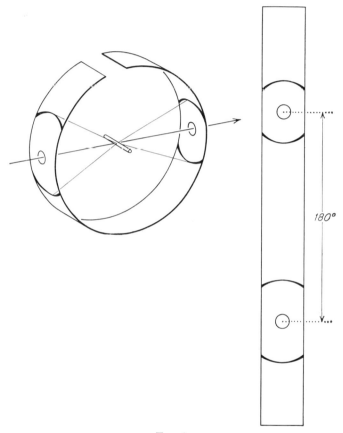

Fig. 6

range. If this is done, the point $2\theta = 180°$ can be located by lines in the front-reflection region.

Direct-beam system. The collimating system of a powder camera was commonly placed in a tube which was external to the camera cylinder. This design increased the source-to-specimen distance by the length of the external collimator. Since the intensity of the x-ray beam falls off as the inverse square of the distance, it is obvious that such design implies an extravagant waste of x-radiation. There is no necessity for collimating the beam outside the camera; this can be adequately provided for within the interior of the camera, thus increasing the available x-ray intensity at the specimen. In the design suggested below, nothing but a small milled nut protrudes beyond the camera cylinder (Figs. 3, 4, and 5).

In some earlier powder-camera designs, the direct beam was prevented from reaching the film by a lead cup placed in front of the film. This scheme has the disadvantage that the back scatter by the inside of the cup reaches the film in a large zone in the back-reflection region. A superior method of getting rid of the beam is to permit it to escape from the camera through a hole. This requires a hole in the film if the film extends to the point $\theta = 0°$. This system has the disadvantage of bringing the source of air scatter very near the film in the region of the hole and thus of producing films fogged by this scatter in the region of the hole. This is corrected in the design to be described. In this design, the x-ray beam is led to the specimen and conducted away from it within tubes which reach nearly to the specimen.

The direct x-ray beam is handled by means of two assemblies (Fig. 5), the entrance-port assembly and the exit-port assembly. The first collimates the beam while the second performs several functions: it removes the direct x-ray beam, at the same time reducing air scatter to the film, and it shows the centering of the beam and specimen on a fluorescent screen. These two assemblies fit into holes on opposite sides of the camera cylinder and are held in place by means of screw threads. Some commercial manufacturers arrange to have entrance-port and exit-port assemblies attached to the camera cylinder by friction fit rather than screw fit. When the film is used Straumanis-fashion so that the beam must enter and leave through holes in the film, both assemblies must be removed and returned in the darkroom when removing or loading a film.

The entrance-port assembly consists essentially of an internal nozzle-shaped tube, pointed toward the specimen. In the tube are mounted two lead disks having central holes which define the direct beam within the camera. The hole toward the x-ray source is the limiting pinhole. It is so located that it lies within the hole in the film and on an extension of the surface of the film cylinder. This location gives rise to well-focused back-reflection lines. The breadth of a line in this region is equal to the diameter of the limiting pinhole. The second pinhole is near the specimen. It merely limits the aperture of the cone originating at the first pinhole, and its size is made only small enough to prevent this cone from spreading beyond the size of the exit hole in the exit-beam screen. A third, larger aperture removes the diffraction originating at the second pinhole.

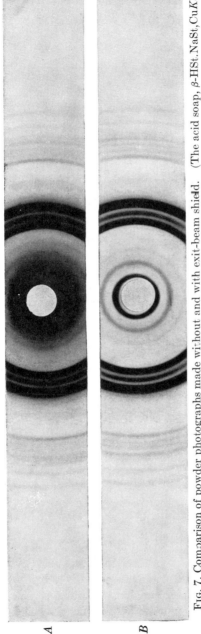

Fig. 7. Comparison of powder photographs made without and with exit-beam shield. (The acid soap, β-HSt.NaSt,CuKα, 1½ hours.)

A. Without exit-beam shield; note darkening about the center due to air scatter near θ = 0.

B. With exit-beam shield; note clean center indicating elimination of air scatter near θ = 0.

After impinging on the specimen, the direct x-ray beam is ushered away from the central part of the camera by an exit-beam shield. This is essentially a tube and is an integral part of the exit-port assembly. Its function is to surround the beam as it traverses the distance between the specimen and the hole in the film and thus shield the film from radiation scattered by the air in the beam path. Cameras provided with such shields produce x-ray photographs which are comparatively clean in the region near $\theta = 0°$ (compare Figs. 7A and 7B). Note that the lines near the exit-port hole are plainly visible in Fig. 7B but are lost in the blackening due to air scatter in Fig. 7A.

The parts of both entrance-port assembly and exit-port assembly which lie within the camera are shaped externally in the form of frustrums of long cones (Figs. 3, 4, and 5). The cones are coaxial and nearly point to point, with their apices slightly to the near sides of the specimen, and their bases outlined by the entrance and exit holes in the film. Shaped this way, these cones do not screen the film from any radiation diffracted by the specimen except that which would fall within the area of the two holes.

The exit-beam shield is an integral part of the exit-port assembly which conducts the direct beam through the exit hole in the film. This assembly (Fig. 5) is essentially a continuous tube which extends well outside the camera cylinder. It is removed, during the loading operation, by unscrewing it from the threaded hole in the camera cylinder.

The hole in the end of the tube is closed by a series of three disks of different material in this succession (Fig. 5): 1. black paper to make the camera interior light-tight; 2. fluorescent screen facing outward, so that the centering of the beam and specimen can be checked at any time; and 3. lead glass, to protect the operator when making such observations, and to prevent the direct beam from leaving the camera and entering the room. These three disks are held against the end of the tube by a friction cap. The whole exit-port assembly acts as a unit and is inserted into the exit hole of the camera cylinder.

Specimen attachment, adjustment, and rotation. The powder sample, prepared in the form of a tiny rod in one of a number of ways, is normally attached to a short section of $\frac{3}{32}$-inch metal rod, which serves as a handle in preparation and mounting. To mount this in the camera, it is merely grasped with small pliers or forceps and thrust into an axial hole in the centering device. Here it is retained in place by pressure from a weak spring made from very fine wire which protrudes into the hole from a longitudinal slot. The pressure is transmitted through a roller.

The powder sample still usually requires a centering adjustment. Centering is accomplished by translating the sample until its axis coincides with the axis about which the specimen may be mechanically rotated. The centering motion is limited to translation by arranging a flat surface, to which the specimen is attached, to slide over another flat surface which is attached to (and normal to) the rotation axis of the camera. When the specimen has been thus mounted in the camera, the eccentricity of the specimen axis is observed through a very simple microscope which is temporarily placed in the normal position of the entrance-port assembly. The microscope consists merely of a lens attached to the end of a small tube. When screwed in place on the camera, the lens is outside the camera and the opposite end of the tube almost touches the specimen. The lens is focused on the specimen; consequently the eye sees the specimen framed in the far end of the tube. When the shaft is rotated, an eccentric specimen appears to execute an up-and-down motion as seen through the microscope. To correct the error, one rotates the shaft with one hand until the specimen is at the top of its motion; then, with the other hand, one operates a pusher which displaces it down to the center of oscillation. The pusher is manipulated by turning a milled head which protrudes from the top of the camera. The screw motion is transmitted as pure translation to a pusher rod keyed to the inside of a tube (Fig. 5). The whole pusher assembly screws into a hole in the top of the camera cylinder. The pusher slides the plate holding the specimen over a plate attached to the shaft. The operation of centering with this device is the work of a moment. When the centering is perfect, the specimen appears unmoved when the shaft is rotated.

The shaft which rotates the specimen works in two gusset-type disk bearings. These are built accurately concentric with the interior cylindrical surface of the camera. This provision reduces eccentricity errors, inherent in other designs, to a negligible amount.

The specimen shaft is rotated through a pulley and rubber-band drive by the smallest electric-clock motor (Fig. 1). An alternative drive is friction coupling through one or two rubber disks. Either method effectively insulates the motor from the camera in the event of a static discharge, which is possible even in commercial x-ray tubes.

The motor is mounted on a rod which is clamped by means of a setscrew to a sleeve in the camera bracket. The position of the pulley is thus easily adjusted by loosening one screw and changing the position of the shaft.

Film cutter and punch. Ordinarily, manufacturers do not supply powder-photograph film cut to size and punched at the proper places for insertion of the entrance and exit ports. The film must be cut from some commercially available x-ray film stock. This is available in rolls of strip film 35 mm wide, and also in standard commercial-sized sheets. If the roll film is used as a source, it must be cut to proper length and punched. If the sheet film is used, a sheet of correct length may be selected, cut in strips of proper width, and punched.

A device for performing the simultaneous cutting and punching of sheet film is illustrated in Figs. 8 and 9. This is a modification of the ordinary photographic-paper trimmer, fitted with an outside stop so that it cuts

Fig. 8

Fig. 9

1-inch strips. Projections on the cutter blade push plungers which perforate the film in the manner of a ticket punch. The cutting ends of the plungers are sharpened in saddle shapes. When a 1-inch strip is being cut, the part of the remainder which will be the next strip to be cut is simultaneously perforated. A universal perforator has three positions.

When a plunger is placed in the central position, a film strip is produced with a hole in the center. When a plunger is placed in each of the outer positions, a film strip is produced for the Straumanis arrangement.

Literature

[1] M. J. Buerger, *The design of x-ray powder cameras*, J. Appl. Phys. **16** (1945) 501–510.

[2] W. Parrish and E. Cisney, *An improved x-ray diffraction camera*, Philips Tech. Rev. **10** (1948) 157–167.

[3] M. Straumanis and A. Ieviņš, *Präzisionsaufnahmen nach dem Verfahren von Debye und Scherrer*, II, Z. Physik **98** (1936) 461–475.

[4] M. J. Buerger, *X-ray crystallography* (John Wiley & Sons, Inc., New York, 1942) 394–396.

[5] A. J. C. Wilson, *Straumanis' method of film-shrinkage correction modified for use without high angle lines*, Rev. Sci. Instr. **20** (1949) 831.

5

Procedures for taking
powder photographs

In Chapter 3 the principles of powder photography were discussed, and in the last chapter the most commonly used powder camera was described in detail. In this chapter the practical procedures for operating the camera are dealt with. This includes attention to the selection of the proper

Table 1
Wavelengths of useful x-radiations and elements which highly absorb them

Element	$K\alpha = \frac{1}{3}(K\alpha_2 + 2K\alpha_1)$	$K\alpha_2$	$K\alpha_1$	$K\beta$	$K\alpha$ is highly absorbed and badly scattered by	$K\beta$ is highly absorbed and badly scattered by
Cr	2.2909 Å	2.29352 Å	2.28962 Å	2.08479 Å	Ti, Sc, Ca	V
Mn	2.1031	2.10570	2.10174	1.91016	V, Ti, Sc	Cr
Fe	1.9373	1.93991	1.93597	1.75654	Cr, V, Ti	Mn
Co	1.7902	1.79279	1.78890	1.62073	Mn, Cr, V	Fe
Ni	1.6591	1.66168	1.65783	1.50008	Fe, Mn, Cr	Co
Cu	1.5418	1.54434	1.54050	1.39217	Co, Fe, Mn	Ni
Zn	1.4364	1.43894	1.43510	1.43510	Ni, Co, Fe	Cu
.	.	.	.	.	.	.
.	.	.	.	.	.	.
.	.	.	.	.	.	.
Mo	0.7107	0.71354	0.70926	0.63225	Y, Sr, Ru	Cb, Zr

x-radiation, how to prepare the specimen, and how to load the camera, take the exposure, and develop the photograph.

While the exact experimental arrangement to be used depends on the particular equipment available and, to some extent, on the nature of the sample, the fundamental principles for taking powder photographs are the same in any case. The discussion is primarily directed, therefore, to the taking of a typical powder photograph of an average specimen with standard equipment. A little attention is given, however, to minor deviations from normal practice.

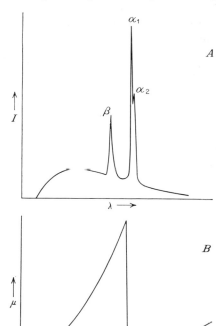

Choice of radiation

Characteristic x-radiation. The x-radiation emitted by an x-ray tube is characteristic of the chemical elements of which its target is composed. In general, the elements 22 (Ti) to 42 (Mo), except for the range 31 (Ga) through 36 (Y), give radiation which is useful for x-ray diffraction investigations.

Table 1 provides some useful data for the characteristic radiations of the elements. Each element, when properly excited, emits several characteristic wavelengths, as illustrated in Fig. 1A. These are called, in order of decreasing wavelength, $K\alpha_2$, $K\alpha_1$, and $K\beta$ radiations. The first two have wavelengths so close together that their reflections constitute a close doublet which is not ordinarily resolved except at large values of θ. It is customary to lump these together when unresolved, and to assign to the combination a wavelength which is the intensity-weighted mean of the two wavelengths, namely,

$$\tfrac{1}{3}(K\alpha_2 + 2K\alpha_1).$$

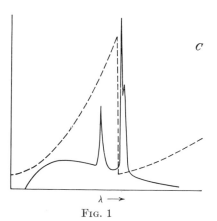

FIG. 1

The $K\beta$ radiation is considerably shorter than the $K\alpha$ and its reflection is always clearly resolved from them under all circumstances. Figure 1A also shows that the characteristic radiation is superposed on a background of *general radiation*, which contains all wavelengths above some minimum value, which is a function of the voltage across the x-ray tube.

In general, the penetrating power of x-radiation increases with decreasing wavelength. Penetrating radiation is sometimes referred to as *hard radiation*, nonpenetrating as *soft radiation*.

Filtered radiation. The variation of the absorption of an element with wavelength is illustrated in Fig. 1*B*. The shape of this absorption curve makes it possible to design a filter for x-radiation which is quite transparent to a given wavelength, but relatively opaque to a somewhat shorter wavelength. This characteristic is utilized in preparing a filter to transmit the $K\alpha$ radiation emitted by a chosen element while screening off its $K\beta$ radiation, thus providing an approximately monochromatic x-ray beam of $K\alpha$ radiation. This is illustrated in Fig. 1*C*. All the elements in the last column of Table 1 can be used as filters to transmit preferentially the $K\alpha$ radiation of the corresponding element in the first column. The most commonly used radiations and suitable filters for separating their α and β components are listed in Table 2.

Table 2
Commercial x-ray tube target wavelengths and suitable filters

Target element	$K\alpha$	Filter element	Density of filter element, g/cm²	Optimum thickness,† mm
Cr	2.2909	V	0.009	0.016
Fe	1.9373	Mn	0.012	0.016
Co	1.7902	Fe	0.014	0.018
Ni	1.6591	Co	0.014	0.013
Cu	1.5418	Ni	0.019	0.021
Mo	0.7107	Zr	0.069	0.108

† This thickness reduces $K\beta/K\alpha$ to 1/600.

Since the $K\beta$ radiation does not add to the information given by the $K\alpha$ radiation it is common practice to use filters to render the x-ray beam approximately monochromatic in this way. Most metals are available in the form of foil, which can be used directly. Certain elements are readily available only as oxides and must be fashioned into suitable filters by affixing the oxide powder to a backing, or by forming thin sheets of the oxide mixed with a plastic binder. It should be remembered that although such filters are particularly opaque to $K\beta$ they also partially absorb $K\alpha$. Optimum filter thicknesses for the densities of material stated are listed in the last column of Table 2.

Since the function of a filter is to absorb $K\beta$ and transmit $K\alpha$ it is immaterial whether the filter is placed between the x-ray tube and the powder specimen or between the specimen and the film. If the first arrangement is used, the filter can be attached either to the window of the x-ray tube† or to the collimator of the camera. Since only the direct beam is intercepted by the filter in this arrangement, pieces of foil about $\frac{1}{2}$ cm across are large enough for this purpose.

A much larger filter is required to intercept all the diffracted beams if it is to be placed between specimen and film. In this case, the filter is usually placed in front of the entire film and about $\frac{1}{8}$ in. ahead of it. In addition to absorbing $K\beta$, the filter, in this arrangement, also absorbs some of the incoherently scattered radiation, thus decreasing the intensity of the background somewhat. Instead of covering the entire film by the filter, a representative half, only, may be covered. The resulting photograph contains a record of diffraction lines due to $K\alpha$ and $K\beta$ on one side, and due to $K\alpha$ alone on the other side. Much shorter exposures can be realized with this arrangement, although care should be exercised not to interpret the absence of a very weak line from the protected part of the film as being necessarily due to absorption of $K\beta$ radiation, since $K\alpha$ is absorbed by the filter also.

Effect of unit-cell size. The locations of diffraction lines on a powder photograph are determined entirely by the dimensions of the unit cell of the crystals in the sample, and by the wavelength of x-radiation used. Bragg's equation shows that for a given spacing the diffraction lines appear at angles directly related to the wavelength of the x-radiation. Thus, if the wavelength used is too short, all visible diffraction lines are crowded closely together in the small-θ region of the photograph. Conversely, a wavelength that is too long disperses the lines to such an extent that not enough lines are recorded. It has been found that, for inorganic crystals with moderate cell dimensions, the characteristic α radiation of copper, called Cu$K\alpha$, which has a wavelength of about 1.54 Å, is ideal. Powder photographs are, therefore, usually made with Cu$K\alpha$ radiation unless there is a specific reason for using other radiation. For example, if an organic crystal with a very large cell is to be studied, it might be advisable to use a radiation having a greater wavelength in order to resolve the lines better.

Effect of specific absorption. The absorption of x-radiation by an element increases with increasing wavelength up to a wavelength called the *absorption edge*, at which a discontinuity occurs (Fig. 1B). On the short-wavelength side of the discontinuity the absorption is very high; on the long-wavelength side it is very low. If the incident radiation has a wavelength lying on the short-wavelength side of the absorption edge of an element, then that element absorbs the wavelength strongly. Such

---

† Do not attempt to attach the filter directly to the window of the x-ray tube, since there is danger that the window may be punctured.

elements are indicated in Table 2 for each wavelength. It will be observed that this element has an atomic number about 2 less than that of the element emitting the characteristic radiation.

The energy absorbed by the element is reemitted in the form of x-rays called *fluorescent radiation*. Since this radiation is emitted in all directions it serves only to increase the background on the film. It is obvious from the above that radiations having wavelengths just on the short-wavelength side of the absorption edge of any element present in the sample should be avoided.

The sample

Effect of sample size.   As the specimen thickness increases absorption cuts down the amount of radiation transmitted through the sample. At the same time, the total amount deviated to a diffraction line on the film increases with the volume of the sample in the beam, and therefore with its thickness. It is possible to compute the optimum thickness that a specimen should have, but this is rarely done in practice since means other than decreasing the size of the sample can be used to decrease the amount of radiation absorbed. The use of shorter-wavelength radiation, for instance, usually decreases the absorption.

The amount by which any substance decreases the intensity of the transmitted beam can be determined from elementary absorption theory. The relation of the intensity of the transmitted beam I to the intensity of the incident beam I_0 is

$$I = I_0\, e^{-\mu t}. \tag{1}$$

The meaning of (1) is that, after radiation passes through a thickness t (in centimeters), the value of the intensity I is equal to the initial intensity I_0 times the exponential factor $e^{-\mu t}$, where e is the base of the natural logarithms and μ is the linear absorption coefficient (in cm^{-1}).

Calculation of linear absorption coefficients.   The linear absorption coefficient can be calculated from a knowledge of the chemical composition of the powder, its specific density, and a table of *mass absorption coefficients*, which are functions of the chemical composition and the wavelength of x-rays employed in the investigation. The mass absorption coefficients of the elements for certain important wavelengths are collected in table form in several reference books, some of which are listed in the literature section at the end of this chapter. These tables list the mass absorption coefficient μ_m whose relation to the linear absorption coefficient μ_l is

$$\mu_m = \frac{\mu_l}{\rho}, \tag{2}$$

where ρ is the density of the element. Thus, the mass absorption coefficient is independent of state of aggregation, chemical combination, etc.

The linear absorption coefficient for use in (1) is readily calculated from the tabulated value of the mass absorption coefficient by the relation

$$\mu = \rho \left(\frac{\mu}{\rho}\right). \tag{3}$$

For a chemical compound, or mixture, ABC . . . , the linear absorption coefficient is calculated from the individual mass absorption coefficients of the elements by the relation

$$\mu = d \sum p \left(\frac{\mu}{\rho}\right)$$
$$= d \left\{ p_A \left(\frac{\mu}{\rho}\right)_A + p_B \left(\frac{\mu}{\rho}\right)_B + p_C \left(\frac{\mu}{\rho}\right)_C + \cdots \right\}, \tag{4}$$

where d is the density of the compound and p is the proportion of the element in the compound. This is, of course, $\left(\frac{\%}{100}\right)_A$, $\left(\frac{\%}{100}\right)_B$, $\left(\frac{\%}{100}\right)_C$, etc.

Preparation of powder mounts. There are essentially only three different ways to prepare a powder mount:

1. By fashioning the sample itself into a suitable shape
2. By coating the outside of a thin fiber with the specimen powder
3. By placing the powder inside a capillary holder

Of the three, the first is preferred since it introduces no material other than the sample in the path of the x-ray beam. Ordinarily the specimen is prepared in the form of a cylindrical rod commonly about $\frac{1}{2}$ mm in diameter. In the case of metal samples this can sometimes be accomplished by machining a piece of the metal to the desired shape. Care should be taken in such instances not to affect the properties of the metal by the machining procedure. Whenever the specimen is naturally plastic the desirable shape can be obtained by packing the specimen into a rigid tube and then extruding it by pushing a piston through one end of the tube. Occasionally a specimen, although not plastic itself, can be formed into a paste by adding a suitable binder, and then be extruded as above. Unfortunately this cannot ordinarily be done with a powdered sample, and some other form of adhesion is required. One adhesive is collodion, which has the advantages of rapid drying and of not introducing spurious lines on the diffraction photograph during normal exposures.

To prepare a collodion mount, the specimen is ground to a uniform size (approximately 200 to 300 mesh) and placed on a glass plate. (If the sample is a metal which cannot be formed readily into a rod or wire, it can be filed carefully with a clean jeweler's file. Any cold-working strains induced in the filings must be removed by annealing at suitable temperatures.) A few milligrams of the sample are then mixed with a drop of collodion until a homogeneous paste results. The amount of powder added to the collodion should not be too large lest the paste become too brittle. The paste is then scooped up on a razor blade and rolled into a rod-shaped specimen with the fingers. A more uniformly shaped cylinder can be obtained by rolling the rod between two pieces of ground glass. The thickness of the rod should be approximately $\frac{1}{2}$ mm and the length 10 mm. After allowing about 10 minutes for the rod to harden, it is affixed to a brass pin by means of ordinary glue. A suitable specimen holder can be made from $\frac{3}{16}$-inch brass rod cut to the desired length and predrilled to allow insertion of the specimen rod partway into the pin. If a highly absorbing sample is being prepared, the total absorption of the final mount can be readily controlled by adjusting the powder-to-collodion ratio. (The absorption of x-rays by collodion is negligible at the wavelengths commonly used.) The collodion contributes a diagram that is too faint to appear during normal exposures.

An alternative way of making powder mounts is first to affix a very thin capillary tube, made of lead-free glass, to a brass pin by means of picein wax. When the wax has hardened, the tube is rotated and pulled through some thinned shellac so that a very thin coating of shellac adheres to the tube walls. The tube thus coated is then rolled in the powder until it is uniformly covered by the specimen.

Substances which are affected by solvents, or are deliquescent, can be packed inside a fine capillary tube. This is accomplished by sealing one end of a lead-free glass capillary and filling it through the open end with the powder. The filling can be facilitated by rubbing the capillary carefully with a file during the filling process. The open end may then be sealed or placed open into the wax as described above. If the sample is highly absorbing it can be mixed with an amorphous material that is less absorbing and does not contribute any spurious lines to the diagram. Some recommended materials are lampblack, casein, and ghatti gum.

It should be noted that these last two methods are not recommended for general use, since the glass tends to absorb x-radiation and also to scatter x-radiation diffusely to the film, causing fogging. These objections can be overcome by using collodion tubes. These can be made by dipping an annealed copper wire in collodion and allowing a thin coating to form on its walls. When the collodion has dried, the copper wire is stretched, thereby decreasing its radius, after which the collodion tube can be readily pulled off. Prior etching of the wire in dilute nitric acid removes any copper oxide on the surface of the wire, facilitating the subsequent withdrawal of the collodion tube. Other materials, such as 10 per cent parlodion in amyl acetate, may be used instead of collodion.

In all the above methods, when the specimen is attached to the brass pin, care should be exercised to maintain parallelism between the specimen and the pin. The brass pin with its affixed powder specimen can best be handled by means of special pliers shown in Fig. 2.

When only a small amount of sample is available, a powder photograph can still be made. In favorable circumstances less than a milligram is sufficient if the sample is properly prepared. For example, in the analysis of the constituents of a rock, one may wish to identify a mineral present in very minute amounts. It is possible to extract this mineral under a microscope by very carefully scratching the rock section with a specially sharpened needle. Alternatively, a thin surface coating can often be removed from the substrate by careful scraping with a clean razor blade.

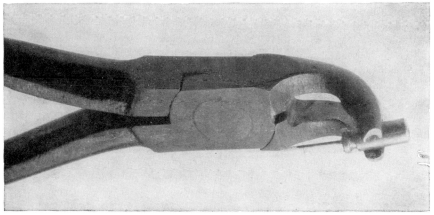

Fig. 2

To prepare a sample for x-ray investigation, the powder thus obtained is first collected on a clean slide. The needle is next dipped in shellac or collodion so that a small drop forms at its tip. The powder is then collected in this drop which is placed on the glass slide and allowed to dry into a small ball. A glass fiber, previously fastened to a brass pin, is then dipped in shellac and the small ball is affixed to it. Since this operation must be performed under the microscope, it is advisable first to practice with some other powder before attempting to fashion such a mount from the sample available only in micro quantities.

Use of the powder camera

Cylindrical powder cameras. The powder cameras most generally used in the United States employ the Straumanis-Ieviņš film arrangement. This arrangement permits simple and accurate measurement of the diffraction angles and is independent of uncertainties in the knowl-

edge of the camera radius.   Although most commercially made cameras
employ the Straumanis-Ieviņš arrangement, they differ in the details of
the placement of the specimen and film in the camera.   Several commercially made cameras are based on the design described in Chapter 4,
and the use of such a camera will be discussed first.

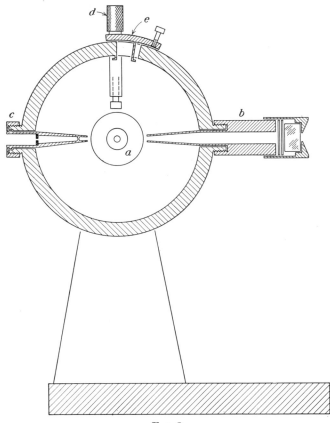

FIG. 3

By using the pair of specially made pliers (Fig. 2) the brass pin, on
which the specimen is mounted, is pushed into a small block a (Fig. 3)
located centrally in the camera.   The direct-beam catcher b is then
removed and a special viewer inserted in its place.   When light is passed
through the collimator c, the mount appears as a horizontal line in the
field of the viewer.   Upon rotation of the external spindle which is connected to the specimen block a, the mount appears to move up and down.
The centering of the mount is accomplished by rotating the spindle until
the mount is in its highest position, and then pushing the block down-

ward by means of the adjusting screw d, until the specimen bisects the field of the viewer. This routine is repeated until the specimen appears motionless during a complete rotation of the spindle. As soon as the specimen is centered, the adjusting screw d should be raised so that it will not interfere with subsequent rotation of the specimen block. The viewer is then removed and the direct-beam catcher and camera lid are replaced.

Before placing the film in the camera, the camera cover, pinhole assembly, and direct-beam catcher must be removed. Before inserting the film, it can be marked with pencil for subsequent identification. The strip of film, cut to size and prepunched to permit subsequent insertion of the collimator and beam-catcher assemblies, is placed flush against the inside wall of the camera. It is important that during the exposure the film be held tightly against the inside wall of the camera. This is assured by pushing the slider e at the top of the camera as far as possible against the film. The collimator, direct-beam-catcher assemblies, and the cover are then replaced. Finally, the camera is placed on its track on the x-ray unit. The external spindle to which the powder mount is affixed is then connected, usually by means of a rubber band, to the small motor.

The powder cameras manufactured by the General Electric Company differ from the above design in that the specimen support and film holder are two separate units. The film holder is an annulus with the direct-beam catcher permanently affixed to it. The film, bearing one hole only, is placed against the outside of the annulus and secured in place by a thin metal cover permanently fastened at one end and secured to the annulus by spring clips at the other end. The collimator is then passed through a hole in the metal strip and the hole in the film. To make the camera light-tight an additional metal cover is placed around the entire assembly. The specimen is attached to a special specimen holder which has a built-in motor for rotating the specimen. A special device for aligning the specimen on the specimen holder is necessary, the centering procedure being similar, in principle, to that described above. An alternative specimen holder is also available which uses a hollow wedge, into which the powder is placed in lieu of making a conventional powder mount. The wedge is centered by placing its edge in the path of the x-ray beam and oscillated through an angle of 20° to assure complete exposure of a maximum number of particles to the x-ray beam. Although the use of a wedge permits shorter exposure and hence is useful for quickly making photographs for comparison with standard films, it does not give as accurate values of the spacings as the conventional methods using cylindrical specimens. The wedge has an additional disadvantage in that it absorbs highly x-radiation in a forward direction. Thus, x-ray lines occurring at small glancing angles may be too weak to appear on the film.

Full rotation of the specimen during the exposure gives the maximum number of crystallites in the specimen an opportunity to contribute to the total diffraction diagram. Failure to rotate the specimen usually results in a diagram with spotty lines. The time necessary for an exposure varies with the type of x-radiation used, the specimen-to-film distance, and the amount of absorption of the x-radiation by the specimen. Ordinarily, an exposure of 1 to 2 hours is adequate for a camera of diameter 57.3 mm.

When the film is fully exposed, it is removed from the camera in a manner reversing the procedures described above. Before development, a clip, preferably numbered, is placed at one end of the film and another clip with a lead weight attached to it is placed at the opposite end. (So-called dental film clips are most suitable.) The function of the lead weight is to prevent the film from curling during the development procedure. The film is developed and fixed according to the manufacturer's instructions. It should be washed for at least $\frac{1}{2}$ hour after fixing and before hanging up the film to dry.

Cylindrical cameras employing somewhat different film arrangements have been produced in Europe. The most popular of these arrangements is one suggested by Bradley and Jay, and modified by Bradley and Bragg. In this modification two strips of film are placed symmetrically about the direct-beam path. The ends of these half films are secured by means of knife-edge film holders. The collimator and direct-beam catcher are inserted in the spaces between the films. The shadows cast by the knife-edge portions of the film holders overlapping the film are used to indicate predetermined values of the diffraction angles. These cameras have a certain advantage in that shorter individual film strips are utilized, particularly if oversized cameras are employed, but they have not gained the popularity of cameras using the Straumanis-Ieviņš arrangement. The principles of operation are quite similar to those of all cylindrical cameras and for this reason they are not further discussed in this book.

Flat-cassette cameras. Occasionally, the specimen is in the form of a thin sheet or a large block or chunk which, for various reasons, cannot be ground down to the powder size required for making cylindrical mounts. In such cases a flat-cassette camera is generally used instead of a cylindrical camera. The dimensions of the flat cassette limit the angular range of the diffracted lines that can be recorded. Two arrangements employing flat cassettes are generally used. One arrangement, the transmission or front-reflection method, places the specimen between a collimator and a flat cassette. The film thus records the x-rays diffracted in the direction of the incident beam. The other arrangement, the back-reflection method, passes the incident x-ray beam through a suitable opening in the film. In this arrangement the film is placed between the x-ray source and the specimen. The film records x-rays diffracted backward, i.e., in a sense opposite to the incident-beam direction. It becomes necessary, therefore, to decide whether the front-reflection region ($0° < \theta < 45°$) or the back-reflection region ($45° < \theta < 90°$) should be photographed. Because of the finite dimensions of the cassette, the angular range over which diffraction lines can be recorded is

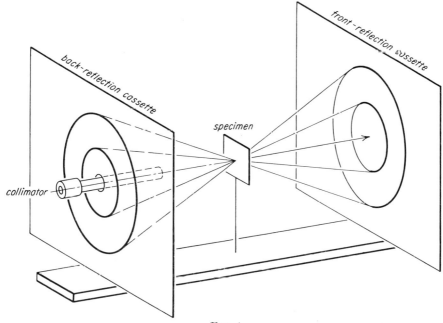

Fig. 4

limited even more than the figures in parentheses indicate. The actual range and, accordingly, the number of diffraction cones intercepted by the film in the flat cassette vary inversely with the specimen-to-film distance.

If the sample is transparent to x-radiation, the front-reflection method or the back-reflection method may be used. If the specimen is opaque to x-rays, then the back-reflection region only can be successfully recorded. The geometry of these arrangements is illustrated in Fig. 4. A front-reflection cassette is usually a rectangular frame with an opaque sheet of paper or metal foil (transparent to the x-radiation used) at the front, and a removable plate at the back. To prevent the direct beam from striking the film, a small lead plug is centrally positioned at the front of the cassette.   The back-reflection cassette differs from the above only in that in place of the lead plug a central opening is provided to transmit the x-ray beam.   In either case, the cassette is loaded by removing the back plate and inserting a precut sheet of film.   It is then placed normal to the x-ray beam.

The specimen is placed at a known specimen-to-film distance in the path of the beam.   Provision for translating the specimen in order to expose different areas of the specimen to the x-rays is sometimes made. In Chapter 3 it was shown that the interpretation of a powder photograph requires the knowledge of specimen-to-film distance.   Since, in the arrangement described above, no means for accurate measurement of specimen-to-film distances are provided, it is advisable to calibrate the film with the aid of a substance whose unit-cell dimensions are known. A very simple trick for doing this is to attach a thin sheet of aluminum foil to the specimen.   The final photograph will then contain the diffraction lines of the unknown in addition to the diffraction lines of aluminum. Aluminum is particularly useful for this purpose since its cell dimensions are very accurately known for a wide range of temperatures.

Literature

Tables listing mass absorption coefficients for x-rays

[1] C. Herman, *Internationale Tabellen zur bestimmung von Kristallstrukturen*, Vol. II (Verlagsbuchhandlung Gebrüder Borntraeger, Berlin, 1935).
[2] Charles D. Hodgman, *Handbook of chemistry and physics* (Chemical Rubber Publishing Co., Cleveland, 1955).

Sample-preparation techniques, general

[3] M. J. Buerger, *An x-ray powder camera*, Am. Mineralogist **21** (1936) 11–17.
[4] M. J. Buerger, *X-ray crystallography* (John Wiley & Sons, Inc., New York, 1942) 182–183.
[5] A. A. Burr, *Specimen holder for powder diffraction samples*, Rev. Sci. Instr. **13** (1942) 127–128.
[6] F. J. Llewellyn, *The X-ray photography of volatile or deliquescent single crystals*, Acta Cryst. **4** (1951) 185.
[7] E. M. Pogainis and E. H. Shaw, Jr., *Dilution and exposure-time of powder samples for optimum x-ray diffraction patterns with copper radiation*, Proc. S. D. Acad. Sci. **31** (1952) 24–30.
[8] André Guinier, *X-ray crystallographic technology*, English translation by T. L. Tippell, edited by Kathleen Lonsdale (Hilger and Watts Ltd., London, 1952).

(Special glass capillaries for x-ray diffraction specimen preparation can be obtained from Paul Raebiger, Simplex Apparate, Franzstrasse 43, Berlin-Spandau, Germany.)

Sample-preparation techniques, capillary mounting

9 K. Lonsdale and H. Smith, *X-ray crystal photography at low temperatures*, J. Sci. Instr. **18** (1941) 133–135.

10 R. Fricke, O. Lohrmann, and W. Schroder, *Capillaren aus Acetylcellulose für Debye-Scherrer-Aufnahmen*, Z. Elektrochem. **47** (1941) 374–379.

11 Jerry M. Waite, *Thin-walled plastic capillaries for x-ray diffraction use*, Rev. Sci. Instr. **17** (1946) 557.

12 Karl E. Beu and Howard H. Claassen, *A rapid method for preparing powder camera specimens with cellulose acetate capillary tubes*, Rev. Sci. Instr. **19** (1948) 179–180.

13 George Gibons and E. J. Bicek, *Filling capillaries for x-ray analysis*, Anal. Chem. **20** (1948) 884.

14 Paul J. Hagelston and Harris W Dunn, *Method for preparing capillary tube samples for x-ray diffraction analysis*, Rev. Sci. Instr. **20** (1949) 373–374.

15 Karl E. Beu, *Improvements in a method for preparing plastic powder sample capillary tubes*, Rev. Sci. Instr. **22** (1951) 62.

Sample-preparation techniques, extrusion

16 Joseph S. Lukesh, *An improved technique for mounting powdered samples for x-ray diffraction*, Rev. Sci. Instr. **11** (1940) 200–201.

17 M. Grotenhuis, G. F. Durst, and A. G. Barkow, *An extruding die for powdered x-ray diffraction specimens*, Non-Destructive Testing **7** (1948) 15–18.

18 L. J. E. Hofer, W. C. Peebles, and P. G. Guest, *Preparing extruded specimens for x-ray diffraction analysis*, Anal. Chem. **22** (1950) 1218.

Sample-preparation techniques, metal samples

19 W. Hume-Rothery and G. V. Raynor, *The application of x-ray methods to the determination of phase boundaries in metallurgical equilibrium diagrams*, J. Sci. Instr. **18** (1941) 74–81.

20 A. Taylor, *An introduction to x-ray metallography* (Chapman & Hall, Ltd., London, 1945) 59–60, 329.

Sample-preparation techniques, microsamples

21 L. K. Frevel and H. C. Anderson, *Powder diffraction patterns from microsamples*, Acta Cryst. **4** (1951) 186.

22 K. H. Olbricht and H. König, *Debye-Scherrer Diagramme dünner Schichten*, Naturwissenschaften **43** (1956) 234.

6

Relation of spacings to cell geometry

The fundamental crystallographic data derivable from powder photographs

A powder photograph shows a series of lines, for each of which a measurement can be made which provides the Bragg angle θ for that line. Each θ can be readily transformed into the interplanar spacing d_{hkl} for the plane responsible for the reflection to that line, as outlined in Chapter 3. Primarily, then, a powder photograph yields a set of numerical values for the various d's of the crystal sample, one d for each line of the powder photograph. A powder photograph is always interpretable up to this point (except for possible complications due to d's so close together that they are unresolved, that is, so that one observed line is really two or more overlapping lines).

One problem whose solution is frequently desired from these fundamental data of powder photography is to find the dimensions of a crystal cell which corresponds to these observed d's. The solution of this problem is not a straightforward one. To make any headway in a solution, the relation between cell dimensions and d's must be known. It is easy to derive this relation for cells whose edges are at right angles to each other, that is, for the orthogonal crystal systems. The derivation is very difficult for cells whose edges make general angles with one another. For this reason the orthogonal case is first discussed in detail.

Relation of spacing to cell edges

An orthogonal cell based upon edges a, b, c is shown in Fig. 1. A stack of planes (hkl) cuts the a axis into h parts, the b axis into k parts, and the c axis into l parts. These planes have a uniform spacing d. Note that one plane cuts the cell at the origin itself. It is convenient to

46

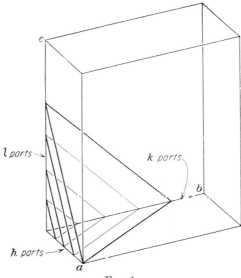

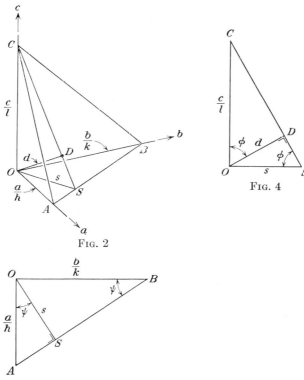

consider the spacing d between this origin plane and the first plane away from the origin. The geometry in the region of the origin is shown enlarged in Fig. 2. The required d is the length of the perpendicular dropped from the origin to this first plane. The first plane intercepts the a, b, and c axis at A, B, and C, respectively, and the distances of these intercept points from the origin are a/h, b/k, and c/l, as indicated in Fig. 2.

To derive the magnitude of d, it is convenient to pass a plane through the c axis and perpendicular to the plane (hkl). This intersects (hkl) in line CS, and plane AOB in line OS. Plane AOB is shown separately in Fig. 3. Since OS is perpendicular to AB, the triangles AOB and ASO are similar; hence

$$\frac{OA}{OS} = \frac{AB}{OB}. \tag{1}$$

Substituting the values of these lengths, there results

$$\frac{\frac{a}{h}}{s} = \frac{\sqrt{\frac{b^2}{k^2} + \frac{a^2}{h^2}}}{b/k} \tag{2}$$

$$\frac{1}{s} = \frac{hk}{ab} \sqrt{\frac{b^2}{k^2} + \frac{a^2}{h^2}} \tag{3}$$

$$= \sqrt{\frac{h^2}{a^2} + \frac{k^2}{b^2}}. \tag{4}$$

Similarly, plane COS is shown separately in Fig. 4. Since OD is perpendicular to SC, the triangles SOC and SDO are similar; hence

$$\frac{OS}{OD} = \frac{SC}{OC}. \tag{5}$$

Substituting the values of these lengths, there results

$$\frac{s}{d} = \frac{\sqrt{\frac{c^2}{l^2} + s^2}}{c/l} \tag{6}$$

$$\frac{1}{d} = \frac{l}{cs} \sqrt{\frac{c^2}{l^2} + s^2} \tag{7}$$

$$= \sqrt{\frac{1}{s^2} + \frac{l^2}{c^2}}. \tag{8}$$

The value of $1/s$ has been determined in (4). If this is substituted in (8), there results

$$\frac{1}{d} = \sqrt{\frac{h^2}{a^2} + \frac{k^2}{b^2} + \frac{l^2}{c^2}}. \tag{9}$$

This is the general relation between spacings and cell edges for orthogonal crystals. The specific values of $1/d^2$ for isometric, tetragonal, and orthorhombic crystals are listed in Table 1.

Table 1
Spacing as a function of cell edges

Crystal system	$\dfrac{1}{d^2_{hkl}}$
Isometric	$\dfrac{1}{a^2}(h^2 + k^2 + l^2)$
Tetragonal	$\dfrac{h^2 + k^2}{a^2} + \dfrac{l^2}{c^2}$
Orthorhombic	$\dfrac{h^2}{a^2} + \dfrac{k^2}{b^2} + \dfrac{l^2}{c^2}$
Hexagonal:	
Hexagonal indices	$\dfrac{4}{3a^2}(h^2 + hk + k^2) + \dfrac{l^2}{c^2}$
Rhombohedral indices	$\dfrac{1}{a^2}\dfrac{(h^2 + k^2 + l^2)\sin^2\alpha + 2(hk + kl + lh)(\cos^2\alpha - \cos\alpha)}{1 + 2\cos^3\alpha - 3\cos^2\alpha}$
Monoclinic	$\dfrac{\dfrac{h^2}{a^2} + \dfrac{k^2}{b^2} - \dfrac{2hk\cos\gamma}{ab}}{\sin^2\gamma} + \dfrac{l^2}{c^2}$ (first setting)
	$\dfrac{\dfrac{h^2}{a^2} + \dfrac{l^2}{c^2} - \dfrac{2hl\cos\beta}{ac}}{\sin^2\beta} + \dfrac{k^2}{b^2}$ (second setting)
Triclinic	$\dfrac{\dfrac{h^2}{a^2}\sin^2\alpha + \dfrac{k^2}{b^2}\sin^2\beta + \dfrac{l^2}{c^2}\sin^2\gamma + \dfrac{2hk}{ab}(\cos\alpha\cos\beta - \cos\gamma)}{1 - \cos^2\alpha - \cos^2\beta - \cos^2\gamma + 2\cos\alpha\cos\beta\cos\gamma}$ $+ \dfrac{\dfrac{2kl}{bc}(\cos\beta\cos\gamma - \cos\alpha) + \dfrac{2lh}{ca}(\cos\gamma\cos\alpha - \cos\beta)}{1 - \cos^2\alpha - \cos^2\beta - \cos^2\gamma + 2\cos\alpha\cos\beta\cos\gamma}$

The derivation of the relation between spacing and cell edges for nonorthogonal crystals is very complicated when attempted in the above manner. It is much simpler when use is made of the relation between d and the geometry of the *reciprocal* cell. This method is used in the next section.

Relation of spacing to reciprocal cell edges

It is rather easy to relate the spacing to the edges of the *reciprocal* cell. The reciprocal lattice and its reciprocal cell are discussed in Chapter 9. This section is best studied after that chapter has been read.

The spacing of any stack of planes (hkl) is represented in reciprocal space by the vector $\boldsymbol{\sigma}_{hkl}$. Specifically,

$$\frac{1}{d_{hkl}} = \boldsymbol{\sigma}_{hkl}. \tag{10}$$

The vector $\mathbf{\delta}_{hkl}$ is the vector sum of multiples of the reciprocal-cell edges,

$$\mathbf{\delta}_{hkl} = h\mathbf{a}^* + k\mathbf{b}^* + l\mathbf{c}^*. \tag{11}$$

If the scalar product of $\mathbf{\delta}_{hkl}$ with itself is formed, there results

$$\begin{aligned}
\mathbf{\delta}_{hkl} \cdot \mathbf{\delta}_{hkl} = \quad & hh\mathbf{a}^* \cdot \mathbf{a}^* + hk\mathbf{a}^* \cdot \mathbf{b}^* + hl\mathbf{a}^* \cdot \mathbf{c}^* \\
& + kh\mathbf{b}^* \cdot \mathbf{a}^* + kk\mathbf{b}^* \cdot \mathbf{b}^* + kl\mathbf{b}^* \cdot \mathbf{c}^* \\
& + lh\mathbf{c}^* \cdot \mathbf{a}^* + lk\mathbf{c}^* \cdot \mathbf{b}^* + ll\mathbf{c}^* \cdot \mathbf{c}^*.
\end{aligned} \tag{12}$$

This reduces to

$$\begin{aligned}
\sigma^2_{hkl} = \quad & h^2 a^{*2} && + hka^*b^* \cos \gamma^* + hla^*c^* \cos \beta^* \\
& + hka^*b^* \cos \gamma^* + k^2 b^{*2} && + klb^*c^* \cos \alpha^* \\
& + hla^*c^* \cos \beta^* + klb^*c^* \cos \alpha^* && + l^2 c^{*2}.
\end{aligned} \tag{13}$$

By collecting like terms and utilizing notation (10), one finds

$$\begin{aligned}
\frac{1}{d^2_{hkl}} = \sigma^2_{hkl} = \quad & h^2 a^{*2} + k^2 b^{*2} + l^2 c^{*2} \\
& + 2hka^*b^* \cos \gamma^* \\
& + 2klb^*c^* \cos \alpha^* \\
& + 2lhc^*a^* \cos \beta^*.
\end{aligned} \tag{14}$$

The specialized forms of (14) for the various symmetries are found in Table 2.

Table 2
Spacing as a function of reciprocal-cell edges

Crystal system	$\dfrac{1}{d^2_{hkl}} = \sigma^2_{hkl} = Q_{hkl}$
Isometric	$(h^2 + k^2 + l^2)a^{*2}$
Tetragonal	$(h^2 + k^2)a^{*2} + l^2 c^{*2}$
Orthorhombic	$h^2 a^{*2} + k^2 b^{*2} + l^2 c^{*2}$

Hexagonal:
   Hexagonal
     indices
   Rhombohedral
     indices
	$(h^2 + hk + k^2)a^{*2} + l^2 c^{*2}$
	
	 $(h^2 + k^2 + l^2)a^{*2} + 2(hk + kl + lh)a^{*2} \cos \alpha^*$

Monoclinic	$h^2 a^{*2} + k^2 b^{*2} + l^2 c^{*2} + 2hka^*b^* \cos \gamma^*$ (first setting) $h^2 a^{*2} + k^2 b^{*2} + l^2 c^{*2} + 2lhc^*a^* \cos \beta^*$ (second setting)
Triclinic	$h^2 a^{*2} + k^2 b^{*2} + l^2 c^{*2} + 2hka^*b^* \cos \gamma^* + 2klb^*c^* \cos \alpha^*$ $+ 2lhc^*a^* \cos \beta^*$

Unfortunately (14) relates d of the direct lattice to the geometry of the reciprocal cell. In Table 1, Chapter 9, are listed relations between the geometry of the reciprocal cell and the direct cell, so that appropriate substitutions can be made in (14). Specific use is made of the following relations:

$$a^* = \frac{bc \sin \alpha}{V}, \tag{15}$$

$$b^* = \frac{ca \sin \beta}{V}, \tag{16}$$

$$c^* = \frac{ab \sin \gamma}{V}, \tag{17}$$

$$\cos \alpha^* = \frac{\cos \beta \cos \gamma - \cos \alpha}{\sin \beta \sin \gamma}, \tag{18}$$

$$\cos \beta^* = \frac{\cos \gamma \cos \alpha - \cos \beta}{\sin \gamma \sin \alpha}, \tag{19}$$

$$\cos \gamma^* = \frac{\cos \alpha \cos \beta - \cos \gamma}{\sin \alpha \sin \beta}, \tag{20}$$

$$V = abc \sqrt{1 - \cos^2 \alpha - \cos^2 \beta - \cos^2 \gamma + 2 \cos \alpha \cos \beta \cos \gamma}. \tag{21}$$

If these are substituted into (14), it becomes

$$\begin{aligned}
\frac{1}{d_{hkl}^2} &= \frac{h^2 b^2 c^2 \sin^2 \alpha}{V^2} + \frac{k^2 a^2 c^2 \sin^2 \beta}{V^2} + \frac{l^2 a^2 b^2 \sin^2 \alpha}{V^2} \\
&+ 2hk \frac{bc \sin \alpha}{V} \cdot \frac{ac \sin \beta}{V} \cdot \frac{\cos \alpha \cos \beta - \cos \gamma}{\sin \alpha \sin \beta} \\
&+ 2kl \frac{ac \sin \beta}{V} \cdot \frac{ab \sin \gamma}{V} \cdot \frac{\cos \beta \cos \gamma - \cos \alpha}{\sin \beta \sin \gamma} \\
&+ 2lh \frac{ab \sin \gamma}{V} \cdot \frac{bc \sin \alpha}{V} \cdot \frac{\cos \gamma \cos \alpha - \cos \beta}{\sin \gamma \sin \alpha} \\
&= h^2 \frac{b^2 c^2 \sin^2 \alpha}{V^2} + k^2 \frac{a^2 c^2 \sin^2 \beta}{V^2} + l^2 \frac{a^2 b^2 \sin^2 \gamma}{V^2} \\
&+ 2hk \frac{abc^2}{V^2} (\cos \alpha \cos \beta - \cos \gamma) \\
&+ 2kl \frac{a^2 bc}{V^2} (\cos \beta \cos \gamma - \cos \alpha) \\
&+ 2lh \frac{ab^2 c}{V^2} (\cos \gamma \cos \alpha - \cos \beta). \tag{22}
\end{aligned}$$

This general form is easily reduced for the crystal systems having symmetries greater than centrosymmetry. The reductions for the several crystal systems are given in the next section. In these reductions, note that, when $\alpha = \beta = \gamma = 90°$, $\cos \alpha = \cos \beta = \cos \gamma = 0$. Therefore, the last three terms of (22) vanish for the otrhogonal systems.

Isometric system

Conditions: $a = b = c,$ $\alpha = \beta = \gamma = 90°,$ $V = a^3.$
Reduction:

$$\frac{1}{d_{hkl}^2} = h^2 \frac{a^2 \cdot a^2 \cdot 1}{a^6} + k^2 \frac{a^2 \cdot a^2 \cdot 1}{a^6} + l^2 \frac{a^2 \cdot a^2 \cdot 1}{a^6} + 0 + 0 + 0;$$

$$\therefore \frac{1}{d_{hkl}^2} = \frac{h^2 + k^2 + l^2}{a^2}. \tag{23}$$

Tetragonal system

Conditions: $a = b \neq c,$ $\alpha = \beta = \gamma = 90°,$ $V = a^2c.$
Reduction:

$$\frac{1}{d_{hkl}^2} = h^2 \frac{a^2 \cdot c^2 \cdot 1}{a^4c^2} + k^2 \frac{a^2 \cdot c^2 \cdot 1}{a^4c^2} + l^2 \frac{a^2 \cdot a^2 \cdot 1}{a^4c^2} + 0 + 0 + 0;$$

$$\therefore \frac{1}{d_{hkl}^2} = \frac{h^2 + k^2}{a^2} + \frac{l^2}{c^2}. \tag{24}$$

Orthorhombic system

Conditions: $a \neq b \neq c,$ $\alpha = \beta = \gamma = 90°,$ $V = abc.$
Reduction:

$$\frac{1}{d_{hkl}^2} = h^2 \frac{b^2c^2 \cdot 1}{a^2b^2c^2} + k^2 \frac{a^2c^2 \cdot 1}{a^2b^2c^2} + l^2 \frac{a^2b^2 \cdot 1}{a^2b^2c^2} + 0 + 0 + 0;$$

$$\therefore \frac{1}{d_{hkl}^2} = \frac{h^2}{a^2} + \frac{k^2}{b^2} + \frac{l^2}{c^2}. \tag{25}$$

Hexagonal system, hexagonal indexing

Conditions:

$$a = b \neq c,\quad \alpha = \beta = 90°,\quad \gamma = 120°,\quad V = a^2c \sin 120°.$$

Reduction:

$$\frac{1}{d_{hkl}^2} = h^2 \frac{a^2c^2 \cdot 1}{a^4c^2 \cdot \frac{3}{4}} + k^2 \frac{a^2c^2 \cdot 1}{a^4c^2 \cdot \frac{3}{4}} + l^2 \frac{a^2a^2 \cdot \frac{3}{4}}{a^4c^2 \cdot \frac{3}{4}}$$

$$+ 2hk \frac{a \cdot a \cdot c}{a^4c^2 \frac{3}{4}} (0 \cdot 0 + \tfrac{1}{2})$$

$$+ 2kl \frac{a^2 \cdot a \cdot c}{a^4c^2 \frac{3}{4}} (0 \cdot [-\tfrac{1}{2}] - 0)$$

$$+ 2lh \frac{a \cdot a^2 \cdot c}{a^4c^2 \frac{3}{4}} ([-\tfrac{1}{2}] \cdot 0 - 0)$$

$$= h^2 \frac{4}{3a^2} + k^2 \frac{4}{3a^2} + l^2 \frac{1}{c^2} + hk \frac{4}{3a^2} + 0 + 0;$$

$$\therefore \frac{1}{d_{hkl}^2} = \frac{4}{3} \frac{h^2 + hk + k^2}{a^2} + \frac{l^2}{c^2}. \tag{26}$$

Hexagonal system, rhombohedral indexing

Conditions:

$$a = b = c, \quad \alpha = \beta = \gamma \neq 90°, \quad V = a^3(1 - 3\cos^2\alpha + 2\cos^3\alpha).$$

Reduction:

$$\frac{1}{d_{hkl}^2} = h^2 \frac{a^2 \cdot a^2 \cdot \sin^2\alpha}{V^2} + k^2 \frac{a^2 \cdot a^2 \cdot \sin^2\alpha}{V^2} + l^2 \frac{a^2 \cdot a^2 \cdot \sin^2\alpha}{V^2}$$

$$+ 2hk \frac{a \cdot a \cdot a^2}{V^2}(\cos\alpha\cos\alpha - \cos\alpha)$$

$$+ 2kl \frac{a^2 \cdot a \cdot a}{V^2}(\cos\alpha\cos\alpha - \cos\alpha)$$

$$+ 2lh \frac{a \cdot a^2 \cdot a}{V^2}(\cos\alpha\cos\alpha - \cos\alpha)$$

$$= \frac{(h^2 + k^2 + l^2)a^4 \sin^2\alpha}{a^6(1 - 3\cos^2\alpha + 2\cos^3\alpha)}$$

$$+ \frac{2(hk + kl + lh)a^4(\cos^2\alpha - \cos\alpha)}{a^6(1 - 3\cos^2\alpha + 2\cos^3\alpha)};$$

$$\therefore \frac{1}{d_{hkl}^2} = \frac{(h^2 + k^2 + l^2)\sin^2\alpha + 2(hk + kl + lh)(\cos^2\alpha - \cos\alpha)}{a^2(1 - 3\cos^2\alpha + 2\cos^3\alpha)}. \quad (27)$$

Monoclinic system

Conditions:

$$a \neq b \neq c, \quad \alpha = \beta = 90° \neq \gamma, \quad V = abc\sin\gamma.$$

Reduction:

$$\frac{1}{d_{hkl}^2} = h^2 \frac{b^2c^2 \cdot 1}{a^2b^2c^2\sin^2\gamma} + k^2 \frac{a^2c^2 \cdot 1}{a^2b^2c^2\sin^2\gamma} + l^2 \frac{a^2b^2\sin^2\gamma}{a^2b^2c^2\sin^2\gamma}$$

$$+ 2hk \frac{abc^2}{a^2b^2c^2\sin^2\gamma}(0 \cdot 0 - \cos\gamma)$$

$$+ 2kl \frac{a^2bc}{a^2b^2c^2\sin^2\gamma}(0 \cdot \cos\gamma - 0)$$

$$+ 2lh \frac{ab^2c}{a^2b^2c^2\sin^2\gamma}(\cos\gamma \cdot 0 - 0)$$

$$= h^2 \frac{1}{a^2\sin^2\gamma} + k^2 \frac{1}{b^2\sin^2\gamma} + l^2 \frac{1}{c^2}$$

$$+ 2hk \frac{1}{ab\sin^2\gamma}(-\cos\gamma) \mid 0 \mid 0;$$

$$\therefore \frac{1}{d_{hkl}^2} = \frac{\dfrac{h^2}{a^2} + \dfrac{k^2}{b^2} + \dfrac{2hk}{ab}\cos\gamma}{\sin^2\gamma} + \frac{l^2}{c^2}. \quad (28)$$

Triclinic system. For triclinic crystals, the general form (22) applies. If the general value of V given by (21) is substituted into (22) there results

$$\frac{1}{d^2_{hkl}} = \frac{\dfrac{h^2}{a^2}\sin^2\alpha + \dfrac{k^2}{b^2}\sin^2\beta + \dfrac{l^2}{c^2}\sin^2\gamma + \dfrac{2hk}{ab}(\cos\alpha\cos\beta - \cos\gamma)}{1 - \cos^2\alpha - \cos^2\beta - \cos^2\gamma + 2\cos\alpha\cos\beta\cos\gamma}$$
$$+ \frac{\dfrac{2kl}{bc}(\cos\beta\cos\gamma - \cos\alpha) + \dfrac{2lh}{ca}(\cos\gamma\cos\alpha - \cos\beta)}{1 - \cos^2\alpha - \cos^2\beta - \cos^2\gamma + 2\cos\alpha\cos\beta\cos\gamma}.$$
$$(29)$$

Thus if $1 - \cos^2\theta$ is substituted for $\sin^2\theta$, and (29) is rearranged, its numerator N can be written

$$N = \frac{h}{a}\left[\frac{h}{a}(1 - \cos^2\alpha) + \frac{k}{b}(\cos\alpha\cos\beta - \cos\gamma) + \frac{l}{c}\right.$$
$$\left.(\cos\alpha\cos\gamma - \cos\beta)\right]$$
$$+ \frac{k}{b}\left[\frac{h}{a}(\cos\alpha\cos\beta - \cos\gamma) + \frac{k}{b}(1 - \cos^2\beta) + \frac{l}{c}\right.$$
$$\left.(\cos\beta\cos\gamma - \cos\alpha)\right]$$
$$+ \frac{l}{c}\left[\frac{h}{a}(\cos\alpha\cos\gamma - \cos\beta) + \frac{k}{b}(\cos\beta\cos\gamma - \cos\alpha) + \frac{l}{c}\right.$$
$$\left.(1 - \cos^2\gamma)\right]. \quad (30)$$

This can be written in determinant form as follows:

$$N = \frac{h}{a}\left\{\frac{h}{a}\begin{vmatrix} 1 & \cos\alpha \\ \cos\alpha & 1 \end{vmatrix} + \frac{k}{b}\begin{vmatrix} \cos\alpha & 1 \\ \cos\gamma & \cos\beta \end{vmatrix} + \frac{l}{c}\begin{vmatrix} \cos\gamma & \cos\beta \\ 1 & \cos\alpha \end{vmatrix}\right\}$$
$$+ \frac{k}{b}\left\{\frac{h}{a}\begin{vmatrix} \cos\alpha & \cos\gamma \\ 1 & \cos\beta \end{vmatrix} + \frac{k}{b}\begin{vmatrix} 1 & \cos\beta \\ \cos\beta & 1 \end{vmatrix} + \frac{l}{c}\begin{vmatrix} \cos\beta & 1 \\ \cos\alpha & \cos\gamma \end{vmatrix}\right\}$$
$$+ \frac{l}{c}\left\{\frac{h}{a}\begin{vmatrix} \cos\gamma & 1 \\ \cos\beta & \cos\alpha \end{vmatrix} + \frac{k}{b}\begin{vmatrix} \cos\beta & \cos\alpha \\ 1 & \cos\gamma \end{vmatrix} + \frac{l}{c}\begin{vmatrix} 1 & \cos\gamma \\ \cos\gamma & 1 \end{vmatrix}\right\}. \quad (31)$$

This can be further condensed to

$$N = \frac{h}{a}\begin{vmatrix} \dfrac{h}{a} & \cos\gamma & \cos\beta \\ \dfrac{k}{b} & 1 & \cos\alpha \\ \dfrac{l}{c} & \cos\alpha & 1 \end{vmatrix} + \frac{k}{b}\begin{vmatrix} 1 & \dfrac{h}{a} & \cos\beta \\ \cos\gamma & \dfrac{k}{b} & \cos\alpha \\ \cos\beta & \dfrac{l}{c} & 1 \end{vmatrix} + \frac{l}{c}\begin{vmatrix} 1 & \cos\gamma & \dfrac{h}{a} \\ \cos\gamma & 1 & \dfrac{k}{b} \\ \cos\beta & \cos\alpha & \dfrac{l}{c} \end{vmatrix}.$$
$$(32)$$

The denominator D of (29) can also be transformed to determinant form as follows:

$$D = 1 - \cos^2 \alpha - \cos^2 \beta - \cos^2 \gamma + 2 \cos \alpha \cos \beta \cos \gamma \tag{33}$$

$$= 1(1 - \cos^2 \alpha) - \cos \gamma(\cos \gamma - \cos \alpha \cos \beta) + \cos \beta(\cos \gamma \cos \alpha - \cos \beta) \tag{34}$$

$$= \begin{vmatrix} 1 & \cos \alpha \\ \cos \alpha & 1 \end{vmatrix} - \cos \gamma \begin{vmatrix} \cos \gamma & \cos \beta \\ \cos \alpha & 1 \end{vmatrix} + \cos \beta \begin{vmatrix} \cos \gamma & \cos \beta \\ 1 & \cos \alpha \end{vmatrix} \tag{35}$$

$$= \begin{vmatrix} 1 & \cos \gamma & \cos \beta \\ \cos \gamma & 1 & \cos \alpha \\ \cos \beta & \cos \alpha & 1 \end{vmatrix}. \tag{36}$$

The relation of spacing to cell dimensions for a triclinic crystal can consequently be expressed by the use of determinants as follows:

$$\frac{1}{d^2_{hkl}} =$$

$$\frac{\begin{vmatrix} \dfrac{h}{a} & \cos \gamma & \cos \beta \\ \dfrac{k}{b} & 1 & \cos \alpha \\ \dfrac{l}{c} & \cos \alpha & 1 \end{vmatrix} + \dfrac{k}{b}\begin{vmatrix} 1 & \dfrac{h}{a} & \cos \beta \\ \cos \gamma & \dfrac{k}{b} & \cos \alpha \\ \cos \beta & \dfrac{l}{c} & 1 \end{vmatrix} + \dfrac{l}{c}\begin{vmatrix} 1 & 1 & \dfrac{h}{a} \\ \cos \alpha & 1 & \dfrac{k}{b} \\ \cos \beta & \cos \alpha & \dfrac{l}{c} \end{vmatrix}}{\begin{vmatrix} 1 & \cos \alpha & \cos \beta \\ \cos \alpha & 1 & \cos \alpha \\ \cos \beta & \cos \alpha & 1 \end{vmatrix}}. \tag{37}$$

where the left determinant is premultiplied by $\dfrac{h}{a}$.

In Table 1 are collected the relations between spacing and cell dimensions for the several crystal systems.

7

Interpretation of powder photographs

The interpretation of a powder photograph requires the identification of all reflections. By measuring the angles subtended by pairs of arcs on a powder photograph, the interplanar spacings of the crystal can be determined. Now, the spacings can also be arrived at in another way, for they are functions of the cell edges a, b, and c, interaxial angles α, β, and γ, and the indices of the reflecting planes (hkl). If the cell dimensions are known, a list of expected d_{hkl} values can be prepared and compared with the measured values. Normally, the cell dimensions are not known, so that the unit-cell dimensions and indices must be determined in some way from the experimental d_{hkl} values. If the crystals belong to the isometric, tetragonal, or hexagonal system, there are graphical procedures for doing this, and these are described in this chapter. Analytical methods for indexing powder photographs of orthorhombic crystals are described in the next chapter. Photographs of crystals that are monoclinic or triclinic can be indexed most readily with the aid of the *reciprocal-lattice* concept described in a subsequent chapter.

Much unnecessary work can be eliminated if, prior to attempting to assign indices to the lines of a powder photograph, something is known about the symmetry of the crystals in the powder. Even an ordinary microscope may be useful in supplying symmetry information by suggesting a crystal system consistent with the crystal habit or cleavage. If the crystals are transparent, valuable information about the symmetry can be obtained by examining the powder with a polarizing microscope. Moreover, a preliminary microscopic examination may disclose the presence of more than one kind of crystal in the powder. If it does turn out that more than one kind of crystal is contained in the powder, it is worthwhile to attempt some means of separating the different constituent crystals prior to photographing such a powder. The presence of more than one kind of unknown crystal in a powder presents a very serious barrier to the successful indexing of the powder photographs.

Measurement of interplanar spacings

In Chapter 3 the relationship between the interplanar spacings of a crystal and the position of the arcs on the photograph was explained. To determine these spacings, it is first necessary to measure the distances

Fig. 1

between arcs of symmetrical pairs. Knowing these separations, the Bragg angles of diffraction can be computed. After the angles are found, Bragg's equation can be used to calculate the interplanar spacings d of the sets of planes giving rise to the arcs.

The measurement of the distance between a pair of arcs is facilitated by the use of a special film-measuring device. Such devices can be obtained from the manufacturers of powder cameras. A typical film-measuring device is illustrated in Fig. 1. When the film is placed in such a device, care should be exercised to ascertain that the long direction of the film is truly parallel to the scale of the measuring device, as shown in Fig. 2.

The readings are made in succession on two corresponding arcs (left and right) by placing the cross hair at the center of each arc and recording the value to 0.1 mm. The sequence of operations can be best understood by preparing a data sheet as shown in Fig. 3. The larger reading

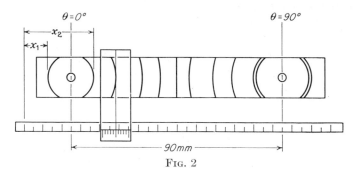

Fig. 2

line no.	x_2	x_1	check x_2+x_1	arc length $S=x_2-x_1$	correction $\frac{P}{100} \times S$	corrected arc length s'	θ $=\frac{S'}{2}$	$\sin \theta$	d $=\frac{\lambda}{2\sin\theta}$	Q $=\frac{1}{d^2}$	hkl

Fig. 3

is placed under the column heading x_2 and the smaller under x_1. The next column contains the sum of x_1 and x_2 which should remain constant with ± 0.1 mm; this serves as a check that the readings have been correctly made.

The relationship of the diffraction angle θ to the measured arc length S is shown in Fig. 4. From this it is seen that

$$4\theta = \frac{S}{R} \qquad\qquad \text{radians,}$$

$$\theta = \left[\frac{1}{4R}\right] S \qquad\qquad \text{radians,} \qquad (1)$$

or

$$\theta = \left[\frac{180}{\pi} \cdot \frac{1}{4R}\right] S \qquad \text{degrees.} \qquad (2)$$

The terms in brackets are constants depending on the radius of the camera used. S and R are normally measured in millimeters. Under S in

Fig. 3 write the difference between x_2 and x_1.   This is the arc length corresponding to the angle of 4θ.

The locations corresponding to $\theta = 0°$ and $\theta = 90°$ (Fig. 2) are next determined by taking a pair of arcs about each hole in the film, and halving the sum of the readings for such a pair.   The difference between these two averages should be 90 mm for cameras of diameter 57.26 mm (or 180 mm for cameras of diameter 114.6 mm).   If it is not, but differs from 90 mm by p per cent, then a corresponding correction of $S \times p/100$ should be added to each value of S recorded.   This correction is tabulated in the column headed "correction," and the corrected value is tabulated under S'.

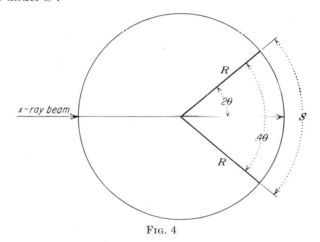

FIG. 4

Substituting for R in equation (2) the assumed radius, $180/\pi$ mm, and for S the corrected arc length S', it follows that

$$\theta = \left[\frac{180}{\pi} \cdot \frac{1}{4} \cdot \frac{\pi}{180}\right] S'$$

$$= \frac{S'}{4} \quad \text{degrees.} \tag{3}$$

From the above it is clear why it is convenient to work with a camera whose radius is a multiple of $180/\pi = 57.3$ (the conversion factor from radians to degrees).   This choice makes it possible to measure the diffraction angle directly with a millimeter scale.

The remaining columns in Fig. 3 are self-explanatory and lead to a solution of Bragg's equation for the value of the spacing d.   These last columns can be eliminated if suitable tables listing θ vs. d for different wavelengths are available.   (See list at end of this chapter.)

It is apparent that a scale can be prepared which shows d mapped on the film itself.   Since d is a function of λ as well as θ, a different scale must be prepared for each wavelength used.   Recently, such scales have been made available commercially for several wavelengths and several variations in the camera radius.   Although these are very useful for rapid determination of all *d* values, they are not as accurate as the direct-measuring technique described above.   The subject of attaining higher accuracy will be discussed more fully in Chapter 14.

Assignment of indices when the crystal system is known

Indexing procedures.   Occasionally the crystal system and the unit cell of the crystalline material in a powder specimen are known before the photograph has been taken, and it is desired to assign indices to the arcs on the photograph.   For instance, if one of the components of a mixture being analyzed by the powder method has been identified, it is usually desirable to index all reflections belonging to that component so that these lines can be eliminated from further consideration.   The procedure usually used in such cases is to compute all possible *d* (or *d²*) values with the aid of the proper equation relating *d* to the indices *hkl*, and to compare them with the *d* values determined from the film.   Alternatively, appropriate graphical methods can be used to reduce greatly the time of computation.   Whichever procedure is used, it is essentially a reversal of the procedures used for assigning indices when the crystal system but not the cell dimensions is known.

When the unit-cell dimensions are not known the assignment of correct indices to the arcs on a powder photograph is more complicated.   If the crystal system is known, however, the procedures described in this chapter can be used to asign the indices.   For crystals belonging to the isometric, hexagonal, or tetragonal systems, either analytical or graphical procedures can be used.   Graphical methods usually are more rapid and therefore are preferred.   For crystals belonging to the orthorhombic, monoclinic, or triclinic systems, however, graphical procedures for assigning indices are not suitable because more than two unknown cell dimensions are involved.   Analytical methods for treating these systems are described in subsequent chapters.

It should be understood that the methods described in this chapter can also be used if the crystal system is not known.   A systematic trial-and-error application of graphical methods can be used to determine whether the crystal can be indexed on an isometric, tetragonal, hexagonal, or rhombohedral frame.   Experience has shown, however, that if no advance information whatever is at hand to indicate the system to which the unknown crystals may belong, the methods outlined in Chapter 10 are more expedient.

Isometric unknown. The relation between the interplanar spacing d and the indices hkl for an isometric crystal is (Table 1, Chapter 6)

$$d_{hkl} = \frac{a}{\sqrt{h^2 + k^2 + l^2}}. \tag{4}$$

Since h, k, and l are integers, $h^2 + k^2 + l^2$ must be equal to an integer. It can be demonstrated that these integers can have all values except $4^p(8n + 7)$ where $p = 0, 1, 2, 3, \ldots$, and $n = 0, 1, 2, 3, \ldots$. The permissible integers are listed in Appendix 1.

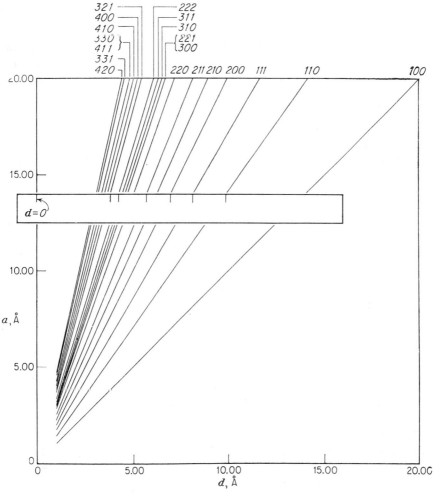

FIG. 5. Indexing chart for isometric crystals.

It can be seen from (4) that for a given plane (*hkl*) the value of d_{hkl} is a function only of the cell edge *a*. Since the possible values of the square root in (4) are the same for any isometric crystal, a list of *d*'s for one isometric crystal is the same as that for any other isometric crystal, except that the entire list is multiplied by a different value of *a* in each case. In other words, the spacings of isometric crystals differ from each other only in scale. This immediately suggests that the assignment of indices for isometric crystals can be achieved quite easily by constructing a chart by plotting *a* vs. *d* in equation (4) for all possible values *a*, say 0 to 20 Å, and all possible combinations of *h*, *k*, *l*. Such a chart is shown in Fig. 5.

The indices are assigned by marking the measured *d* values on a strip of paper, using the same scale as in the chart, and sliding the paper strip up and down, keeping the strip horizontal and keeping the point *d* = 0 on the *Y* axis, until a location is found for which all the lines on the strip correspond to lines on the graph. For greater convenience it is suggested that the chart used in the laboratory be drawn to a scale of 20 by 20 inches.

Other graphical methods for indexing powder photographs of isometric crystals have been described as early as 1918 by Scherrer.[9] Shortly after Hull and Davey published their logarithmic charts (see next section) for indexing tetragonal, hexagonal, and rhombohedral crystals, Schiebold[16] published analogous logarithmic charts for the isometric system. Somewhat different isometric gnomograms were described by Eulitz[19] in 1930. More recently, a novel indexing procedure was developed by Straumanis as a consequence of a graphical attempt to ascertain the best wavelength to use for precision determination of the lattice constant *a* (see Chapter 14). The simplicity of the graphical method first described, however, makes a detailed description of these other methods superfluous.

Tetragonal unknown. For tetragonal crystals, the relation between *d* and *hkl* is (Table 1, Chapter 6)

$$d_{hkl} = \frac{1}{\sqrt{\dfrac{h^2 + k^2}{a^2} + \dfrac{l^2}{c^2}}}. \tag{5}$$

Although it is possible to use (5) with the analytical methods described in the next chapter to determine the lattice constants *a* and *c*, graphical methods are much simpler. Hull and Davey[11] were the first to describe such a graphical method. Equation (5) can be recast as follows:

$$d_{hkl}^2 = \frac{a^2}{(h^2 + k^2) + (l^2)\dfrac{1}{c^2/a^2}}. \tag{6}$$

If logarithms of both sides are taken there results

$$2 \log d_{hkl} = 2 \log a - \log \left[(h^2 + k^2) + (l^2) \frac{1}{c^2/a^2} \right]. \qquad (7)$$

If all possible values of the ratio $C = c/a$ are considered, the specific values of c and a are not important. For simplicity, therefore, let $a = 1$ so that $log\ a = 0$. Equation (7) then becomes

$$2 \log d_{hkl} = - \log \left[(h^2 + k^2) + (l^2) \frac{1}{C^2} \right]. \qquad (8)$$

Hull and Davey used equation (8) to construct charts by plotting log d_{hkl} vs. C on semilogarithmic paper for each possible combination of the indices hkl. Such a chart is illustrated in Fig. 6.

To use this chart for determining the indices of a tetragonal crystal, the observed log d values are plotted to the scale of the chart on a strip of paper. This is accomplished by placing a strip of paper against the scale found at the bottom of such charts and drawing on it lines for each observed value of d. The strip is then translated parallel to itself on the chart, care being taken to maintain the strip parallel to the X axis, until a match is obtained. An acceptable match occurs when every line on the strip corresponds to some line on the chart. A fortuitous match can be readily obtained for high-order reflections due to the dense accumulation of lines to the left on the chart. Accordingly, in searching for a match, the first few lines on the strip (large d values) must correspond to low-order reflections. When such a match has been obtained the $C = c/a$ value is read off the ordinate. The a value can then be determined by substituting a d_{hk0} value and corresponding $hk0$ value in equation (5). Similarly the c value is found by substituting the d_{00l} value and corresponding $00l$ value in (5).

Since (8) is not a linear equation, the construction of Hull-Davey charts involves a considerable expenditure of effort. A chart that is much simpler to construct has been described by Bjurström.[21] Equation (6) can be written in the form

$$\frac{1}{d_{hkl}^2} = \frac{h^2 + k^2}{a^2} + \frac{l^2}{c^2}. \qquad (9)$$

Subtracting and adding l^2/a^2 on the right-hand side of (9), it becomes

$$\frac{1}{d_{hkl}^2} = \frac{h^2 + k^2}{a^2} - \frac{l^2}{a^2} + \frac{l^2}{a^2} + \frac{l^2}{c^2} \qquad (10)$$

$$= (h^2 + k^2 - l^2) \frac{1}{a^2} + l^2 \left(\frac{1}{a^2} + \frac{1}{c^2} \right). \qquad (11)$$

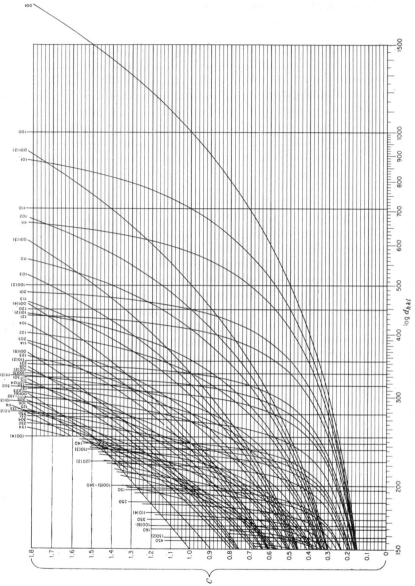

Fig. 6. Hull-Davey chart for tetragonal crystals.

If specialized units for measuring a, c, and d are chosen such that the following two conditions are satisfied,

$$\frac{1}{a^2} + \frac{1}{c^2} = 1, \quad \text{and} \quad \begin{cases} 0 \leq \dfrac{1}{a^2} \leq 1 \\ 1 \geq \dfrac{1}{c^2} \geq 0 \end{cases}, \qquad (12)$$

then (11) assumes the form

$$y = k_1 x + k_2,$$

which is the equation for a family of straight lines.

According to (12), $x = 1/a^2$ can have all values from 0 to 1. Since it is necessary to know only two points to plot a straight line, consider the limiting values of $x = 1/a^2 = 1$, and $x = 1/a^2 = 0$. When $1/a^2 = 1$, then according to (12), $1/c^2 = 0$, and equation (11) becomes

$$\frac{1}{d^2} = (h^2 + k^2)\frac{1}{a^2}$$
$$= (h^2 + k^2).$$

Similarly, when $1/a^2 = 0$, then $1/c^2 = 1$, and equation (11) becomes

$$\frac{1}{d^2} = l^2\frac{1}{c^2}$$
$$= l^2.$$

The Bjurström chart (Fig. 7) is prepared by plotting $1/d^2$ vs. $1/a^2$ on an arbitrary scale. It is most easily constructed by plotting along the vertical line corresponding to $1/a^2 = 1$, all possible values of $1/d^2 = h^2 + k^2$, and along the vertical line corresponding to $1/a^2 = 0$, all possible values of $1/d^2 = l^2$. If all the points on the left side are joined with all the points on the right side by straight lines, the resulting chart is simply a plot of (11) for all selected values of h, k, and l.

Since the horizontal line in Fig. 7 represents a plot of the values of $1/a^2$, from $1/a^2 = 0$ to $1/a^2 = 1$, (12) can be used to determine the corresponding c/a ratio. Thus:

When $1/a^2 = 1$, then $1/c^2 = 0$, or $c^2 = \infty$, and $c/a = \infty$.

When $1/a^2 = 0$, or $a^2 = \infty$, then $c/a = 0$.

When $1/a^2 = \frac{1}{2}$, then $1/c^2 = \frac{1}{2}$ and $c/a = 1$.

Similarly, the other values of the ratio c/a can be determined and noted along this line.

The indexing of a set of lines on a powder photograph using the Bjurström chart requires, simply, the plotting of measured $1/d^2$ values on a strip of paper and a search for a match on the chart, similarly to the procedure used in the Hull-Davey method. There is one difficulty, however, in doing this, namely, the lack of knowledge of the scale on which to plot the $1/d^2$ values. Since the Bjurström chart was prepared by plotting a^2, c^2, and, therefore, d^2, in arbitrary units, an additional degree of freedom must be introduced to relate the $1/d^2$ values of a given crystal to the $1/d^2$ values in the Bjurström chart. This is most easily done by preparing an additional graph on which $1/d^2$ is plotted on several scales. A continuously varying scale of $1/d^2$ can be prepared in the form

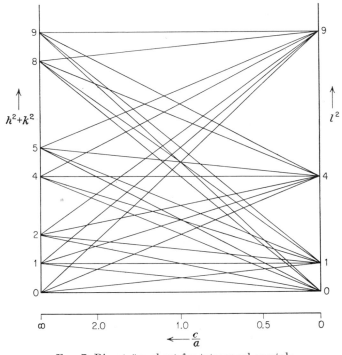

FIG. 7. Bjurström chart for tetragonal crystals.

of a graph, by plotting all values of $1/d^2$ on the ordinate and joining these points, by straight lines, to a common point on the abscissa. The resulting "fan" diagram (Fig. 8) is drawn on transparent paper and superimposed on Fig. 7. A trial-and-error attempt is then made to match the two sets of lines. A match occurs when all the lines in Fig. 8 cross the lines of Fig. 7 along any vertical line. The vertical line then determines the c/a ratio and the indices corresponding to the measured lines. The cumbersomeness of this procedure can be alleviated somewhat by constructing a parallelogram-shaped mechanical aid described in Bjurström's original paper.

The one great advantage of the Bjurström charts is that they are very easily constructed. They have a great disadvantage in that the superposition of two sets of lines can be very confusing. This difficulty can be overcome by plotting log $(h^2 + k^2)$ along one ordinate and log l^2 along the other, then joining these points by logarithmic curves, that is, by plotting c/a on a logarithmic scale also. This variation was suggested by Bunn,[24] and one of his charts is reproduced in Fig. 9. The use of Bunn's chart is entirely analogous to the use of the Hull-Davey chart.

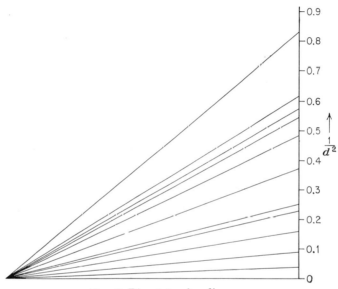

FIG. 8. Bjurström fan diagram.

A similar chart prepared by Bond (unpublished) plots log $(h^2 + k^2)$ and log l^2 against c/a directly. This variation has the advantage that such charts are easier to prepare. Unfortunately, at very large or very small values of c/a the curves are nearly vertical, making them more difficult to use in these ranges. It is questionable, however, whether such very small (or very large) values of the axial ratio are often encountered in practice. A Bond chart for the tetragonal system is shown in Fig. 10.

A somewhat different procedure was suggested by Harrington.[23] Two special cases of (9) are

$$\frac{1}{d_{hk0}^2} = \frac{h^2 + k^2}{a^2}, \tag{13}$$

$$\frac{1}{d_{00l}^2} = \frac{l^2}{c^2}. \tag{14}$$

If these are added, one obtains

$$\frac{1}{d_{hk0}^2} + \frac{1}{d_{00l}^2} = \frac{h^2 + k^2}{a^2} + \frac{l^2}{c^2}. \tag{15}$$

The right side of (15) is the same as the right side of (9). It therefore follows that

$$\frac{1}{d_{hkl}^2} = \frac{1}{d_{hk0}^2} + \frac{1}{d_{00l}^2}. \tag{16}$$

With the aid of (15) and (16) it is easy to see that the interplanar spacing of a pyramid (hkl) plane is asymptotic to the interplanar spacing of the corresponding prism $(hk0)$ plane or pinacoid $(00l)$ plane as either a^2

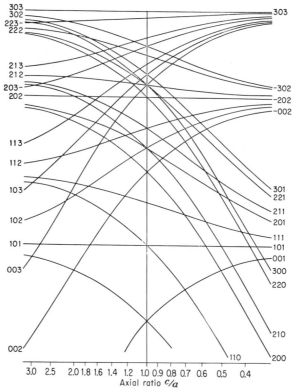

Fig. 9. Bunn chart for tetragonal crystals.

or c^2 becomes very large. This means that, if both the $(hk0)$ curves and $(00l)$ curves are drawn as families of parallel straight lines, the (hkl) curves, being asymptotic to them, always have the same shape. This is so since, according to (16), they are independent of the specific values of *hkl*. Harrington realized this condition by plotting the axial ratio on a logarithmic scale. When this is done it turns out that all (*hk*0) curves have a slope of −2, and all (00*l*) curves a slope of +2. The corresponding (*hkl*) curves are therefore symmetrical about their intersection as

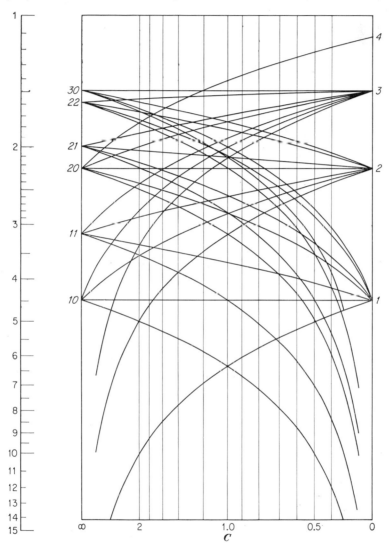

FIG. 10. Bond chart for tetragonal crystals.

shown in Fig. 11. Since the curves of all pyramid planes have the same shape, Harrington's charts can be constructed very easily with the aid of a single template. The chart is used in an analogous manner to the procedures described for Hull-Davey charts.

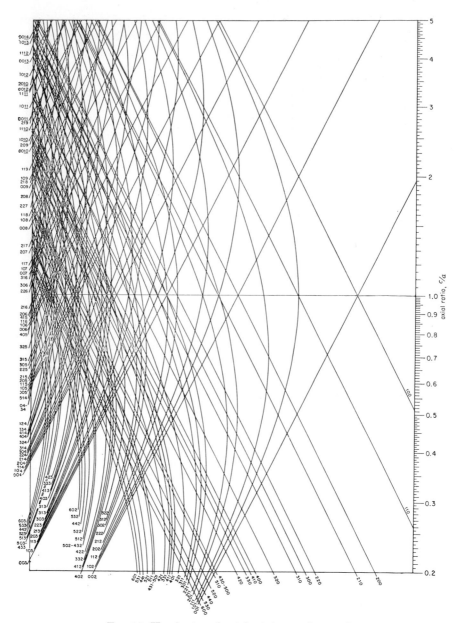

Fig. 11. Harrington chart for tetragonal crystals.

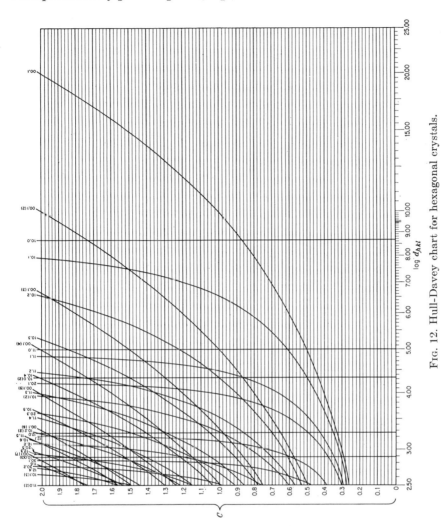

Fig. 12. Hull-Davey chart for hexagonal crystals.

Hexagonal unknown.   The relation between d^2_{hkl} and the indices hkl for the hexagonal system is (Table 1, Chapter 6)

$$d_{hkl} = \frac{1}{\sqrt{\dfrac{4}{3}\dfrac{(h^2 + hk + k^2)}{a^2} + \dfrac{l^2}{c^2}}}, \tag{17}$$

or

$$\frac{1}{d^2_{hkl}} = \frac{4}{3}(h^2 + hk + k^2)\frac{1}{a^2} + l^2\frac{1}{c^2}. \tag{18}$$

Although analytical interpretations of equation (17) or (18) can be attempted, they are generally not so fruitful as graphical methods.

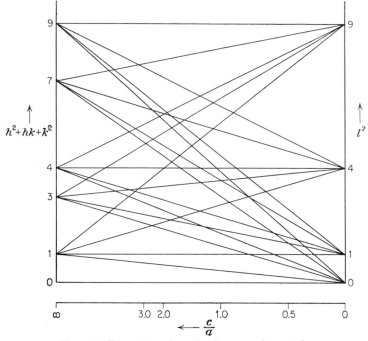

FIG. 13. Bjurström chart for hexagonal crystals.

Because of the obvious similarity between equations (17) and (5), charts very similar to the ones described above for the tetragonal system have been devised for the hexagonal system by Hull and Davey[11] (Fig. 12), Bjurström[21] (Fig. 13), Bunn,[24] Bond† (Fig. 14), and Harrington[23] (Fig. 15).

These charts can equally well be used if the crystal is based upon a rhombohedral lattice.   The unit cell and indices must, of course, be those referred to a hexagonal cell, but these can be readily transformed later to

† Unpublished.

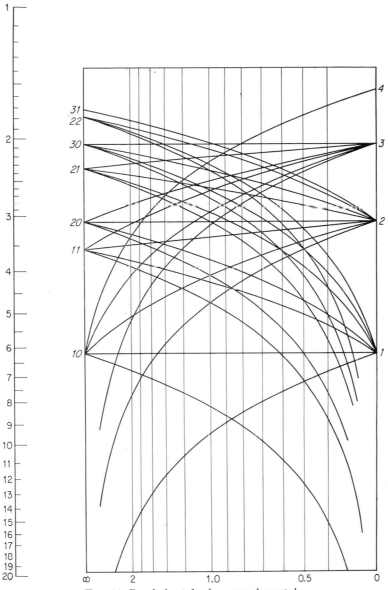

Fig. 14. Bond chart for hexagonal crystals.

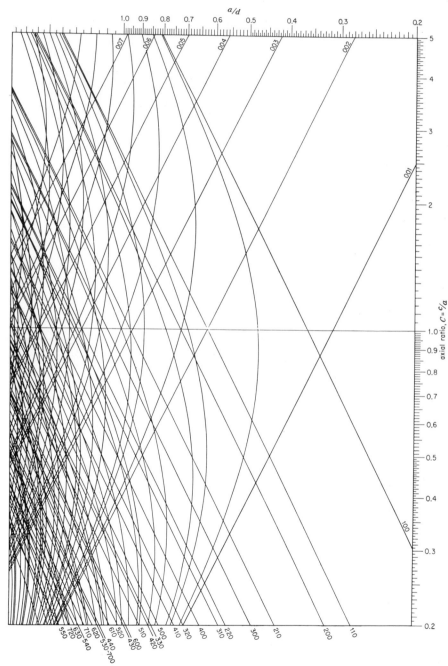

Fig. 15. Harrington chart for hexagonal crystals.

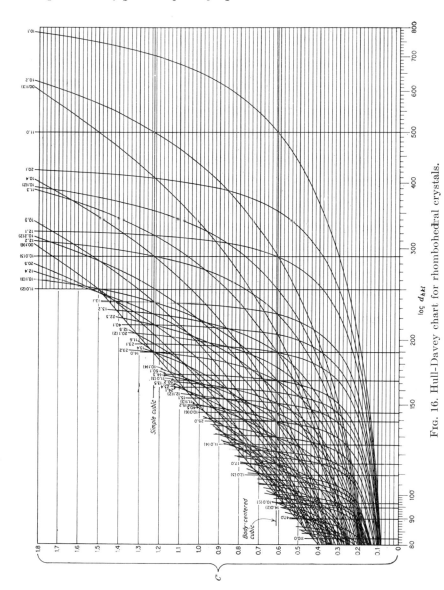

FIG. 16. Hull-Davey chart for rhombohedral crystals.

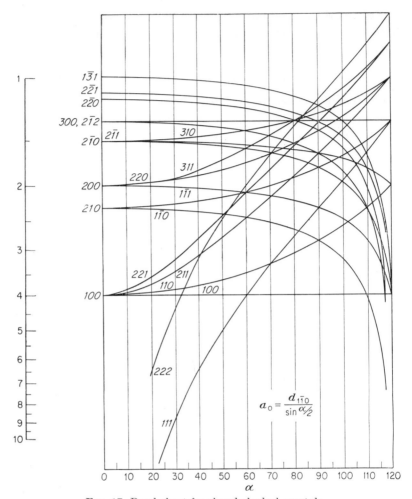

Fɪɢ. 17. Bond chart for rhombohedral crystals.

those of a rhombohedral cell, if so desired. A crystal with a rhombohedral lattice, but referred to hexagonal axes, is readily recognized as such, by the absence of all lines not obeying the condition

$$-h + k + l = 3n, \qquad n = 0, 1, 2, 3, \ldots \qquad (19)$$

The relation between d_{hkl} in rhombohedral indexing is

$$d_{hkl} = \cfrac{1}{\cfrac{1}{a} \sqrt{\cfrac{(h^2 + k^2 + l^2)\ \sin^2 \alpha + 2(hk + kl + lh)(\cos^2 \alpha - \cos \alpha)}{1 - 3 \cos^2 \alpha + 2 \cos^3 \alpha}}}. \qquad (20)$$

Charts giving rhombohedral indices directly have been prepared by Hull and Davey[11] and by Bond (unpublished). These charts are reproduced in Figs. 16 and 17.

Literature

Scales for reading *d* directly

[1] A. Weissberger, *Physical methods of organic chemistry* (Interscience Publishers, Inc., New York, 1945). Chapter on *x*-ray diffraction by I. Fankuchen.
[2] P. Terpstra, *Graphical methods for determining interplanar spacings*, Am. Mineralogist **39** (1954) 149–150.
[3] Christian Hoffroge and Hermann Weyerer, *Abstandsmessung von Röntgeninterferenzen. Ein neues Längenmessgerät*, Z. angew. Phys. **6** (1954) 419–420.

(Transparent scales for Co, Cr, Cu, Fe, or Mo radiation for 57.3-mm, 143.2-mm cameras can be obtained from N. P. Nies, 1495 Collidge Ave., Pasadena 7, Calif.)

Tables listing *d* vs. *θ* (or *2θ*)

[4] C. Herman, *Internationale Tabellen zur bestimmung von Kristallstrukturen*, Vol. II (Verlagsbuchhandlung Gebrüder Borntraeger, Berlin, 1935).
[5] *Tables for conversion of x-ray diffraction angles to interplanar spacings*, N.B.S. Applied Mathematics Series, Vol. 10 (Government Printing Office, Washington, D.C., 1950).
[6] *Tables of interplanar spacings for angle 2θ* (General Electric Co., Milwaukee, Wis.).
[7] W. Parrish and B. W. Erwin, *Data for x-ray analysis, Vol. I, Charts for solution of Bragg's equation* (Phillips Technische Bibliotheek, Eindhoven, 1953).

Graphical indexing procedures for isometric, tetragonal, and hexagonal crystals

[8] C. Runge, *Die bestimmung eines Kristallsystems durch Röntgenstrahlen*, Physik. Z. **18** (1917) 509–515.
[9] P. Scherrer, *Das Raumgitter des Aluminums*, Physik. Z. **19** (1918) 23–27.
[10] A. Johnsen and O. Toeplitz, *Über die mathematische Auswertung der Debye-Scherrerschen Röntgen-Spektrogramme*, Physik. Z. **19** (1918) 47–55.
[11] Albert W. Hull and Wheeler P. Davey, *Graphical determination of hexagonal and tetragonal crystal structures from x-ray data*, Phys. Rev. **17** (1921) 549–570.
[12] Wheeler P. Davey, *A new x-ray diffraction apparatus*, J. Opt. Soc. Amer. **5** (1921) 479–493.
[13] Wheeler P. Davey, *A new x-ray diffraction apparatus*, Gen. Elec. Rev. **25** (1922) 565–580.
[14] O. Pauli, *Die Debye-Scherrer-Methode zur untersuchung von Kristallstrukturen*, Z. Krist. **56** (1921–1922) 591–609.
[15] E. A. Owen and G. D. Preston, *Modification of the powder method of determining the structure of metal crystals*, Proc. Phys. Soc. (London) **35** (1923) 101–108.
[16] E. Schiebold, *Über graphische Auswertung von Röntgenphotogrammen*, Z. Physik. **28** (1924) 355–370.

[17] Gustav Kettman, *Beiträge zur Auswertung von Debye-Scherrer-Aufnahmen*, Z. Physik. **53** (1929) 198–209.

[18] Gustav Kettman, *Berichtung zu der Arbeit von Gustav Kettman: Beiträge zur Auswertung von Debye-Scherrer-Aufnahmen*, Z. Physik. **54** (1929) 596.

[19] Werner Eulitz, *Ein einfaches graphisches Verfahren zur Auswertung von Debye-Scherrer-Diagrammen*, Z. Physik. **64** (1930) 452–457.

[20] Nelson W. Taylor, *Die Kristallstrukturen der Verbindungen Zn_2TiO_4, Zn_2Sno_4, Ni_2SiO_4, und $NiTiO_3$*, Phys. Chemie, Abteilung B **9** (1930) 241–264.

[21] T. Bjurström, *Graphishe Methoden zum Aufsuchen der quadratischen form aus röntgenographischen Pulverphotogrammen*, Z. Physik. **69** (1931) 346–355.

[22] Fritz Ebert, *Graphische hzw. maschinelle Auswertung von Debye-Scherrer-Diagrammen Kubischer, Tetragonaler, Hexagonaler und Rhombischer Symmetrie (mit anwendungsbeispielen: WC, PdF_2, und $HgCl_2$)*, Z. Krist. **78** (1931) 489–495.

[23] Robert A. Harrington, *A simplified construction of Hull-Davey charts*, Rev. Sci. Instr. **9** (1938) 429–430.

[24] Charles William Bunn, *Chemical crystallography* (Oxford University Press, New York, 1946), especially 379–382.

[25] G. Homés and J. Gouzou, *Recherche d'une méthode nouvelle de déchiffrement des diagrammes de Debye-Scherrer*, Rev. mét. **51** (1954) 749–757.

(A complete set of laboratory-size Bunn's charts for tetragonal and hexagonal crystals can be obtained from Polycrystal Book Service, 99 Livingston St., Brooklyn 1, N.Y.)

Graphical indexing procedures for orthorhombic crystals

[26] J. O. Wilhelm, *An extension of the graphical method for determining crystal structures to the orthorhombic system*, Trans. Roy. Soc. Canada **21** (1927) 41–43.

[27] C. W. Jacob and B. E. Warren, *The crystalline structure of uranium*, J. Am. Chem. Soc. **59** (1937) 2588—2591.

[28] R. J. Wasilewski, *The solubility of oxygen in, and the oxidates of, tantalum*, J. Am. Chem. Soc. **75** (1953) 1001–1002.

(Laboratory-size charts for indexing powder photographs of orthorhombic crystals, based on Harrington's[23] construction, can be obtained from Batelle Memorial Institute, Columbus, Ohio.)

8

Analytical methods for indexing powder photographs

The graphical methods of the previous chapter can be extended to include the orthorhombic system also. The interplanar spacing for the orthorhombic system is

$$d_{hkl} = \frac{1}{\sqrt{\dfrac{h^2}{a^2} + \dfrac{k^2}{b^2} + \dfrac{l^2}{c^2}}}, \tag{1}$$

or

$$\frac{1}{d_{hkl}^2} = \frac{h^2}{a^2} + \frac{k^2}{b^2} + \frac{l^2}{c^2}. \tag{2}$$

Multiplying both sides of (2) by c^2,

$$\frac{c^2}{d_{hkl}^2} = \frac{h^2}{a^2/c^2} + \frac{k^2}{b^2/c^2} + \frac{l^2}{c^2/c^2}$$

$$= \frac{h^2}{A^2} + \frac{k^2}{B^2} + \frac{l^2}{1}, \tag{3}$$

where $A = a/c$, $B = b/c$. Taking logarithms of both sides, (3) becomes

$$2 \log c - 2 \log d_{hkl} = \log \left(\frac{h^2}{A^2} + \frac{k^2}{B^2} + \frac{l^2}{1} \right). \tag{4}$$

Assigning successive values to $A = a/c$, (4) can be used to construct charts for finding $B = b/c$ with the aid of the observed values of d_{hkl}. Such charts can be prepared in a way similar to the procedures described in the previous chapter. Of course, a separate chart must be prepared for each assigned value of A. A set of such charts for the orthorhombic systems, based on Harrington's charts, has been recently published by Bell and Austin.† Although similar graphical methods have been used

† See note at end of literature list of foregoing chapter.

in the past by several investigators, they are less accurate and more time-consuming than the analytical procedures described in this chapter.

The first to publish a straightforward analytical method was Hesse. Shortly thereafter, Lipson published an identical procedure which he had independently discovered at about the same time.　An extension of this method by Stosick, who uses linear Diophantine equations to systematize the method further, appears useful for tetragonal and hexagonal crystals but would become laborious if extended to the orthorhombic system.

A different approach was used by Vand to index powder photographs of organic compounds having one large spacing.　In the powder photographs of crystals having one large spacing, the lines are grouped into bands.　The separation between these bands is used to determine the length of the large spacing, reducing the problem thereby to two dimensions.　The two-dimensional problem then can be solved analytically or graphically.

As mentioned previously, it is possible to use analytical procedures to index powder photographs of isometric, tetragonal, and hexagonal crystals also.　Usually these procedures will prove more time-consuming than the graphical procedures already described.　For completeness, however, these methods are briefly described in this chapter.

Isometric system

The interplanar spacing equation

$$d_{hkl} = \frac{a}{\sqrt{h^2 + k^2 + l^2}} \tag{5}$$

can be written alternatively

$$(h^2 + k^2 + l^2)d_{hkl}^2 = a^2. \tag{6}$$

The sum $h^2 + k^2 + l^2$ must be equal to an integer, since h, k, and l are integers.　To assign suitable indices to lines, therefore, the measured d^2's can be multiplied by successive permissible integers, 1, 2, 3, . . . and if the products all have the same value, the appropriate indices can be assigned to these planes.　Let $\Sigma = (h^2 + k^2 + l^2)$.　Then

$$\Sigma \times d_{hkl}^2 = a^2 \qquad \text{where} \qquad \Sigma = h^2 + k^2 + l^2,$$

specifically,

$$
\begin{aligned}
1 \times d_{100}^2 &= a^2 \qquad \text{since} \qquad 1 = 1^2 + 0^2 + 0^2, \\
2 \times d_{110}^2 &= a^2 \qquad \text{since} \qquad 2 = 1^2 + 1^2 + 0^2, \\
3 \times d_{111}^2 &= a^2 \qquad \text{since} \qquad 3 = 1^2 + 1^2 + 1^2, \\
4 \times d_{200}^2 &= a^2 \qquad \text{since} \qquad 4 = 2^2 + 0^2 + 0^2, \\
5 \times d_{210}^2 &= a^2 \qquad \text{since} \qquad 5 = 2^2 + 1^2 + 0^2, \text{ etc.}
\end{aligned}
\tag{7}
$$

It should be borne in mind that certain values of $h^2 + k^2 + l^2$ may be missing if the corresponding planes (hkl) do not reflect x-rays. Such absences may be purely fortuitous, or they may be due to special conditions of symmetry. For example, systematic absences occur because of *lattice extinctions*. Reflections for all combinations of h, k, and l are possible if the lattice is primitive. On the other hand, if the lattice is body-centered, only those reflections can occur for which the sum of h, k, and l is an even number. Similarly, if a lattice is face-centered, only those reflections can occur whose indices are all odd or all even. This is illustrated diagrammatically in Fig. 1.

In addition to such absences, certain integers cannot be expressed as the sum of the squares of three numbers, and reflections corresponding to

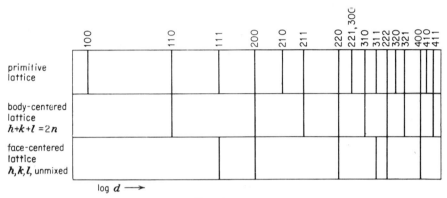

Fig. 1

these are therefore missing. Such numbers as 7, 15, 23, or more generally, any number of the form $4^p(8q + 7)$, cannot be expressed as the sum of the squares of three integers, and so must be omitted from consideration. A list of permissible values of the sum $h^2 + k^2 + l^2$ up to 200 is given in Appendix 1.

The above procedure can be somewhat simplified by recasting equation (6) in the form

$$\frac{a^2}{d_{hkl}^2} = h^2 + k^2 + l^2. \tag{8}$$

It is then only necessary to assume values of a^2 such that the ratio a^2/d^2 for all lines is an integer. This can be done rather quickly on a slide rule. The A (or B) scale on a slide rule consists of two decades and the C (or D) scale consists of one decade. Remove the slider and invert it so that the C scale is opposite the A scale and their senses are respectively inverted, that is, the A scale increases from left to right and the C scale increases from right to left. Now, the observed d values are noted on the C scale on the slider and it is moved until all the d values on the C scale are opposite integers on the A scale. This procedure is the equivalent of the normal division operation on a slide rule. In this case, a appears on the C scale opposite the integer 1 on the A scale, and the other possible quotients of (8) are the integers on the A scale that match the d values noted on the C scale. In actual practice it may be found that these quotients only approximate integers, that is, they may have values like 0.95, 1.04, 1.95, etc.

Tetragonal system

The interplanar spacing for the tetragonal system can be written

$$d_{hkl}^2 = \frac{1}{\dfrac{h^2 + k^2}{a^2} + \dfrac{l^2}{c^2}}. \tag{9}$$

If $l = 0$, equation (9) becomes

$$d_{hk0}^2 = \frac{a^2}{h^2 + k^2}. \tag{10}$$

The similarity with equation (5) is obvious. If the possible combinations of h and k are considered, the value of a^2 can be deduced in a manner similar to the procedure used for the isometric case. For example:

$$
\begin{array}{llll}
1 \times d_{100}^2 = a^2 & \text{since} & 1 = 1^2 + 0^2, & \\
2 \times d_{110}^2 = a^2 & \text{since} & 2 = 1^2 + 1^2, & \\
4 \times d_{200}^2 = a^2 & \text{since} & 4 = 2^2 + 0^2, & (11) \\
5 \times d_{210}^2 = a^2 & \text{since} & 5 = 2^2 + 1^2, & \\
8 \times d_{220}^2 = a^2 & \text{since} & 8 = 2^2 + 2^2, \text{ etc.} &
\end{array}
$$

If a set of experimentally observed d's can be found that satisfies the above relations, then the length of the a axis can be determined. Knowing a, (9) can be used to determine c by iteration, and to assign indices to the remaining lines on the photograph.

Hexagonal system

The analytical procedure for the hexagonal system is analogous to the procedure described above for the tetragonal system. When $l = 0$ the interplanar spacing is

$$d_{hk0}^2 = \frac{3}{4}\left(\frac{a^2}{h^2 + hk + h^2}\right). \tag{12}$$

This can be used to determine the value of a. After a is known, the entire equation

$$d_{hkl}^2 = \frac{1}{\dfrac{4(h^2 + hk + k^2)}{3a^2} + \dfrac{l^2}{c^2}} \tag{13}$$

can be used to determine the value of c.

The interplanar-spacing equation for the rhombohedral subsystem is too unwieldly to be used directly. It is possible, however, to use (12) and (13) to assign hexagonal indices to a powder photograph of a rhombohedral crystal.

Orthorhombic system

Hesse-Lipson procedure. The interplanar spacing for the orthorhombic system can be written

$$\frac{1}{d_{hkl}} = \sqrt{\frac{h^2}{a^2} + \frac{k^2}{b^2} + \frac{l^2}{c^2}}. \tag{14}$$

On the other hand, the Bragg equation can be written in the form

$$\frac{1}{d_{hkl}} = \frac{2 \sin \theta_{hkl}}{\lambda}. \tag{15}$$

If equations (14) and (15) are squared and combined, the following relation is obtained:

$$\frac{4 \sin^2 \theta_{hkl}}{\lambda^2} = \frac{1}{d_{hkl}^2}$$
$$= h^2 \frac{1}{a^2} + k^2 \frac{1}{b^2} + l^2 \frac{1}{c^2}, \tag{16}$$

or
$$\sin^2 \theta_{hkl} = h^2 \frac{\lambda^2}{4a^2} + k^2 \frac{\lambda^2}{4b^2} + l^2 \frac{\lambda^2}{4c^2}$$
$$= h^2 A + k^2 B + l^2 C, \tag{17}$$

where
$$A = \frac{\lambda^2}{4a^2}$$
$$B = \frac{\lambda^2}{4b^2},$$
$$C = \frac{\lambda^2}{4c^2}.$$

An interesting characteristic of (17) is that $\sin^2 \theta$ is a simple sum of several parts, each part depending on its index h, k, or l. Thus, since

$$\sin^2 \theta_{h00} = h^2 A \tag{18}$$

and
$$\sin^2 \theta_{0k0} = k^2 B \tag{19}$$

it follows that

$$\sin^2 \theta_{h00} + \sin^2 \theta_{0k0} = h^2 A + k^2 B. \tag{20}$$

But this has exactly the same value as

$$\sin^2 \theta_{hk0} = h^2 A + k^2 B. \tag{21}$$

Consequently, relationships of the following type occur:

$$\begin{aligned}
\sin^2 \theta_{h_1 k_1 0} &= \sin^2 \theta_{h_1 00} &&+ \sin^2 \theta_{0k_1 0}, \\
\sin^2 \theta_{h_1 k_1 l_1} &= \sin^2 \theta_{00l_1} &&+ \sin^2 \theta_{h_1 k_1 0}, \\
\sin^2 \theta_{h_1 0 l_1} &= \sin^2 \theta_{h_1 00} &&+ \sin^2 \theta_{00l_1}, \text{ etc.}
\end{aligned} \tag{22}$$

Difference relations also occur, for example,

$$\begin{aligned}
\sin^2 \theta_{0k_1 0} &= \sin^2 \theta_{h_1 k_1 0} &&- \sin^2 \theta_{h_1 00}, \\
\sin^2 \theta_{00l_1} &= \sin^2 \theta_{h_1 k_1 l_1} &&- \sin^2 \theta_{h_1 k_1 0}, \\
\sin^2 \theta_{00l_1} &= \sin^2 \theta_{h_2 k_2 l_1} &&- \sin^2 \theta_{h_2 k_2 0}, \text{ etc.}
\end{aligned} \tag{23}$$

The Hesse-Lipson[3,5] method is based on the assumption that, if all possible combinations of $\sin^2 \theta$ values are used to form equations like those shown above, then the $\sin^2 \theta$ values appearing most frequently in these equations are those of the pinacoid reflections $h00$, $0k0$, and $00l$. The practical application of the Hesse-Lipson method is illustrated with the aid of a set of $\sin^2 \theta$ values obtained from a powder photograph of a compound having the marcasite structure. The twenty-eight $\sin^2 \theta$ values obtained are listed in Table 1. Table 2 contains a list of differences between the $\sin^2 \theta$ values of Table 1. A systematic tabulation is obtained by first subtracting $\sin^2 \theta_1$ from $\sin^2 \theta_2$ through $\sin^2 \theta_{28}$ and entering the

Table 1
Observed $\sin^2 \theta$ values of an unknown orthorhombic crystal

Line	Sin² θ	Line	Sin² θ
1	.0865	15	.4956
2	.1396	16	.5208
3	.1406	17	.5472
4	.1755	18	.5624
5	.1912	19	.5973
6	.2413	20	.6930
7	.2804	21	.7020
8	.3308	22	.7748
9	.3460	23	.8538
10	.3564	24	.8675
11	.3658	25	.8875
12	.4031	26	.9241
13	.4350	27	.9616
14	.4547	28	1.0131

Table 2
Partial list of differences between sin² θ values given in Table 1
(decimal point omitted)

No.	Sin² θ	1	2	3	4	5	6	7	8	9	10	11	12	13	14	15	16
1	.0865																
2	.1396	0531															
3	.1406	0541	0010														
4	.1755	0890	0359	0349													
5	.1912	1047	0516	0506	0157												
6	.2413	1548	1017	1007	0658	0501											
7	.2804	1939	1408	1398	1049	0892	0391										
8	.3308	2443	1912	1902	1553	1396	0895	0504									
9	.3460	2595	2064	2054	1705	1548	1047	0656	0152								
10	.3564	2699	2168	2158	1809	1652	1151	0760	0256	0104							
11	.3658	2793	2262	2252	1903	1746	1245	0854	0350	0198	0094						
12	.4031	3166	2635	2625	2276	2119	1618	1227	0723	0571	0567	0473					
13	.4350	3485	2954	2944	2595	2438	1937	1546	1044	0892	0788	0694	0319				
14	.4547	3682	3151	3141	2792	2635	2134	1743	1239	1087	0983	0889	0516	0197			
15	.4956		3560	3550	3201	3044	2543	2152	1648	1496	1392	1298	0925	0606	0409		
16	.5208			3802	3453	3296	2795	2404	1900	1748	1644	1550	1177	0858	0661	0252	
17	.5472				3717	3560	3056	2668	2164	2012	1908	1814	1441	1122	0925	0616	0264

differences in column 1. Next, sin² θ_2 is subtracted from sin² θ_3 through sin² θ_{28} and the differences are entered in column 2, etc., until all possible differences are obtained.

After such a table of differences is prepared, it is examined to determine the frequency of occurrence of the differences. This can be best accomplished by plotting all the differences on a chart. Hesse used a bar

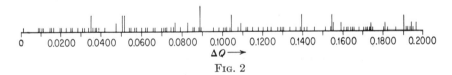

Fig. 2

graph (Fig. 2) in which each difference is indicated by a vertical line on the abscissa. If a given difference occurs more than once, the height of the line is correspondingly increased. The distribution of these lines is examined by observing the number of lines occurring within a region whose width is the range of the experimental error, in this case estimated as $\Delta \sin^2 \theta \approx 0.0004$.

Alternatively, Lipson plotted the differences along horizontal lines, one for each column of Table 2, in a graph such as the one shown in Fig. 3. Each difference is given a finite width on the chart, the width representing the assumed experimental error. The chart is studied to ascertain the locations of concentrations of plotted points along vertical lines, such as those shown in Fig. 3.

If either Fig. 2 or Fig. 3 is examined, the following distribution of differences is found:

.0890 occurs 7 times,
.1548 occurs 6 times,
.1045 occurs 5 times,
.1396 occurs 5 times, (24)
.0350 occurs 4 times,
.0505 occurs 4 times,
.0515 occurs 4 times, etc.

The problem now is to select from these numbers the values of $\sin^2 \theta_{100}$, $\sin^2 \theta_{010}$, and $\sin^2 \theta_{001}$ according to (23). Two points should be borne in mind when making the selections. The first is to check whether higher-order reflections are present, that is, after $\sin^2 \theta_{100}$ is selected, $\sin^2 \theta_{200}$,

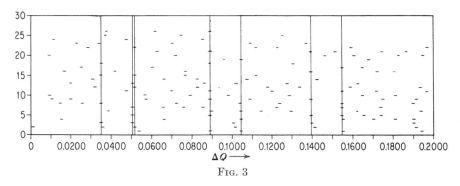

FIG. 3

$\sin^2 \theta_{300}$, etc., should be computed and compared with the list of observed $\sin^2 \theta$ values to see if any of the observed $\sin^2 \theta$ values correspond to the computed value of higher orders. The second point to remember is that one or more of the $\sin^2 \theta$ values of a pinacoid reflection must be equal to, or smaller than, the smallest value of the observed $\sin^2 \theta$ values.

Thus, in the most frequently occurring difference given above in (24), if .0890 is chosen as $\sin^2 \theta_{100}$, then $\sin^2 \theta_{200} = 2^2 \sin^2 \theta_{100} = .3560$. A check of Table 1 shows that $\sin^2 \theta_{10} = .3564$, which makes this selection acceptable.

The second point mentioned above states that at least one of the other pinacoid reflections must have a $\sin^2 \theta$ smaller than .0865 ($\sin^2 \theta_1$ in Table 1). If the list of differences in (24) is examined, it becomes apparent that, if .0350 (occurring four times) is chosen as $\sin^2 \theta_{010}$, then $\sin^2 \theta_{020} = .1400$. Since .1396 appears five times in this list (note also that $\sin^2 \theta_2 = .1396$ in Table 1) this selection also appears acceptable. Equation (4) can now be used to compute the values of $\sin^2 \theta_{hk0}$ for different values of h and k. Thus:

$$\sin^2 \theta_{100} = .0890$$
$$\sin^2 \theta_{200} = \sin^2 \theta_{10} \text{ in Table 1} = .3560$$
$$\sin^2 \theta_{010} = .0350$$
$$\sin^2 \theta_{020} = \sin^2 \theta_2 \text{ in Table 1} = .1400 \tag{25}$$
$$\sin^2 \theta_{110} = .1240$$
$$\sin^2 \theta_{220} = \sin^2 \theta_{15} \text{ in Table 1} = .4960, \text{ etc.}$$

A check of Table 1 shows that the smallest observed $\sin^2 \theta$ value (.0865) is not accounted for as yet. Since this value did not appear among the most frequent differences listed in (20), it is not likely to be the value of $\sin^2 \theta_{001}$. It could, however, be the value of $\sin^2 \theta_{011}$, for example. If this assumption is made, then the value of $\sin^2 \theta_{001}$ can be readily computed, since

$$\sin^2 \theta_{001} = \sin^2 \theta_{011} - \sin^2 \theta_{010}$$
$$= .0865 - .0350$$
$$= .0515.$$

This value occurs four times in the list of differences in (24). Moreover, if the above values of $\sin^2 \theta_{100}$, $\sin^2 \theta_{010}$, and $\sin^2 \theta_{001}$ are used to compute $\sin^2 \theta_{hkl}$ for different values of hkl, it becomes possible to index *all* the lines in Table 1. This is the final confirmation that the $\sin^2 \theta$ values for the pinacoid reflections were correctly selected. By substituting the values of $\sin^2 \theta_{100}$, $\sin^2 \theta_{010}$, and $\sin^2 \theta_{001}$ in equation (16) the values of a, b, and c can be determined.

When applying the Hesse-Lipson method it should be remembered that space-group extinctions may prevent certain reflections from appearing. In such cases less direct methods of trial and error, similar to the method used in finding $\sin^2 \theta_{001}$ above, must be resorted to.

Vand's method for crystals having one large spacing. Some organic substances, such as certain forms of soaps, fats, and fatty acids, have unit cells for which one dimension is much larger than the other two. To distinguish such cases from the more general ones of long-chain compounds whose three unit-cell edges have more nearly equal lengths, Vand[2,4] proposed the term "long-spacing compound."

If the Hesse-Lipson notation,

$$\sin^2 \theta_{hkl} = h^2 A + k^2 B + l^2 C, \tag{26}$$

is adopted, it is easy to prove that one very long axis, say c, has the effect of grouping all reflections that have the same h and k values closely together.

Consider the case:

$$c \gg a, b \quad \text{or} \quad C \ll A, B.$$

Since C is much smaller than either A or B, the product l^2C is much smaller than the sum of h^2A and k^2B in (26). As a consequence of this, all reflections having the same h and k values but different l values have very nearly the same $\sin^2 \theta_{hkl}$ values. As a matter of fact, if c is sufficiently longer than both a and b, the lines on a powder photograph corresponding to such reflections may be so close together as to be not well resolved and may constitute a band. Barring this difficulty, however, equations like (23) can be used to determine the length of c from the lines comprising any such bands. Since all the lines within one band have the same hk indices, the following relations apply:

$$\sin^2 \theta_{h_1k_1l_2} - \sin^2 \theta_{h_1k_1l_1} = (h_1^2A + k_1^2B + {}_2^2C) - (h_1^2A + k_1^2B + l_1^2C)$$
$$= (l_2^2 - l_1^2)C. \tag{27}$$

Equation (27) can be used, therefore, to determine c from successive pairs of reflections within a band. (Since l_1^2 and l_2^2 can only have the values 1, 4, 9, 16, etc., their differences can only have the values 1, 3, 5, 7, 8, 12, etc.) As a check, the same computations can be tried in another band since the h and k indices within any one band have the same values. Remembering that

$$\frac{1}{d_{hkl}^2} = \frac{h^2}{a^2} + \frac{k^2}{b^2} + \frac{l^2}{c^2}, \tag{28}$$

and
$$\frac{1}{d_{hk0}^2} = \frac{h^2}{a^2} + \frac{k^2}{b^2}, \tag{29}$$

it is easy to see that

$$\frac{1}{d_{hk0}^2} = \frac{1}{d_{hkl}^2} - \frac{l^2}{c^2}. \tag{30}$$

Also, it is clear that (30) can be rewritten

$$d_{hk0}^2 = \frac{1}{\dfrac{h^2}{a^2} + \dfrac{k^2}{b^2}}$$
$$= \frac{a^2}{h^2 + \dfrac{k^2}{b^2/a^2}}. \tag{31}$$

The similarity between (31) and equation (6) of Chapter 7 immediately suggests that the graphical methods of Chapter 7 can be used to determine a and b as well as the indices h and k. This is easily accomplished as follows: After the value of c has been determined by the procedure described above, the l^2 values of each reflection are determined with the aid of relation (27). The appropriate value of l^2/c^2 can then be subtracted from the $1/d^2$ value of each observed reflection. The result is a set of $1/d^2$ values for each band.   The appropriate values of h and k, as well as a and b, can then be determined by graphical methods.

The details of this procedure are not given here since both graphical and analytical methods for solving this kind of problem have already been described.   It should be noted here, however, that Vand's original procedures for indexing powder photographs of long-spacing compounds were designed for crystals of any symmetry, i.e., monoclinic and triclinic crystals also.   The procedures in such cases become much more involved since more than three unknowns must be determined.   Although the details of this essentially trial-and-error procedure are beyond the scope of this book, the special case of orthorhombic symmetry, described above, illustrates the principles underlying Vand's method.

Monoclinic system

The interplanar-spacing relation for the monoclinic system is (Table 1, Chapter 6)

$$d_{hkl} = \cfrac{1}{\sqrt{\cfrac{\dfrac{h^2}{a^2} + \dfrac{l^2}{c^2} + \dfrac{2hl}{ac}\cos\beta}{\sin^2\beta} + \dfrac{k^2}{b^2}}}. \tag{32}$$

This equation has four unknowns, a, b, c, β, in addition to hkl.   It is not possible to index all the reflections from crystals in this system by any of the methods described so far.   Since b is perpendicular to a and c, it is possible to use procedures similar to the Hesse-Lipson method to determine the value of b from $hk0$ or $0kl$ reflections.   Consider the form of equation (32) when $l = 0$,

$$d_{hk0} = \cfrac{1}{\sqrt{\dfrac{h^2}{a^2\sin^2\beta} + \dfrac{k^2}{b^2}}}. \tag{33}$$

By analogy to equations (14), (15), (16), and (17), equation (33) can be alternatively written

$$\sin^2\theta_{hk0} = h^2A + k^2B, \tag{34}$$

where

$$\left.\begin{array}{l} A = \dfrac{\lambda^2}{4a^2\sin^2\beta} \\[2ex] B = \dfrac{\lambda^2}{4b^2}. \end{array}\right\} \tag{35}$$

and

Similarly,

$$\sin^2\theta_{0kl} = k^2B + l^2C, \tag{36}$$

where

$$\left.\begin{array}{l} B = \dfrac{\lambda^2}{4b^2} \\[2ex] C = \dfrac{\lambda^2}{4c^2\sin^2\beta}. \end{array}\right\} \tag{37}$$

and

A procedure identical to the one described for the orthorhombic system can now be used to determine the values of $\sin^2 \theta_{h00}$, $\sin^2 \theta_{0k0}$, and $\sin^2 \theta_{00l}$ which, in turn, can be used to identify $hk0$ and $0kl$ reflections. Using this procedure, the length of b can be determined from either (34) or (36).

The angle β can be determined only from $h0l$ or hkl reflections, i.e., the remaining unindexed reflections. Equation (32) can be rewritten

$$\frac{1}{d_{hkl}^2} = \frac{h^2}{a^2 \sin^2 \beta} + \frac{l^2}{c^2 \sin^2 \beta} + \frac{2hl}{ac} \frac{\cos \beta}{\sin^2 \beta} + \frac{k^2}{b^2}$$

or, in the Hesse-Lipson notation,

$$\sin^2 \theta_{hkl} = h^2 A + l^2 C + 2hl \sqrt{AC} \cos \beta + k^2 B, \tag{38}$$

where A, B, and C have the same meaning as in (35) and (37). If the values of A, B, and C are known from the procedure described above using $hk0$ and $0kl$ reflections, (38) can be used by trial-and-error procedures to determine β and, consequently, a and c. Unfortunately, this procedure is not too easily implemented in practice. The inability to distinguish $hk0$ and $0kl$ reflections from the more general hkl reflections in the preliminary stages complicates the application of (34) and (36). Consequently, procedures using the *reciprocal-lattice* concept are recommended·

Triclinic system

In the triclinic system the relation between d and (hkl) is given by the last entry in Table 1, Chapter 6.

The number of unit-cell constants requiring definition increases from four in the monoclinic system to six in the triclinic. It is not surprising, therefore, that the methods described above cannot be used to advantage to determine the values of these constants. A workable method for indexing photographs of triclinic crystals is described in Chapter 10, where procedures for indexing crystals regardless of their symmetry are discussed.

Literature

Analytical indexing procedures

[1] D. E. Thomas, *A slide rule for x-ray diffraction by cubic crystals*, J. Sci. Instr. **18** (1941) 205.

[2] Vladimir Vand, *Indexing method of powder photographs of long-spacing compounds*, Acta Cryst. **1** (1948) 109–115.

[3] R. Hesse, *Indexing powder photographs of tetragonal, hexagonal and orthorhombic crystals*, Acta Cryst. **1** (1948) 200–207.

[4] Vladimir Vand, *A third graphical method of indexing powder photographs of long-spacing compounds*, Acta Cryst. **1** (1948) 290–291.

[5] H. Lipson, *Indexing powder photographs of orthorhombic crystals*, Acta Cryst. **2** (1949) 43–45.

[6] A. J. Stosick, *A method for indexing powder photographs, using linear diophantine equations, and some tests for crystal classes*, Acta Cryst. **2** (1949) 271–277.

[7] N. F. M. Henry, H. Lipson, and W. A. Wooster, *The interpretation of x-ray diffraction photographs* (D. Van Nostrand Company, Inc., Amsterdam, 1951), especially 181–184.

[8] Riccardo Ferro, *Differential crystallography of powders. I. Interpretation of Debye-Scherrer reflections*, Atti reale accad. nazl. Lincei **15** (1953) 285–289.

[9] Riccardo Ferro, *Differential crystallography of powders. II. First approximation of the differential analysis*, Atti reale acad. nazl. Lincei **15** (1953) 404–414.

[10] D. R. Das Guptai, *An x-ray study of $Na_2SO_4\,III$*, Acta Cryst. **7** (1954) 275–276.

9

The reciprocal lattice

Reciprocal representation of planes

In the introductory part of this book, the diffraction of x-rays by a crystal was described as reflection from individual planes in the crystal according to Bragg's law. Although it is easy to visualize such reflection by any given plane, it becomes increasingly more difficult to retain such an accurate picture as more and more planes are considered. A two-dimensional plane can be equally well represented by its normal, which has only one dimension. The direction of the normal, of course, specifies the orientation of the plane. If the length assigned to each normal is proportional to the reciprocal of the interplanar spacing of that plane, the points at the end of the normals (drawn from a common origin) form a lattice which is called the *reciprocal lattice*.

The reciprocal lattice is an invaluable aid to the interpretation of the diffraction of x-rays by single crystals. In the powder method it has had limited application because an individual reciprocal lattice must be associated with each individual crystal in the powder. Nevertheless, the concept can be used to simplify greatly the indexing procedures for a powder photograph, particularly when the crystal belongs to a system of low symmetry, such as the monoclinic or triclinic systems.

Graphical demonstration of the reciprocal lattice in two dimensions

An understanding of the relation of the reciprocal lattice to the direct lattice can be easily acquired by considering how a zone of planes can be represented by points on their normals. Since the planes to be considered are parallel to a common line (the *zone axis*), the normals to the planes are confined to a plane perpendicular to this line. Such a situation is pictured in Fig. 1, which shows the unit cell of a monoclinic crystal looking along its unique axis, here designated b. The cell edges seen in Fig. 1 are accordingly a and c. The illustration also shows four $(h0l)$ planes, namely (100), (101), (102), and (001). Since all these planes are parallel to b, their normals lie in the plane of the paper.

Now consider finding the points which represent these planes. The rules for finding these points are as follows:

1. From a common origin, erect a normal to each plane.
2. Place a point on the normal to each plane (hkl) at a distance from the origin equal to $1/d_{hkl}$.

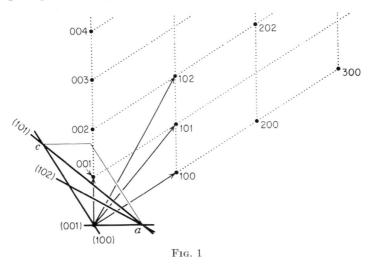

FIG. 1

Each of the points preserves all the important characteristics of the stack of planes it represents. The direction of the point from the origin preserves the orientation of the plane, and the distance of the point from the origin preserves the interplanar spacing of that stack of planes. The student should perform the actual construction diagramed in Fig. 1. When this is done it becomes apparent that the points labeled 100, 101, and 102 lie along a straight line which is perpendicular to the crystal plane (001). In a similar way, if points are located for each possible plane $(h0l)$, the set of points is found to lie on a two-dimensional lattice.[†]

Vector-algebraic discussion of the reciprocal lattice

The reciprocal lattice has just been demonstrated by direct construction. It will now be shown that the collection of points in three dimen-

---

† A simple geometrical proof that the points *must* lie on a lattice is given in M. J. Buerger, *X-ray crystallography* (John Wiley & Sons, Inc., New York, 1942) 108–116.

sions conforming to conditions 1 and 2, above, is indeed a lattice. As an aid to this demonstration, the statement describing the reciprocal lattice given in the previous section can be rephrased as follows: to each plane (*hkl*) of the crystal, a normal is drawn of length

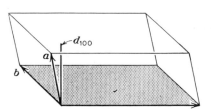

$$\sigma_{hkl} = \frac{1}{d_{hkl}}. \tag{1}$$

The sum total of points at the ends of these normals must be shown to lie on a lattice.

Fig. 2

Figure 2 shows the three axes *a*, *b*, *c* of a primitive crystal lattice. The volume of a unit cell in this lattice is clearly equal to the area of the shaded parallelogram whose sides are *b* and *c*, times the altitude, which is d_{100}:

$$V = \text{area} \cdot d_{100},$$

so
$$\frac{1}{d_{100}} = \frac{\text{area}}{V}. \tag{2}$$

In vector notation, the normal to a plane is represented by the unit vector **n**, so that (1) can be written

$$\mathbf{\delta}_{hkl} = \frac{1}{d_{hkl}} \mathbf{n}. \tag{3}$$

An area is represented by the vector product of its sides, so that (2) can be written

$$\frac{1}{d_{100}} \mathbf{n} = \frac{\mathbf{b} \times \mathbf{c}}{V}. \tag{4}$$

Combining (3) and (4), and expressing the volume in vector form, one obtains

$$\mathbf{\delta}_{100} = \frac{1}{d_{100}} \mathbf{n}$$
$$= \frac{\mathbf{b} \times \mathbf{c}}{\mathbf{a} \cdot \mathbf{b} \times \mathbf{c}}. \tag{5}$$

By analogy, similar expressions can be written for $\mathbf{\delta}_{010}$ and $\mathbf{\delta}_{001}$. These three vectors are chosen as the three reciprocal axes,†

---

† It is customary to indicate that a geometrical feature pertains to the reciprocal lattice by adding an asterisk to its symbol. The asterisk is usually called a "star." Thus a^* is pronounced "*a*-star."

$$\mathbf{a}^* \equiv \boldsymbol{\delta}_{100} = \frac{\mathbf{b} \times \mathbf{c}}{\mathbf{a} \cdot \mathbf{b} \times \mathbf{c}},$$

$$\mathbf{b}^* \equiv \boldsymbol{\delta}_{010} = \frac{\mathbf{c} \times \mathbf{a}}{\mathbf{a} \cdot \mathbf{b} \times \mathbf{c}}, \qquad (6)$$

$$\mathbf{c}^* \equiv \boldsymbol{\delta}_{001} = \frac{\mathbf{a} \times \mathbf{b}}{\mathbf{a} \cdot \mathbf{b} \times \mathbf{c}}.$$

The reciprocal axes bear a simple relationship to the crystal axes which follows from the vector notation of (6):

$$\begin{aligned} &\mathbf{a}^* \text{ is normal to } \mathbf{b} \text{ and } \mathbf{c},\\ &\mathbf{b}^* \text{ is normal to } \mathbf{c} \text{ and } \mathbf{a},\\ &\mathbf{c}^* \text{ is normal to } \mathbf{a} \text{ and } \mathbf{b}. \end{aligned} \qquad (7)$$

These conditions require the following vector relations:

$$\begin{array}{ll} \mathbf{a}^* \cdot \mathbf{b} = 0, & \mathbf{a}^* \cdot \mathbf{c} = 0\\ \mathbf{b}^* \cdot \mathbf{c} = 0, & \mathbf{b}^* \cdot \mathbf{a} = 0\\ \mathbf{c}^* \cdot \mathbf{a} = 0, & \mathbf{c}^* \cdot \mathbf{b} = 0. \end{array} \qquad (8)$$

These relations can also be derived by forming the scalar product of both sides of the first relation of (6) with $\mathbf{b}$ and with $\mathbf{c}$, etc. A corollary to the three conditions in (8) are the three relations

$$\begin{aligned} \mathbf{a}^* \cdot \mathbf{a} &= 1\\ \mathbf{b}^* \cdot \mathbf{b} &= 1\\ \mathbf{c}^* \cdot \mathbf{c} &= 1. \end{aligned} \qquad (9)$$

These relations can also be derived by forming the scalar product of both sides of the first relation of (6) with $\mathbf{a}$, etc.

If a lattice is constructed using the reciprocal-lattice vectors of (6) it follows that successive points in the $\mathbf{a}^*$ direction represent successive submultiples h of the spacing of (100); in the $\mathbf{b}^*$ direction, successive submultiples k of the spacing of (010); and in the $\mathbf{c}^*$ direction, successive submultiples l of the spacing of (001). That this is so is clearly evident from (3) since

$$\begin{aligned} \mathbf{a}^* = \ \ \boldsymbol{\delta}_{100} &= \frac{1}{d_{100}} \mathbf{n}\\[4pt] 2\mathbf{a}^* = 2\boldsymbol{\delta}_{100} &= \frac{2}{d_{100}} \mathbf{n}\\[4pt] = \ \ \boldsymbol{\delta}_{200} &= \frac{1}{d_{200}} \mathbf{n}\\[4pt] 3\mathbf{a}^* = 3\boldsymbol{\delta}_{100} &= \frac{3}{d_{100}} \mathbf{n}\\[4pt] = \ \ \boldsymbol{\delta}_{300} &= \frac{1}{d_{300}} \mathbf{n}. \end{aligned} \qquad (10)$$

To reach any reciprocal-lattice point *hkl* one goes *h* units along $\mathbf{a}^*$, *k* units along $\mathbf{b}^*$, and *l* units along $\mathbf{c}^*$. Accordingly, the reciprocal-lattice vector $\boldsymbol{\sigma}_{hkl}$ can be written in vector notation

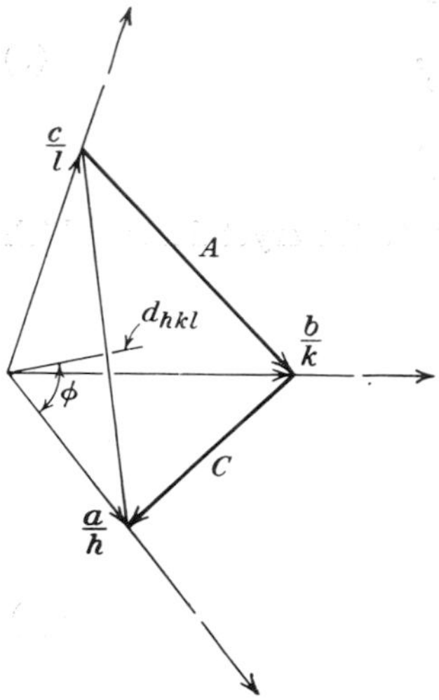

$$\boldsymbol{\sigma}_{hkl} = h\mathbf{a}^* + k\mathbf{b}^* + l\mathbf{c}^*. \tag{11}$$

It can now be shown that the set of points at the ends of the collection of vectors $\boldsymbol{\sigma}_{hkl}$ conforms to conditions 1 and 2. The simplest way to prove that $\boldsymbol{\sigma}_{hkl}$ is normal to the crystal plane (*hkl*) is to show that the scalar products of $\boldsymbol{\sigma}_{hkl}$ and vectors lying in the (*hkl*) plane vanish. The plane (*hkl*), shown in Fig. 3, intercepts $\mathbf{a}$ at $\mathbf{a}/h$, $\mathbf{b}$ at $\mathbf{b}/k$, $\mathbf{c}$ at $\mathbf{c}/l$. Consider the vector

$$\mathbf{C} = \frac{\mathbf{a}}{h} - \frac{\mathbf{b}}{k} \tag{12}$$

Fig. 3

lying in this plane. The scalar product of $\mathbf{C}$ in (12) with $\boldsymbol{\sigma}_{hkl}$ in (11) is

$$
\begin{aligned}
\mathbf{C} \cdot \boldsymbol{\sigma}_{hkl} &= \left(\frac{\mathbf{a}}{h} - \frac{\mathbf{b}}{k}\right) \cdot (h\mathbf{a}^* + k\mathbf{b}^* + l\mathbf{c}^*) \\
&= \frac{\mathbf{a}}{h}(h\mathbf{a}^* + k\mathbf{b}^* + l\mathbf{c}^*) - \frac{\mathbf{b}}{k}(h\mathbf{a}^* + k\mathbf{b}^* + l\mathbf{c}^*) \\
&= \left(\frac{h}{h} + 0 + 0\right) - \left(0 + \frac{k}{k} + 0\right) \\
&= 1 - 1 \\
&= 0. \tag{13}
\end{aligned}
$$

Similarly, the scalar product of the vector $\mathbf{A} = \mathbf{b}/k - \mathbf{c}/l$ with $\boldsymbol{\sigma}_{hkl}$ vanishes also:

$$
\begin{aligned}
\mathbf{A} \cdot \boldsymbol{\sigma}_{hkl} &= \left(\frac{\mathbf{b}}{k} - \frac{\mathbf{c}}{l}\right) \cdot (h\mathbf{a}^* + k\mathbf{b}^* + l\mathbf{c}^*) \\
&= \left(0 + \frac{k}{k} + 0\right) - \left(0 + 0 + \frac{l}{l}\right) \\
&= 1 - 1 \\
&= 0. \tag{14}
\end{aligned}
$$

Since, according to (13) and (14), $\boldsymbol{\sigma}_{hkl}$ is normal to $\mathbf{C}$ and $\mathbf{A}$, it is normal to the plane containing $\mathbf{C}$ and $\mathbf{A}$, that is, to the plane (*hkl*). In view of this, $\mathbf{n}$, the unit vector normal to (*hkl*), is parallel to $\boldsymbol{\sigma}_{hkl}$, so that (11) can be written

$$|\boldsymbol{\sigma}_{hkl}|\mathbf{n} = h\mathbf{a}^* + k\mathbf{b}^* + l\mathbf{c}^* \tag{15}$$

from which it follows that

$$\mathbf{n} = \frac{(h\mathbf{a}^* + k\mathbf{b}^* + l\mathbf{c}^*)}{|\boldsymbol{\delta}_{hkl}|}. \tag{16}$$

The length of the interplanar spacing of the plane shown in Fig. 3 is obviously

$$d_{hkl} = \frac{a}{h} \cos \varphi$$

$$= \frac{\mathbf{a}}{h} \cdot \mathbf{n}$$

$$= \frac{\mathbf{a}}{h} \cdot \frac{(h\mathbf{a}^* + k\mathbf{b}^* + l\mathbf{c}^*)}{|\boldsymbol{\delta}_{hkl}|}$$

$$= \frac{1}{|\boldsymbol{\delta}_{hkl}|}, \tag{17}$$

verifying relation (1).

It is a relatively simple matter to show that expressions similar to (6) can be used to express the crystal-lattice axes in terms of the reciprocal-lattice axes. Consider the reciprocal of the reciprocal-lattice vector

$$(\mathbf{a}^*)^* = \frac{\mathbf{b}^* \times \mathbf{c}^*}{\mathbf{a}^* \cdot \mathbf{b}^* \times \mathbf{c}^*}. \tag{18}$$

Multiplying the right member by $\mathbf{a} \cdot \mathbf{a}^*$ which, according to (8), is unity:

$$(\mathbf{a}^*)^* = \mathbf{a} \cdot \mathbf{a}^* \cdot \frac{\mathbf{b}^* \times \mathbf{c}^*}{\mathbf{a}^* \cdot \mathbf{b}^* \times \mathbf{c}^*}$$

$$= \mathbf{a} \cdot \frac{\mathbf{a}^* \cdot \mathbf{b}^* \times \mathbf{c}^*}{\mathbf{a}^* \cdot \mathbf{b}^* \times \mathbf{c}^*}$$

$$= \mathbf{a}. \tag{19}$$

Combining (18) and (19), it follows that

$$\mathbf{a} = \frac{\mathbf{b}^* \times \mathbf{c}^*}{\mathbf{a}^* \cdot \mathbf{b}^* \times \mathbf{c}^*}$$

and similarly,
$$\mathbf{b} = \frac{\mathbf{c}^* \times \mathbf{a}^*}{\mathbf{a}^* \cdot \mathbf{b}^* \times \mathbf{c}^*}, \tag{20}$$

$$\mathbf{c} = \frac{\mathbf{a}^* \times \mathbf{b}^*}{\mathbf{a}^* \cdot \mathbf{b}^* \times \mathbf{c}^*}$$

There are other useful relationships between the elements of the crystal lattice and their counterparts in the reciprocal lattice. A detailed derivation of them is beyond the scope of this book. Since some of these relations may be necessary in the chapters that follow, Table 1 lists these relations expressed for the most general lattice. The relations for more symmetrical lattices are easily derived from these by substitution of

Table 1
Relations between the dimensions of the direct and reciprocal cells

Angular parameters

$$\cos \alpha^* = \frac{\cos \beta \cos \gamma - \cos \alpha}{\sin \beta \sin \gamma}$$

$$\cos \beta^* = \frac{\cos \gamma \cos \alpha - \cos \beta}{\sin \gamma \sin \alpha}$$

$$\cos \gamma^* = \frac{\cos \alpha \cos \beta - \cos \gamma}{\sin \alpha \sin \beta}$$

$$\cos \alpha = \frac{\cos \beta^* \cos \gamma^* - \cos \alpha^*}{\sin \beta^* \sin \gamma^*}$$

$$\cos \beta = \frac{\cos \gamma^* \cos \alpha^* - \cos \beta^*}{\sin \gamma^* \sin \alpha^*}$$

$$\cos \gamma = \frac{\cos \alpha^* \cos \beta^* - \cos \gamma^*}{\sin \alpha^* \sin \beta^*}$$

Linear parameters

$$a^* = \frac{bc \sin \alpha}{V}$$

$$b^* = \frac{ca \sin \beta}{V}$$

$$c^* = \frac{ab \sin \gamma}{V}$$

$$a = \frac{b^*c^* \sin \alpha^*}{V^*}$$

$$b = \frac{c^*a^* \sin \beta^*}{V^*}$$

$$c = \frac{a^*b^* \sin \gamma^*}{V^*}$$

Volume

$$V^* = a^*b^*c^* \sqrt{1 - \cos^2 \alpha^* - \cos^2 \beta^* - \cos^2 \gamma^* + 2 \cos \alpha^* \cos \beta^* \cos \gamma^*}$$

$$V = abc \sqrt{1 - \cos^2 \alpha - \cos^2 \beta - \cos^2 \gamma + 2 \cos \alpha \cos \beta \cos \gamma}$$

appropriate values for α, β, and γ.   An example of an actual set of computations based upon the relations in Table 1 is presented at the end of this chapter.

Interpretation of Bragg's law in terms of the reciprocal lattice

The reciprocal lattice provides a convenient way of dealing with x-ray diffraction phenomena.[1,2]   The Bragg condition for the diffraction of x-rays by a stack of planes (hkl) in a crystal can be written to give the glancing angle θ_{hkl} in terms of the variables involved,

$$\sin \theta_{hkl} = \frac{\lambda/2}{d_{hkl}} = \frac{1/d_{hkl}}{2/\lambda}. \tag{21}$$

A direct geometrical interpretation of (21) is given in Fig. 4.   Here θ is the angle between the diameter of a circle of radius $1/\lambda$ and the line

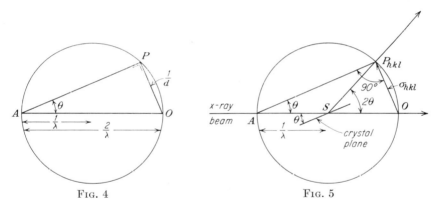

FIG. 4   FIG. 5

drawn to the end of the line OP representing the reciprocal-lattice vector of length $1/d_{hkl}$.   With this construction, a physical interpretation of x-ray diffraction can be given as follows:

1. Let AO be taken not only as a length, but also as the direction of the x-ray beam; then, since AP makes the angle θ with AO, AP is the slope of the reflecting crystal plane.

2. OP is the normal to the reflecting plane ($\angle APO = 90°$) and hence it has the direction of the reciprocal-lattice vector σ_{hkl} drawn from the origin to the reciprocal-lattice point P.   Its length is also $1/d_{hkl} = \sigma_{hkl}$.

3. The above two conditions are indicated in Fig. 5.   It is also clear from this figure that $\angle OSP = 2\angle OAP = 2\theta$; hence the vector from the center of the circle S to the reciprocal-lattice point P_{hkl} represents the direction of the x-ray reflection.

The following can now be said about the meaning of Fig. 5:

1. The crystal can be pictured as located at the center of the circle S.

2. The point O where the direct beam leaves the circle is the origin of the reciprocal lattice, which is oriented so that every σ_{hkl} is normal to plane (hkl).

3. Whenever a plane (hkl) of the crystal makes the angle θ with the direct-beam direction, the reciprocal-lattice point hkl at the end of σ_{hkl} lies on the circumference of the circle, and the reflected x-ray beam passes through this point.

4. Diffraction can only occur, therefore, when a reciprocal-lattice point touches the circle. (This placement of the reciprocal-lattice point is accomplished by proper orientation of the crystal and its associated reciprocal lattice.)

In three dimensions, of course, the circle of Fig. 5 becomes a sphere, appropriately called the *sphere of reflection*. The construction of Fig. 5, therefore, represents a diametral plane of the sphere.

The collection of reciprocal lattices of the crystals in a powder

The reciprocal-lattice relations developed in the last section can be used in the solution of problems arising in the powder method. Each crystal may be envisaged as a reciprocal lattice. Since the several crystals in a powder are randomly oriented, the reciprocal lattices associated with them are randomly oriented also. The origins of all these lattices, however, lie at the point where the direct beam leaves the sphere of reflection.

Consider the reciprocal-lattice point hkl at the end of the vector σ_{hkl} in Fig. 5. If there are an infinite number of crystals in the powder, there must be an infinite number of such points hkl, all lying at a vector distance σ_{hkl} from the origin. Since these vectors are randomly directed in space, the reciprocal-lattice points hkl must lie on a sphere centered at the origin. This is obviously true of any reciprocal-lattice point hkl. The sum total of all reciprocal-lattice points, therefore, comprises a set of concentric spheres of radii σ_{hkl} centered at an origin which lies on the sphere of reflection. These reciprocal-lattice spheres, therefore, intersect the sphere of reflection in small circles. Since, as was shown in the last section, a diffracted beam develops whenever a reciprocal-lattice point intersects the sphere of reflection, the diffracted beams form cones emanating from the center of the sphere of reflection, as illustrated in Fig. 6 for one such cone.

It is clear from Fig. 5 that one can determine all the angles of reflection 2θ for crystals whose reciprocal lattices are known. From this it might appear, offhand, that the converse is true, that is, that once all angles of reflection are known, the reciprocal lattice could be reconstructed from this knowledge. Actually, only the magnitude of σ_{hkl} can be determined experimentally. The directions in reciprocal space of the *set of vectors* σ_{hkl} collectively for all the crystals in the powder can be determined. It is not possible, however, to resolve a particular σ_{hkl} for any one crystal. Thus, it would appear at first sight that the reciprocal lattice is of limited value in the interpretation of experimentally determined diffraction directions, and cannot be used for assigning appropriate indices to the lines in a powder photograph.

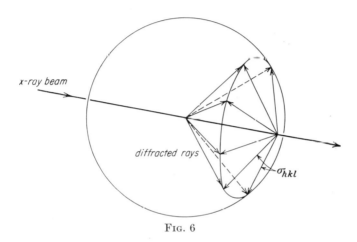

x-ray beam

diffracted rays

σ_{hkl}

FIG. 6

It is possible, however, to solve this problem indirectly. A reciprocal lattice whose vectors σ_{hkl} have magnitudes identical to the experimentally observed vector magnitudes is first postulated. A reciprocal unit cell that exhibits the true symmetry of the lattice can then be selected. This makes it possible to determine the correct unit cell in *crystal space* and to assign indices to the lines in a powder photograph. A systematic method for doing this is described in the next chapter.

Determination of the reciprocal lattice of crystals in a powder

In the previous section, it was shown how the lengths of the reciprocal-lattice vectors σ_{hkl} are recorded on a powder photograph. Having measured these lengths, it is possible, in principle, to determine the correct reciprocal lattice from this information.[3] To understand the principle, consider the two-dimensional analogue (Fig. 7). In two dimensions, concentric circles are drawn whose radii are the observed absolute values of the reciprocal-lattice vectors σ_{hkl}. After one of these radii is selected as the length of a trial axis a^*, its direction is fixed by drawing an arbitrary (say vertical) line through the origin. The values of ha^* are marked off as distances along this line. These are reciprocal-lattice points which correspond to multiples of the fundamental reciprocal translation a^*; they should fall on one or more of the prescribed circles. Next, the radius of another circle is selected as the length of a second trial axis. To do this one proceeds as before, except that the direction of this line is not known since the interaxial angle is unknown. To determine this angle, a second trial reciprocal-lattice row is rotated about the origin

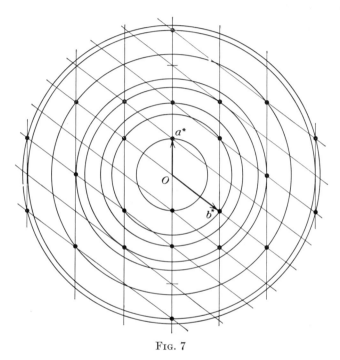

Fig. 7

until the points of the lattice defined by the two selected axes fall on the circles. A lattice thus determined is illustrated in Fig. 7, which shows the coincidence of the points of the proposed lattice and the circles of the observed absolute values of the reciprocal-lattice vectors. As can be seen in this figure, the unit cell selected defines an acceptable lattice, although an alternative unit cell may better describe the symmetry of the reciprocal lattice.

In three dimensions the circles become spheres, and the problem of locating three vectors to define the reciprocal lattice becomes impracticable. This problem, however, can be treated analytically and can be solved uniquely in all cases where accurately measured spacings are available. This analytical procedure is described in the next chapter.

Example of transformation computation

It has just been shown that a diffraction record, such as a powder photograph, supplies information about the geometry of the reciprocal lattice. In the next chapter the way this geometry is ferreted out of a powder photograph is discussed in detail. The result of the work, however, is a knowledge of the reciprocal-cell dimensions. To put this information on a more useful basis, the reciprocal-cell dimensions must be transformed into direct-cell dimensions. This is customarily done with the relations in Table 1.

To illustrate the practical nature of these relations, an actual transformation computation is presented below in detail. The reciprocal-cell data are from (51) of Chapter 10, which are results obtained from an analysis of a powder photograph of $MgWO_4$. These are as follows:

$$
\begin{aligned}
a^{*2} &= .0310 & a^* &= .1761 & \alpha^* &= 89.5° \\
b^{*2} &= .0454 & b^* &= .2131 & \beta^* &= 49.0° \\
c^{*2} &= .0721 & c^* &= .2685 & \gamma^* &= 90.0°.
\end{aligned}
\tag{22}
$$

The sines and cosines of α^*, β^*, and γ^* are needed for some of the relations in Table 1, so that the initial data for computing these relations are

$$
\begin{aligned}
a^* &= .1761 & \sin \alpha^* &= 1.0000 & \cos \alpha^* &= .0087 \\
b^* &= .2131 & \sin \beta^* &= .7547 & \cos \beta^* &= .6561 \\
c^* &= .2685 & \sin \gamma^* &= 1.0000 & \cos \gamma^* &= 0.
\end{aligned}
\tag{23}
$$

The direct-cell angles are functions of the reciprocal-cell angles only, and are computed as follows:

$$
\begin{aligned}
\cos \alpha &= \frac{\cos \beta^* \cos \gamma^* - \cos \alpha^*}{\sin \beta^* \sin \gamma^*} \\
&= \frac{0 - .0087}{(.7547)(1.0000)} \\
&= -.0115 \\
\alpha &= 90°40'
\end{aligned}
\tag{24}
$$

$$
\begin{aligned}
\cos \beta &= \frac{\cos \gamma^* \cos \alpha^* - \cos \beta^*}{\sin \gamma^* \sin \alpha^*} \\
&= \frac{0 - .6561}{(1.0000)(1.0000)} \\
&= -.6561 \\
\beta &= 131°0'
\end{aligned}
\tag{25}
$$

$$\cos \gamma = \frac{\cos \alpha^* \cos \beta^* - \cos \gamma^*}{\sin \alpha^* \sin \beta^*}$$

$$= \frac{(.0087)(.6561) - 0}{(1.0000)(.7547)}$$

$$= +.0076$$

$$\gamma = 89°34'. \tag{26}$$

In order to compute the direct-cell edges, the reciprocal volume must first be computed. Although this involves the product $a^*b^*c^*$, substitution of numerical values for these quantities is not made, since pairs of them will cancel in later substitution. The reciprocal volume is given by

$$V^* = a^*b^*c^* \sqrt{1 - \cos^2 \alpha^* - \cos^2 \beta^* - \cos^2 \gamma^* + 2 \cos \alpha^* \cos \beta^* \cos \gamma^*}$$

$$= a^*b^*c^* \sqrt{1 - .0001 - .4305 - 0 + 0}$$

$$= a^*b^*c^* \sqrt{.5694}$$

$$= .7546a^*b^*c^*. \tag{27}$$

Using this value of V^*, the direct-cell edges can be computed:

$$a = \frac{b^*c^* \sin \alpha^*}{V^*}$$

$$= \frac{b^*c^* \sin \alpha^*}{.7546a^*b^*c^*}$$

$$= \frac{\sin \alpha^*}{.7546a^*}$$

$$= \frac{1.0000}{.7546(.1761)}$$

$$= 7.524 \text{ Å}. \tag{28}$$

Similarly,

$$b = \frac{\sin \beta^*}{.7546b^*}$$

$$= \frac{.7547}{.7546(.2131)}$$

$$= 4.693 \text{ Å}, \tag{29}$$

and

$$c = \frac{\sin \gamma^*}{.7546c^*}$$

$$= \frac{1.0000}{.7546(.2685)}$$

$$= 4.936 \text{ Å}. \tag{30}$$

The direct-cell dimensions resulting from this transformation can now be assembled as follows:

$$a = 7.524 \text{ Å} \qquad \alpha = 90°40'$$
$$b = 4.693 \qquad\quad \beta = 131°0' \tag{31}$$
$$c = 4.936 \qquad\quad \gamma = 89°34'.$$

Literature

[1] P. P. Ewald, *Das "reziproke Gitter" in der Strukturtheorie*, Z. Krist. (A) **56** (1921) 148–150.

[2] J. D. Bernal, *On the interpretation of x-ray single-crystal rotation photographs* Proc. Roy. Soc. London (A) **113** (1926) 120–123.

[3] T. Ito, *X-ray studies on polymorphism* (Maruzen Co., Ltd., Tokyo, 1950) 210–214.

[4] W. L. Bond, *Nomographs for triclinic cell computations*, Am. Mineralogist **35** (1950) 239–244.

10

Indexing powder photographs with the aid of the reciprocal lattice

Methods for assigning indices to powder photographs of crystals belonging to the isometric, tetragonal, hexagonal, and orthorhombic systems have been described in Chapter 8. For all but isometric crystals, these methods are iterative since more than one unknown is involved. Moreover, these methods become increasingly more difficult to use as the number of unknowns increases from two to six.

A method for indexing powder photographs regardless of symmetry was presented by Ito[1] in 1948. Ito's method is based on the fact that each reflection of a powder photograph corresponds to a vector in reciprocal space. If three noncoplanar vectors are selected, the edges of a unit cell are defined, and if three more are selected, their interaxial angles are fixed. This corresponds to the selection of six appropriate lines on the powder photograph. Ito's method is concerned with the strategy of selecting these six lines. Once an appropriate unit cell is selected, it becomes possible to index all the lines on the powder photograph.

Experience has shown that, whenever the symmetry of the crystal is not known, Ito's method provides a comparatively rapid procedure for indexing a powder photograph, particularly if the crystal is monoclinic or triclinic. As part of the procedure of applying Ito's method, it is advisable to make successive tests to determine whether the crystal is isometric, tetragonal, or hexagonal. After the experimental measurements have been expressed in a suitable form, these tests can be used to limit the attendant computations. In this chapter, Ito's method is first described and then there follows an outline of a practical procedure for indexing a powder photograph whenever the symmetry is not known.

Determination of the correct reciprocal lattice

Notation. In the general case, the reciprocal-lattice vector $\boldsymbol{\sigma}_{hkl}$ is related to the unit-cell dimensions a, b, c, α, β, γ in crystal space by the complicated expression shown in Table 1, Chapter 6. On the other hand, the reciprocal-lattice vector $\boldsymbol{\sigma}_{hkl}$ is related to the reciprocal-cell dimensions a^*, b^*, c^*, α^*, β^*, γ^* by the relatively simpler expression

$$
\begin{aligned}
\sigma_{hkl}^2 = {} & h^2 a^{*2} + k^2 b^{*2} + l^2 c^{*2} \\
& + 2hka^*b^* \cos \gamma^* \\
& + 2klb^*c^* \cos \alpha^* \\
& + 2lhc^*a^* \cos \beta^*.
\end{aligned}
\tag{1}
$$

Since most manipulations of powder-photograph data in reciprocal space involve terms σ_{hkl}^2, it is customary to rewrite this as Q_{hkl}, so that

$$
\begin{aligned}
Q_{hkl} \equiv {} & \sigma_{hkl}^2 \\
= {} & h^2 a^{*2} + k^2 b^{*2} + l^2 c^{*2} \\
& + 2hka^*b^* \cos \gamma^* \\
& + 2klb^*c^* \cos \alpha^* \\
& + 2lhc^*a^* \cos \beta^*.
\end{aligned}
\tag{2}
$$

By virtue of (17) of Chapter 9, it follows that

$$
Q_{hkl} = \frac{1}{d^2_{hkl}}.
\tag{3}
$$

If this is combined with (4), Chapter 2, Bragg's law can be written in the form

$$
Q_{hkl} = \frac{4 \sin^2 \theta_{hkl}}{\lambda^2}.
\tag{4}
$$

Since study and manipulation of powder-photograph data in reciprocal space involve Q's, it is convenient to transform such data to a list of Q's. A table for converting d's to Q's is given in Appendix 2. The indices corresponding to each Q are not known in advance; so these Q's are listed in order of increasing value, as shown in Table 2. Until indices are assigned, the nth Q value in the list is temporarily designated Q_n.

Determination of the reciprocal-cell edges. Any three noncoplanar vectors in reciprocal space can be chosen as the edges of a reciprocal unit cell. It does not follow that the unit cell determined by three arbitrarily chosen vectors is necessarily primitive, but such a set of vectors, if noncoplanar, does define a unit cell. The chances that the unit cell is primitive are enhanced if the vectors are short.

According to the last section of the last chapter, each powder line corresponds to a reciprocal-lattice vector. If three such lines are chosen, three reciprocal-cell vectors are defined, and this implies that a unit cell is defined if the vectors turn out to be noncoplanar. The first step in interpreting a powder photograph, therefore, is to select from the list of Q's three having small values to represent the reciprocal vectors $\mathbf{a}^*$, $\mathbf{b}^*$, and $\mathbf{c}^*$. Later tests will prove whether the vectors corresponding to the Q's chosen are noncoplanar and, if so, whether they define a primitive cell.

Determination of the reciprocal-cell interaxial angles. The way the interaxial angles may be determined is suggested by Fig. 1, which

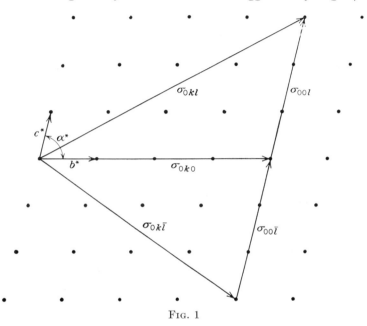

Fig. 1

shows the plane of the reciprocal lattice containing the vectors $\mathbf{b}^*$ and $\mathbf{c}^*$. In this diagram the vector $\boldsymbol{\sigma}_{0kl}$ can be decomposed as follows:

$$\boldsymbol{\sigma}_{0kl} = \boldsymbol{\sigma}_{0k0} + \boldsymbol{\sigma}_{00l} \tag{5}$$
$$= k\mathbf{b}^* + l\mathbf{c}^*. \tag{6}$$

The length of $\boldsymbol{\sigma}_{0kl}$ is given by

$$\sigma_{0kl}^2 = k^2 b^{*2} + l^2 c^{*2} + 2kl b^* c^* \cos \alpha^*. \tag{7}$$

If the σ^2's are written as Q's, according to (2), this relation can be rewritten

$$Q_{0kl} = k^2 Q_{010} + l^2 Q_{001} + 2kl b^* c^* \cos \alpha^*. \tag{8}$$

A straightforward way of determining α^* would be to solve (8) for $\cos \alpha^*$:

$$\cos \alpha^* = \frac{Q_{0kl} - k^2 Q_{010} - l^2 Q_{001}}{2kl b^* c^*}. \tag{9}$$

A related way of determining α^* makes use of $\delta_{0k\bar{l}}$: From Fig. 1, this vector is given by

$$\delta_{0k\bar{l}} = \delta_{0k0} - \delta_{00l}. \tag{10}$$

From a sequence of steps similar to (5), (6), (7), and (8), this can be transformed into

$$Q_{0k\bar{l}} = k^2 Q_{010} + l^2 Q_{001} - 2klb^*c^* \cos \alpha^*. \tag{11}$$

If (8) and (11) are added and subtracted, there results

adding: $\qquad\qquad Q_{0kl} + Q_{0k\bar{l}} = 2k^2 Q_{010} + 2l^2 Q_{001} \qquad\qquad (12)$

subtracting: $\qquad Q_{0kl} - Q_{0k\bar{l}} = 4klb^*c^* \cos \alpha^*. \qquad\qquad (13)$

The last relation can be solved for $\cos \alpha^*$:

$$\cos \alpha^* = \frac{Q_{0kl} - Q_{0k\bar{l}}}{4klb^*c^*}. \tag{14}$$

Both these methods of determining $\cos \alpha^*$ require the identification of Q_{0kl} or $Q_{0k\bar{l}}$, or both, from a list of unindexed Q's. Ito's method provides a means of identifying these two required Q's. According to (12), the average value of these two Q's is

$$Q'_{0kl} \equiv \frac{Q_{0kl} + Q_{0k\bar{l}}}{2} = k^2 Q_{010} + l^2 Q_{001}. \tag{15}$$

To solve (14), therefore, one must first search among the Q's to find two whose average value is $k^2 Q_{010} + l^2 Q_{001}$. For example, if the reflections 011 and 01$\bar{1}$ are not absent, then the average value of these Q's is $b^{*2} + c^{*2}$.

Similar relations hold for finding β^* and γ^*. A list of the useful relations is given in Table 1.

Verification of the reciprocal unit cell. Having selected a^*, b^*, c^* arbitrarily from the list of Q's of a powder photograph, and having determined α^*, β^*, γ^* for these axes, a reciprocal unit cell is defined and, therefore, the reciprocal lattice is determined. That the lattice determined by this unit cell is indeed the correct lattice for the powder under investigation can be verified by computing the complete set of Q_{hkl}'s according to equation (2) and comparing this set with the experimentally determined Q's. If the comparison is acceptable, this test is sufficient to prove the correctness of the assigned reciprocal lattice. However, the unit cell selected may not display the symmetry of the lattice. Procedures for transforming this cell into another cell which does display the desired symmetry are described in Chapters 11 and 12.

The successful application of Ito's method requires that the reciprocal-lattice translations selected as Q_{100}, Q_{010}, and Q_{001} be conjugate translations of the lattice, i.e., that the unit cell formed by these transla-

Table 1
Relations useful for finding α^*, β^*, and γ^*

		α^*		β^*		γ^*	
From general reflections	Inter-axial angle	$\cos\alpha^* = \dfrac{Q_{0kl} - Q_{0\bar{k}l}}{4klb^*c^*}$	(16A)	$\cos\beta^* = \dfrac{Q_{h0l} - Q_{\bar{h}0l}}{4lhc^*a^*}$	(16B)	$\cos\gamma^* = \dfrac{Q_{hk0} - Q_{h\bar{k}0}}{4hka^*b^*}$	(16C)
	Q'	$\overline{Q_{0kl} + Q_{0\bar{k}l}} = k^2 Q_{010} + l^2 Q_{001}$ $\equiv Q'_{0kl}$	(17A) (18A)	$\overline{Q_{h0l} + Q_{\bar{h}0l}} = l^2 Q_{001} + h^2 Q_{100}$ $\equiv Q'_{h0l}$	(17B) (18B)	$\overline{Q_{hk0} + Q_{\bar{h}k0}} = h^2 Q_{100} + k^2 Q_{010}$ $\equiv Q'_{hk0}$	(17C) (18C)
From special reflections	Inter-axial angle	$\cos\alpha^* = \dfrac{Q_{011} - Q_{01\bar{1}}}{4b^*c^*}$	(19A)	$\cos\beta^* = \dfrac{Q_{101} - Q_{\bar{1}01}}{4c^*a^*}$	(19B)	$\cos\gamma^* = \dfrac{Q_{110} - Q_{\bar{1}10}}{4a^*b^*}$	(19C)
	Q'	$\overline{Q_{011} + Q_{01\bar{1}}} = b^{*2} + c^{*2}$ $\equiv Q'_{011}$	(20A) (21A)	$\overline{Q_{101} + Q_{\bar{1}01}} = c^{*2} + a^{*2}$ $\equiv Q'_{101}$	(20B) (21B)	$\overline{Q_{110} + Q_{\bar{1}10}} = a^{*2} + b^{*2}$ $\equiv Q'_{110}$	(20C) (21C)

tions does not enclose a lattice point within the cell. If a nonprimitive cell is used to index the lines of a powder photograph, the list of computed indices contains only $1/n$th of the number of reflections that the observed list contains, where n is the multiplicity of the nonprimitive cell. If all reflections cannot be indexed, therefore, it must be assumed that a nonprimitive reciprocal cell has been chosen, and a new selection of reflections for 100, 010, and 001 must be made.

Practical aspects of Ito's method

Before proceeding to illustrate the application of Ito's method to an actual case, some practical aspects of applying this method will be considered. The difficulties encountered are twofold. In the first place, reciprocal-lattice points corresponding to weak or extinguished reflections are not recorded. The consequence of this is that, after Q_{100}, Q_{010}, Q_{001} have been originally selected from the observed Q's, and appropriate combinations of these Q's have been used to compute different Q_{hk0}, Q_{h0l}, Q_{0kl} values, the computed Q's may not account for all the observed values. When this is the case, it is because, with some of these Q's absent from the experimentally observed set, the ones selected and labeled Q_{010}, Q_{010}, and Q_{001} were not *all* correctly selected. In this event, it becomes necessary to consider other possible values for the pinacoidal Q's. The procedure for doing this is illustrated by actual examples in the following sections.

The other difficulty arises from the inaccuracy with which the Q's have been determined. It turns out, in practice, that the Q's must be known to three, or preferably to four, figures in order for this method to work readily. Since this implies that the interplanar spacings must be known to at least three places to the right of the decimal point, the necessity for accurate experimental measurements is obvious.

Illustration of the use of Ito's method

The data for $MgWO_4$ have been selected to illustrate the application of Ito's method. The interplanar spacings and corresponding Q values are listed in Table 2. The first three lines in this table are selected as reasonable choices for Q_{100}, Q_{010}, and Q_{001}. Before proceeding further it is advisable to compute, according to (3), the higher orders of these lines, i.e., Q_{200}, Q_{300}, Q_{400}, . . . , Q_{020}, Q_{040}, etc. If corresponding values are present among the observed Q's, the higher-order reflections can be used to correct the Q values of the first-order reflections selected, since the accuracy in measuring Q increases as θ increases. The Q's of lines 1, 2, and 3 in Table 2 and their higher orders are listed in Table 3.

Table 2
Observed Q values for MgWO$_4$†

Line	d	$Q = 1/d^2$	Line	d	$Q = 1/d^2$
1	5.68	.0310	21	1.754	.3429
2	4.68	.0457	22	1.735	.3322
3	3.70	.0730	23	1.724	.3364
4	3.607	.0769	24	1.708	.3428
5	2.928	.1165	25	1.702	.3451
6	2.902	.1187	26	1.689	.3505
7	2.841	.1239	27	1.652	.3664
8	2.462	.1649	28	1.639	.3723
9	2.426	.1699	29	1.617	.3824
10	2.346	.1816	30	1.578	.4016
11	2.260	.1957	31	1.565	.4083
12	2.194	.2077	32	1.502	.4432
13	2.170	.2123	33	1.499	.4450
14	2.047	.2386	34	1.491	.4498
15	2.026	.2436	35	1.473	.4608
16	1.993	.2517	36	1.465	.4659
17	1.975	.2563	37	1.448	.4768
18	1.892	.2793	38	1.434	.4863
19	1.862	.2884	39	1.426	.4918
20	1.806	.3065	40	1.423	.4938
			41	1.364	.5376

† Prepared from data in *Standard x-ray diffraction powder patterns*, Natl. Bur. Standards Circ. 539 (1953) 85.

Table 3
Selection of Q_{100}, Q_{010}, and Q_{001} from Table 2

Q_{hkl}	Computed	Observed	Error in Q_{h00}
Q_{100}	. . . . .	.0310	
Q_{200}	.1240	.1239	$-\dfrac{.0001}{4} = 0$
Q_{300}	.2790	.2793	$+\dfrac{.0003}{9} = 0$
Q_{010}	. . . . .	.0457	
Q_{020}	.1828	.1816	$-\dfrac{.0012}{4} = -.0003$
Q_{030}	.4113	.4083	$-\dfrac{.0030}{9} = -.0003$
Q_{001}	. . . . .	.0730	
Q_{002}	.2920	.2884	$-\dfrac{.0036}{4} = -.0009$
Q_{003}	.6570		

From Table 3, it is concluded that:

$$Q_{100} = .0310 \pm .0000 = .0310 = a^{*2}, \qquad a^* = .1761,$$
$$Q_{010} = .0457 - .0003 = .0454 = b^{*2}, \qquad b^* = .2131, \qquad (22)$$
$$Q_{001} = .0730 - .0009 = .0721 = c^{*2}, \qquad c^* = .2685.$$

For these particular values of Q_{100} and Q_{010} equation (20C) becomes

$$Q'_{110} = .0310 + .0454 = .0764. \qquad (23)$$

According to (21C) there should exist a pair of Q's in Table 3, namely, Q_{110} and $Q_{1\bar{1}0}$, which are respectively greater and less than Q'_{110} by the same amount. A check of Table 2 for a pair of Q's symmetrically disposed about Q'_{110} that satisfies equation (21C) reveals that line 4 has a Q value of .0769. According to equation (21C), this would indicate that $Q'_{110} = Q_{110}$, i.e., $\cos \gamma^* = 0$, or $\gamma^* = 90°$. To verify this conclusion, several other values of Q_{hk0} are computed and are compared with the Q values listed in Table 2. The agreement of the observed and computed values is shown in Table 4.

Table 4
Agreement of computed and observed Q_{hk0}'s for MgWO$_4$

Q_{hk0}	Computed	Observed	Difference
Q_{110}	.0764	.0769	+.0005
Q_{120}	.2126	.2123	−.0003
Q_{210}	.1694	.1699	+.0005
Q_{220}	.3056	.3065	+.0009
Q_{130}	.4396		
Q_{310}	.3244	.3249	+.0005
Q_{230}	.5326		
Q_{320}	.4606	.4608	+.0002

A rapid trial-and-error check of Table 2 and Table 4 shows that the agreement between the observed and computed Q's cannot be appreciably improved by changing the values of either a^{*2} or b^{*2}. It is concluded from this that:

$$\begin{aligned} a^{*2} &= .0310 & a^* &= .1761 \\ b^{*2} &= .0454 & b^* &= .2131 \end{aligned} \qquad \gamma^* = 90°,$$

Substituting Q_{010} and Q_{001} in (20A) it becomes

$$Q'_{011} = .0454 + .0721 = .1175. \qquad (24)$$

A check of Table 2 for a pair of Q's symmetrically disposed about Q'_{011} shows that lines 5 and 6 come closest to satisfying the relation of equation (21A). This requires

$$Q_{011} - Q'_{011} = Q'_{011} - Q_{01\bar{1}}. \qquad (25)$$

Substituting the Q value of line 6 for Q_{011},

$$Q_{011} - Q'_{011} = Q_6 - Q'_{011}$$
$$= .1187 - .1175$$
$$= .0012. \tag{26}$$

Similarly, substituting for $Q_{01\bar{1}}$ the Q value of line 5,

$$Q'_{011} - Q_{01\bar{1}} = Q'_{011} - Q_5$$
$$= .1175 - .1165$$
$$= .0010. \tag{27}$$

The slight discrepancy between the two differences above can be attributed to errors in the measured values of Q_{011} and $Q_{01\bar{1}}$ or in the computation of Q'_{011}. Since b^{*2} appears to be correctly known from the previous calculations, the assumed value of c^{*2} and, therefore, Q'_{011} should be modified. Thus, with the new values of $c^{*2} = .0722$ and $Q'_{011} = .1176$, the two sides of equation (25) become

$$Q_{011} - Q'_{011} = Q'_{011} - Q_{011}$$
$$.1187 - .1176 = .1176 - .1165$$
$$.0011 = .0011. \tag{28}$$

To determine α^*, Q_5 and Q_6 are substituted in equation (19A):

$$\alpha^* = \cos^{-1}\left(\frac{Q_{011} - Q_{01\bar{1}}}{4b^*c^*}\right)$$
$$= \cos^{-1}\left(\frac{.1187 - .1165}{4(.2131 \times .2687)}\right)$$
$$= 89.5°. \tag{29}$$

As a check, equations (8) and (11) can be used to compute Q_{0kl} and $Q_{0k\bar{l}}$ for several combinations of k and l. The computed and observed values are listed in Table 5.

Table 5
Agreement of computed and observed Q_{0kl}'s for MgWO$_4$

Q_{0kl}	Computed	Observed	Difference
Q_{011}	.1187	.1187	.0000
$Q_{01\bar{1}}$	.1165	.1165	+.0000
Q_{012}	.3364	.3364	.0000
$Q_{01\bar{2}}$	.3320	.3322	+.0002
Q_{021}	.2560	.2563	+.0003
$Q_{02\bar{1}}$	.2516	.2517	+.0001
Q_{022}	.4748	.4768	+.0020
$Q_{02\bar{2}}$	.4660	.4659	−.0001
Q_{013}	.6985		
Q_{031}	.4841	.4863	+.0022
$Q_{03\bar{1}}$	.4775	.4768	−.0007

A check of Table 5 shows that the agreement between the observed and computed values of Q_{okl} is satisfactory. The following parameters, therefore, have been determined so far:

$$
\begin{aligned}
a^{*2} &= .0310 & a^* &= .1761 & \alpha^* &= 89.5° \\
b^{*2} &= .0454 & b^* &= .2131 & & \\
c^{*2} &= .0722 & c^* &= .2687 & \gamma^* &= 90°.
\end{aligned}
\tag{30}
$$

The only remaining unknown angle, β^*, can now be determined with the help of equations (19B), (20B), and (21B). Substituting the values of Q_{100} and Q_{001} in (21B), it becomes

$$
\begin{aligned}
Q'_{101} &= Q_{100} + Q_{001} \\
&= .0310 + .0722 \\
&= .1032.
\end{aligned}
\tag{31}
$$

An inspection of Table 2 fails to yield two lines that will satisfy the condition of (20B). This means that either Q_{101} or $Q_{10\bar{1}}$ has not been observed. Accordingly, another order of Q_{00l} is selected to substitute in (18B), e.g., Q_{002}. This gives

$$
\begin{aligned}
Q'_{102} &= Q_{100} + Q_{002} \\
&= .0310 + (2^2)\,.0722 \\
&= .3198.
\end{aligned}
\tag{32}
$$

A check of Table 2 shows that lines 11 and 32 come closest to satisfying the condition of (17B) since

$$
\begin{aligned}
Q_{102} - Q'_{102} &= Q_{32} - Q'_{102} \\
&= .4432 - .3198 \\
&= .1234,
\end{aligned}
\tag{33}
$$

and

$$
\begin{aligned}
Q'_{102} - Q_{10\bar{2}} &= Q'_{102} - Q_{11} \\
&= .3198 - .1957 \\
&= .1241.
\end{aligned}
\tag{34}
$$

As before, the discrepancy between the two differences may be attributed to errors in the measured values of Q_{102} and $Q_{10\bar{2}}$ or in the computation of Q'_{102}. Postponing the resolution of this difficulty until later, β^* can now be computed from (16B) by assuming that a^* and c^* are correctly known. On this assumption, the value of β^* becomes

$$
\begin{aligned}
\beta^* &= \cos^{-1}\left(\frac{Q_{102} - Q_{10\bar{2}}}{4 \cdot 1 \cdot 2 \cdot a^* c^*}\right) \\
&= \cos^{-1}\left(\frac{.4432 - .1957}{8 \times .1716 \times .2687}\right) \\
&= 47.9°.
\end{aligned}
\tag{35}
$$

As a check, equation (2) can be used to compute Q_{h0l} and $Q_{h0\bar{l}}$ for several values of h and l. The computed and observed values are listed in Table 6.

Table 6
Agreement of computed and observed Q_{h0l}'s for MgWO$_4$

Q_{h0l}	Computed	Observed	Difference
Q_{101}	.1645	.1649	+.0004
$Q_{10\bar{1}}$	.0419		
Q_{102}	.4424	.4450	+.0026
$Q_{10\bar{2}}$	.1972	.1957	−.0015
Q_{201}	.3188		
$Q_{20\bar{1}}$	.0736	.0730	−.0006
Q_{202}	.6890		
$Q_{20\bar{2}}$	.1986		
Q_{103}	.8647		
$Q_{10\bar{3}}$	.4969	.4938	−.0031
Q_{301}	.5351	.5376	+.0025
$Q_{30\bar{1}}$	.1673		
$Q_{30\bar{2}}$	.2000		
$Q_{40\bar{1}}$	.3230	.3249	+.0019
$Q_{40\bar{2}}$	.2944		

A casual inspection of Table 6 would indicate that the discrepancy between the observed and computed values of Q_{h0l} is very large. It should be borne in mind, however, that the "observed" values are merely the values of Q in Table 2 that come nearest to the computed values and need not, necessarily, be the actual corresponding Q values. They could be, for instance, the Q values of hkl rather than $h0l$ reflections. A somewhat better agreement can be obtained, however, by changing the value of c^{*2} back to its original value and adjusting the value of β^* accordingly. The new dimensions of the reciprocal cell then become

$$
\begin{aligned}
a^{*2} &= .0310 & a^* &= .1761 & \alpha^* &= 89.5° \\
b^{*2} &= .0454 & b^* &= .2131 & \beta^* &= 49.0° \\
c^{*2} &= .0721 & c^* &= .2685 & \gamma^* &= 90.0°.
\end{aligned} \tag{36}
$$

As a final check, enough values of Q_{hkl} should be computed from (2) to index every line in the observed powder diagram. The complete set of observed and computed Q values is listed in Table 7. The comparison in this table shows that the reciprocal cell described by the above parameters satisfactorily accounts for the observed reflections of MgWO$_4$.

The primitive cell of MgWO$_4$ determined in this section is one of the many possible primitive cells in the correct reciprocal lattice of MgWO$_4$.

Table 7

Final agreement of observed and computed Q_{hkl}'s for MgWO$_4$

Powder-diagram line	$Q_{obs.}$	$Q_{comp.}$	hkl	Powder-diagram line	$Q_{obs.}$	$Q_{comp.}$	hkl
1	.0310	.0310	100	21	.3249	.3244	310
2	.0457	.0454	010	22	.3322	.3320	$01\bar{2}$
3	.0730	.0721	001	23	.3364	.3364	012
4	.0769	.0764	110	24	.3428	.3420	$22\bar{2}$
5	.1165	.1166	$01\bar{1}$	25	.3451	.3444	$\bar{1}2\bar{1}$
6	.1187	.1187	011	26	.3505	.3508	$\bar{2}2\bar{2}$
7	.1239	.1240	200	27	.3664	.3664	211
		.1648	$20\bar{2}$	28	.3723	.3708	$30\bar{3}$
8	.1649			29	.3824	.3818	$1\bar{2}\bar{2}$
		.1650	101	30	.4016	.4015	$20\bar{3}$
9	.1699	.1694	210	31	.4083	.4084	030
10	.1816	.1818	020	32	.4432	.4432	102
11	.1957	.1956	$10\bar{2}$	33	.4450	.4436	$21\bar{3}$
12	.2077	.2083	$21\bar{2}$	34	.4498	.4502	$2\bar{1}\bar{3}$
13	.2123	.2126	120	35	.4608	.4606	320
14	.2386	.2388	$11\bar{2}$	36	.4659	.4662	$02\bar{2}$
15	.2436	.2432	$1\bar{1}\bar{2}$	37	.4768	.4774	$03\bar{1}$
16	.2517	.2515	$02\bar{1}$	38	.4863	.4842	031
17	.2563	.2559	021	39	.4918	.4908	112
18	.2793	.2790	300	40	.4938	.4942	$10\bar{3}$
19	.2884	.2884	002	41	.5376	.5326	230
20	.3065	.3056	220				

Another cell, however, may better display the symmetry of the lattice. Methods for determining such a cell are described in Chapters 11 and 12.

Systematic procedure for indexing any powder photograph

The method described above clearly shows that the reciprocal-lattice concept can be used to index a powder photograph when the crystal symmetry is not known. Two difficulties were also mentioned, namely, the absence of certain reflections from the photograph and the need for accurately measured Q values. It should be realized, however, that these difficulties are not unique to this method. They are present and equally troublesome in the methods described in Chapters 7 and 8 also, although they are not so apparent there. The advantage which the methods of this chapter have lies in their ability to detect the absence of certain reflections immediately. This ability can be used to test quickly whether the crystal is isometric, tetragonal, or hexagonal, as discussed below. If these tests indicate one of these high symmetries, much computational labor can be saved. In any event, the same procedures can be extended to assign indices if the crystals are orthorhombic. If the crystal does not prove to be orthorhombic, the Ito method can be used to determine the reciprocal lattice and, hence, the indices of the powder-photograph lines.

Test for isometric system. After the experimentally determined Q's have been tabulated, a rapid test can be made to determine whether

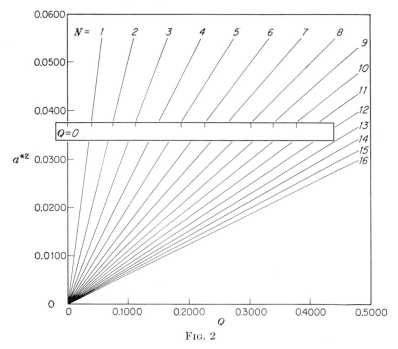

Fig. 2

the crystal is isometric. This is a consequence of the simple form of equation (2) in the isometric system:

$$Q_{hkl} = (h^2 + k^2 + l^2)a^{*2}$$
$$= N_C a^{*2}. \tag{37}$$

The only values that $N_C = (h^2 + k^2 + l^2)$ can have are 1, 2, 3, 4, etc. (Appendix 1). Accordingly, if the first value of Q observed is assumed to be $Q_{100} = (1)a^{*2}$, the other possible Q's can be rapidly computed and compared with the observed list. If the set of Q's computed by this method fails to match the observed set, the first value of Q observed is next assumed to be $Q_{110} = (2)a^{*2}$, and the above procedure is repeated. This can be done most conveniently with a slide rule.

Alternatively, (37) can be solved graphically. Figure 2 is a plot of equation (37) for different values of N_C and a^{*2}. Its use is analogous to the similar chart shown in Fig. 5 of Chapter 7.

Tests for tetragonal and hexagonal systems. The interplanar-spacing equations for the tetragonal and hexagonal systems can be written in terms of the reciprocal-lattice constants $a^* = 1/d_{100}$ and $c^* = 1/d_{001}$ as follows in the tetragonal system:

$$Q_{hkl} = (h^2 + k^2)a^{*2} + l^2c^{*2}$$
$$= N_T a^{*2} + l^2c^{*2}, \tag{38}$$

and in the hexagonal system:

$$Q_{hkl} = (h^2 + hk + k^2)a^{*2} + l^2c^{*2}$$
$$= N_H a^{*2} + l^2c^{*2}. \tag{39}$$

The coefficients N_T and N_H can have the following values (Appendix 1):

$$N_T = 1, 2, 4, 5, 8, 9, 10, 13, 16, \ldots$$
$$N_H = 1, 3, 4, 7, 9, 12, 13, 16, 19, \ldots \tag{40}$$

If every value of N_T is multiplied by 2, another permissible value of N_T is obtained. Similarly, if every value of N_H is multiplied by 3 another permissible value of N_H is obtained. To a limited extent, therefore, as an indication of the probable crystal system of an unknown, it is possible to examine a set of observed Q values to determine which factor is present. If the presence of one of these factors is ascertained, equation (38) or (39) can be used directly to determine the magnitudes of the reciprocal-lattice constants and hence the indices. It should be realized, of course, that these factors apply only to $hk0$ reflections and that usually they are obscured by the presence of hkl reflections.

It is possible, however, to take advantage of the coefficients N_T and N_H in another way. Equations (38) and (39) suggest the existence of relations of the type:

$$Q_{h_1 k_1 l_1} - Q_{h_2 k_2 l_1} = (N_1 a^{*2} + l_1^2 c^{*2}) - (N_2 a^{*2} + l_1^2 c^{*2})$$
$$= \Delta N a^{*2}, \tag{41}$$

where ΔN simply denotes the difference between the coefficients N. (ΔN can have any integral value from zero to infinity.) The above relations can be stated: *Whenever two reflections have the same l index, the difference between their respective Q values is simply $\Delta N a^{*2}$.*

There are only two other types of relations that can occur if differences between Q values are considered. These occur when the hk values of two reflections are the same but the l values are not, for example,

$$Q_{h_1 k_1 l_1} - Q_{h_1 k_1 l_2} = (N_1 a^{*2} + l_1^2 c^{*2}) - (N_1 a^{*2} + l_2^2 c^{*2})$$
$$= \Delta l^2 c^{*2}, \tag{42}$$

and when neither hk nor l have the same values, for example,

$$Q_{h_1k_1l_1} - Q_{h_2k_2l_2} = (N_1a^{*2} + l_1^2c^{*2}) - (N_2a^{*2} + l_2^2c^{*2})$$
$$= \Delta Na^{*2} + \Delta l^2c^{*2}, \tag{43}$$

where Δl^2 can have the values 1, 3, 4, 5, 7, 8, 9, 12, 15, 16, 21, etc. From the above, it is clear that a list of differences between all pairs of observed Q values contains certain differences that can be used to determine the magnitudes of a^{*2} and c^{*2}.

Since more different relations are possible in (41) than in (42), it follows that, if differences between all possible values of Q are considered, ΔNa^{*2} values should recur more often than Δl^2c^{*2} values. On the other hand, recurrences of specific differences in (43) are purely fortuitous. It is possible, therefore, to use the most frequently recurring differences to determine the magnitudes of a^{*2} and c^{*2}. The procedure for doing this is illustrated by an actual problem below.

The Q values, obtained from a powder photograph of GeO_2, are listed in Table 8. Differences between all pairs of observed Q values are first

Table 8
Observed Q values for GeO_2

Q_1	.0536	Q_{11}	.3396	Q_{21}	.6567
Q_2	.0850	Q_{12}	.3750	Q_{22}	.6599
Q_3	.1606	Q_{13}	.4067	Q_{23}	.6729
Q_4	.1786	Q_{14}	.4440	Q_{24}	.6968
Q_5	.1919	Q_{15}	.4959	Q_{25}	.7281
Q_6	.2145	Q_{16}	.5002	Q_{26}	.7668
Q_7	.2456	Q_{17}	.5139	Q_{27}	.7817
Q_8	.2817	Q_{18}	.5544	Q_{28}	.8220
Q_9	.2860	Q_{19}	.6075	Q_{29}	.8356
Q_{10}	.3357	Q_{20}	.6431	Q_{30}	.8767

computed. The differences for the first twelve Q's are plotted on the bar graph shown in Fig. 3. The graph is used to determine the differences that recur more than twice. From Fig. 3, it can be seen that

$$\begin{aligned}
&.0314 \text{ recurs 4 times,} \\
&.0536 \text{ recurs 4 times,} \\
&.1070 \text{ recurs 3 times,} \\
&.1252 \text{ recurs 4 times,} \\
&.1608 \text{ recurs 8 times,} \\
&.2145 \text{ recurs 6 times,} \\
&.2281 \text{ recurs 3 times,} \\
&.2682 \text{ recurs 4 times,} \\
&.3217 \text{ recurs 7 times,} \\
&.3750 \text{ recurs 3 times.}
\end{aligned} \tag{44}$$

The differences tabulated above are assumed to represent the left side of equation (41). The right side of this equation is identical with the right side of the parallel equation (37) for the isometric system:

$$\begin{aligned}
(37): \quad Q_{hkl} &= (h^2 + k^2 + l^2)a^{*2} \\
&= N_C a^{*2}.
\end{aligned}$$

Since equation (41) contains only one unknown, a^{*2}, the graphical-indexing procedure commonly used for the isometric system can be used to determine the magnitude of a^{*2}, as shown in Fig. 2. The differences in (44) are marked on a strip of paper which is moved up and down on the chart until an acceptable match is obtained. A match for all but three lines occurs when $a^{*2} = .0536$.

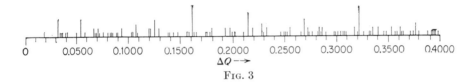

FIG. 3

After the magnitude a^{*2} is known, equation (38) or (39) can be first used to assign indices to all observed $hk0$ reflections. In doing this, the permissible values of either N_T or N_H can be used; this part of the procedure, therefore, determines the crystal system. Obviously, the $hk0$ indices can be found conveniently with the help of the graph shown in Fig. 2. When the crystal system is known, the appropriate relation can be used to determine the magnitude of c^{*2} from hkl reflections, by trial-and-error methods. In the case of GeO_2, two of the three lines (.0314 and .1252) not indexed in Fig. 2 can be assumed to represent the left side of equation (42) and used to determine c^{*2}. By carrying out the above procedure, the crystal system of GeO_2 is determined as hexagonal with $a^{*2} = .0536$ and $c^{*2} = .0314$. All the observed lines can now be indexed using equation (39).

As a further demonstration of the usefulness of the procedure described above, the most frequently recurring differences between the first twelve observed Q values of several other compounds are given in Table 9. The reader can use these values in combination with Fig. 2 to find the unit-cell constants.

The values given in Table 9 for calcite (for which $a^{*2} = .0536$, $c^{*2} = .0034$) show up the complication occasionally encountered with this method. If c is very much larger than a (so that $c^{*2} \ll a^{*2}$), then relations (42) and (43) become more frequent than relations (41). The recurrence of differences, however, is still much greater in (42) than in

(43).   Although this complicates the analysis, it in no way decreases the validity of the method.   Fortunately such cases are quite rare.

Table 9
The most frequently occurring differences between Q values for several crystals

Rutile	Zr	Red PbO	Calcite
.0475 (3 times)	.1280 (4 times)	.0643 (4 times)	.0560 (4 times)
.0945 (5 times)	.2555 (3 times)	.1268 (6 times)	.1044 (3 times)
.1422 (4 times)	.3830 (7 times)	.2534 (6 times)	.1079 (4 times)
.1615 (3 times)	.4530 (3 times)		.1607 (7 times)
.1899 (6 times)	.5100 (4 times)		.2695 (5 times)
.3792 (3 times)			.3214 (6 times)

Procedures when tests for high symmetry are negative.   Failure to index a powder photograph by the above procedures indicates that the crystal is probably not isometric, tetragonal, or hexagonal.   The next most reasonable possibility is the orthorhombic system.   It will be recalled from Chapter 8 that the Hesse-Lipson analytical procedure employs differences between the observed $\sin^2 \theta_{hkl}$ values to identify the pinacoid reflections with whose aid a, b, and c are determined.   It is a simple matter to transpose the equations of that chapter to reciprocal-lattice coordinates with the aid of (4).   If equations (17) through (38) of Chapter 8 are multiplied by $4/\lambda^2$, the parallelism with the equations of this chapter is obvious.   After completing the tests described above, it is possible, therefore, to proceed directly with the application of the Hesse-Lipson procedure for indexing powder photographs of orthorhombic crystals as described in Chapter 8.

Some discretion, however, should be used at this point.   The tests for isometric, tetragonal, and hexagonal symmetry are fast and simple.   On the other hand, the Hesse-Lipson procedure is relatively more time-consuming.   If it should fail in any given case, it is then necessary to use the even more lengthy Ito procedure.   Since the data are already in the form of a list of Q's, would it not be more expedient at this point to proceed directly to the Ito method, which is equally capable of indexing the powder photograph should the crystals turn out to be orthorhombic?   The investigator's own judgment can best answer this question.   In the tests for tetragonal or hexagonal symmetry of the crystals, a list of ΔQ's was prepared.   If, upon inspection of such a list, it appears that the crystal is probably orthorhombic (the key is the number of frequently recurring ΔQ values), then the Hesse-Lipson procedure should be used.   If there is serious doubt whether or not the crystal is orthorhombic, then the Ito method should be used.   Herein lies the advantage of the tests of the preceding section. After completion of these tests, the powder-photograph data are in such a form that either procedure can be utilized without further preparatory labor. Moreover, with just a casual familiarity with both procedures, the investigator can intelligently surmise the most rewarding procedure to follow.

Literature

[1] T. Ito, *X-ray studies on polymorphism* (Maruzen Co., Ltd., Tokyo, 1950) 187–228.

11

Reduced cells and their application

In the last chapter it was shown that it is possible to analyze a powder photograph in such a way that a cell† is found on the basis of which every line of the powder photograph can be assigned an index. This cell defines a lattice which is the lattice of the crystal. A specific cell defines one and only one lattice, but the converse is not true, for any lattice can be defined by an infinite number of cells.‡

A crystal is characterized (among other things) by the dimensional aspects of its lattice, so that the crystal could be identified by reference to suitable tables of dimensions. To make such a scheme practical, it is vital that one of the infinite number of cells which characterize a lattice be accepted as standard. Fortunately such a standard cell, known as the *reduced cell*, is available for this purpose.

The reduced cell is a special one that has been recognized since the earliest days of crystallography. It has been thoroughly discussed from a crystallographic viewpoint[1-7] in the literature. The choice of this cell as a standard cell is far from arbitrary, since it corresponds to the *reduced ternary quadratic form* of mathematics, upon which much has been written.[8-11]

More than one Bravais lattice type occurs in each crystal system other than the triclinic, where only the primitive occurs. To each of these fourteen Bravais lattice types, there corresponds one or more specific kinds of reduced cells. Conversely, if one can find that a crystal has a specific kind of reduced cell, the Bravais lattice type and the crystal system are immediately identified.

---

† It is assumed in this chapter that a primitive cell has been found.

‡ See M. J. Buerger, *Elementary crystallography* (John Wiley & Sons, Inc., New York, 1956) 15–16.

124

Theory of reduced cells

The reduced cell in two dimensions. The idea of a reduced cell is inherent in Bravais' work[1] of 1850. Beginning with a two-dimensional lattice, Bravais defined the *principal triangle* of the net as that triangle which has as its base the smallest lattice translation and the base angles

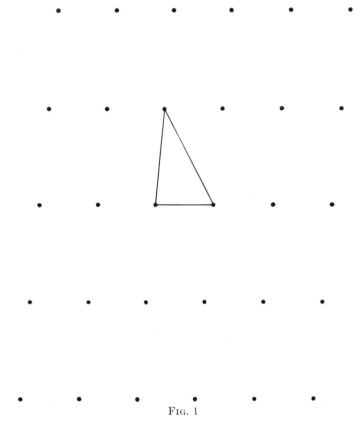

Fig. 1

of which are acute (one of which might, exceptionally, be a right angle). This is the triangle shown in Fig. 1. It is the only triangle whose three angles are acute. A most important property is that it contains the three shortest translations of the net. Secondarily, its three angles are limited as follows:

The smallest angles lies in the range 0 to 60°.
The middle angle lies in the range 45 to 90°.
The largest angle lies in the range 60 to 90°.

Two principal triangles can be placed edge to edge in three ways only, depending on which of the three edges is common to the two triangles (Fig. 2). Each such pair forms a parallelogram, which is a possible unit cell of the net. Johnsen[2] recognized three categories of cells, depending

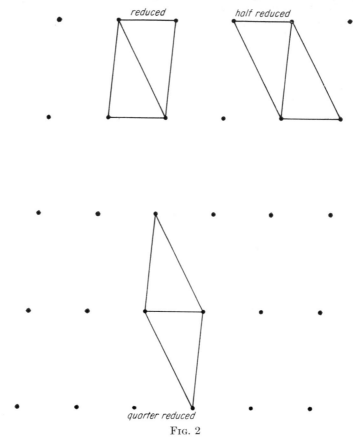

Fɪɢ. 2

on which pair of the three shortest vectors became the sides of the parallelogram. These are illustrated in Fig. 2 and are defined as follows:

Sides of parallelogram	*Name of cell*
Shortest and second shortest..........	Reduced
Shortest and third shortest............	Half-reduced
Second and third shortest.............	Quarter-reduced

The reduced cell in three dimensions. Bravais defined as an *elementary tetrahedron* any tetrahedron three of whose edges are the conjugate translations† radiating from the same lattice point. The volume

† i.e., translations defining a primitive cell. See M. J. Buerger, *Elementary crystallography* (John Wiley & Sons, Inc., New York, 1956) 15–16.

of such a tetrahedron is one-sixth that of the primitive cell. The *principal tetrahedron* was defined as the one whose base has the two shortest translations of the lattice (that is, the principal triangle of the densest net), and whose sides all make acute dihedral angles with the base. (Two of these dihedral angles may exceptionally be right angles.) This can be interpreted with the aid of Fig. 3. The three lattice points of a principal triangle of the densest net are shown as dots at the triangle corners. The dot representing the fourth point of the principal tetrahedron must be placed so that each of the three sides it forms with the edges of the base triangle makes an acute angle with the base. To make the angles acute, the sides must slope upward in the direction of the

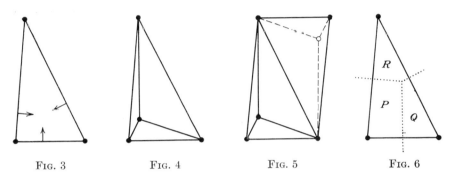

FIG. 3   FIG. 4   FIG. 5   FIG. 6

arrows in Fig. 3. The fourth dot, therefore, must fall above the triangle but project within it, as shown in Fig. 4. Since every lattice is centro-symmetrical, a corresponding point shown as an open circle in Fig. 5 occurs *below* the adjacent triangle, so that the tetrahedra occur in pairs.

Since the apical lattice point in Fig. 4 projects within the triangle, it is evident that every plane angle of the principal tetrahedron is acute. Exceptionally, if the upper lattice point projects exactly over a corner of the base triangle, two angles can be right angles. If the base triangle is a right triangle but the upper lattice point projects within the triangle, one plane angle of the principal tetrahedron is a right angle. If both conditions occur, three plane angles of the principal tetrahedron can be right angles.

The most important characteristic of the principal tetrahedron is that it contains the two smallest translations of the lattice, and also the smallest translation lying outside the plane of these two.† This can be

† This third translation is not necessarily the third shortest translation of the lattice. To see this, let the principal triangle contain very short translations *a* and *b*. Let the next plane of the lattice parallel to this net be at a considerable distance from it. Then not only are *a* and *b* the first two shortest translations, but many linear vector sums of *a* and *b*, such as $\mathbf{a} + \mathbf{b}$ and $\mathbf{a} - \mathbf{b}$, may be shorter than the shortest translation **c** which relates the two net planes.

succinctly stated by saying that the principal tetrahedron contains the three shortest noncoplanar translations of the lattice. The principal tetrahedron also contains the four densest net planes of the lattice.

A parallelepiped whose edges are three edges of any elementary tetrahedron is a primitive unit cell of the lattice. The particular parallelepiped whose edges are the three shortest edges of the principal tetrahedron has been called the *reduced cell* by Johnsen[2] and by Niggli.[3] Briefly, *the reduced cell is characterized by having as edges the three shortest noncoplanar translations of the lattice.*

The reduced plane cell of the densest net plane (that is, that plane containing the two shortest translations) is one of the faces of the reduced cell of the three-dimensional lattice. Several cases arise depending upon

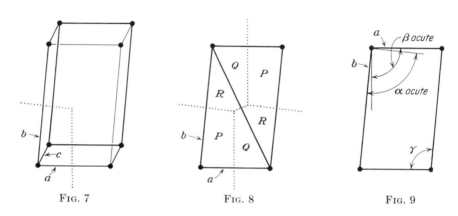

FIG. 7   FIG. 8   FIG. 9

which region of the principal triangle contains the projection of the apex of the principal tetrahedron, shown in Fig. 4. This depends upon which of the three edges from the base to the apex is the shortest. The several cases are easily studied by partitioning the principal triangle by the perpendicular bisectors of its sides (Fig. 6). The projection of the apex must fall into one of the regions P, Q, or R. Whichever region it falls into, the shortest translation outside the principal triangle is from the apex to the lattice point in that partition.

If the projection of the apex falls in the partition P of the least acute angle of the principal triangle, then the resulting cell, shown in Fig. 7, has all interaxial angles acute. If it falls in partition Q or R, these two can be considered together by considering the adjacent principal triangles of the reduced plane parallelogram (Fig. 8). Consider only the upper pair of partitions R and Q (Fig. 8). Draw perpendiculars to the two sides of the parallelogram at one lattice point. Then the regions in which α and β are acute are limited by the perpendiculars as shown in

Fig. 9. The projection of c can fall into three different possible regions. Since $\gamma = a \wedge b$ is obtuse, as drawn, these three possibilities give rise to angular combinations which can be represented in the following way: Let an acute angle be characterized as $+$, and an obtuse angle as $-$ (according to their cosines). Then the three sets are:

α	β	γ
$+$	$-$	$-$
$+$	$+$	$-$
$-$	$+$	$-$

Together with the set $+ + +$, which was discovered in Fig. 7, there appear, offhand, to be four kinds of reduced cells, with respect to the combination of angles. Not all these sets are distinct, however, as discussed in the next section.

Axial representations of a cell. At each lattice point, eight cells meet. Since all the cells are identical, they can be represented by the edges and interaxial angles of any of these eight cells which meet at the common lattice point. Each such set of three directed edges and their three included angles may be taken as a *representation* of the cell. Since every lattice point is a center of symmetry of the lattice, four of these eight are centrosymmetrical equivalents of the other four, so that four may be regarded as right-handed representations and four as left-handed representations.

One can transform from one representation to another by reversing the direction of one, two, or three axes. To study this in detail, let the acute or obtuse nature of an interaxial angle be called its *character*. These may be symbolized by the signs $+$ for acute, and $-$ for obtuse. If one reverses the direction of an axis, it changes the character of the angle that axis makes with any line crossing it. Thus, if the direction of a is reversed, β transforms to $\pi - \beta$ and γ transforms to $\pi - \gamma$. This is succinctly described by saying that, when a is reversed, β and γ change character. (From this it can be readily shown that, if two axes are reversed, the characters of the two angles corresponding to the reversed axes are changed, whereas if all three axes are reversed, the characters of no angles are changed.)

Now consider the eight character representations of a particular cell, one of whose character representations is $+ + +$, i.e., the three interaxial angles α, β, and γ are acute. Starting with the character representation $+ + +$ and applying the simple rule of transformation, the character representations given in Table 1 can be derived. It will be

Table 1
Derivation of other representations from axial character
representation + + +

Axes reversed from original representation	Character representation
	+ + +
a	+ − −
b	− + −
c	− − +
a, b	− − +
a, c	− + −
b, c	+ − −
a, b, c	+ + +

observed that there are four distinct character representations, each duplicated.   The duplicates are right- and left-handed equivalents.

Now it is remarkable that the character representation − − − is missing from Table 1.   Furthermore, study of Table 1 shows that all character representations are described by an odd number of + characters and
an even number of − characters, while those described by the reversed
condition, to which − − − belongs, are all missing.   Since − − − represents a cell one of whose corners is all-obtuse, which is certainly a possible parallelepiped, Table 1 certainly does not contain axial character
representations of all possible kinds of cells.   Some more character representations can evidently be obtained by transforming the character
representation − − −.   These transformations are derived in Table 2.

Table 2
Derivation of other representations from axial character
representation − − −

Axes reversed from original representation	Character representation
	− − −
a	− + +
b	+ − +
c	+ + −
a, b	+ + −
a, c	+ − +
b, c	− + +
a, b, c	− − −

All the character representations in Table 2 are described by an odd
number of − characters and an even number of + characters.   These
character representations are, therefore, exactly complementary to those
of Table 1.   Furthermore, the eight distinct character representations in
Tables 1 and 2 together are the only ways in which the two signs + and

— can be distributed over three positions. This discussion implies that there are two distinct types of parallelepipeds, or cells. They can be designated types I and II. Each can be represented in four ways on right-handed axes and four corresponding ways on left-handed axes. The simplest representation is one in which all angles have the same character; these are called *normal representations*. Thus the normal character representation for type I cell is $+ + +$, while that of type II cell is $- - -$.

The discussion just given is perfectly general. It implies that any given cell must fall into the category of either type I or type II. In any lattice, each of the infinite number of cells which can be selected must fall into type I or type II. Conversely, both type I and type II cells can be found in any lattice. Since, however, a specific lattice can have only one reduced cell, this must necessarily be *either* type I *or* type II.

Methods of finding the reduced cell

Since the dimensions of the reduced cell are to be used for tabulation of cell data for purposes of identification, it is essential to be able to start with any primitive cell, and transform it to the reduced cell. Fortunately, it is quite easy to do this.

Now, one can start at any point of a lattice and reach any other lattice point by a translation which is a vector sum of multiples of edges of any primitive cell. That is, by using a translation

$$\mathbf{t} = u\mathbf{a} + v\mathbf{b} + w\mathbf{c} \tag{1}$$

(where u, v, and w are integers), every point in the lattice can be reached from any other. The problem of finding the reduced cell is simply that of finding the three shortest noncoplanar $\mathbf{t}$'s, given any set of $\mathbf{a}$, $\mathbf{b}$, $\mathbf{c}$. These three $\mathbf{t}$'s then become the edges of the reduced cell. This problem can be solved graphically or analytically.

Graphical determination of the reduced cell. The most straightforward method of finding the reduced cell is to make a simple drawing of the lattice and select the reduced cell by inspection. This method has been followed by crystallographers for many years. In case the shortest translations of the lattice are not obvious by inspection, a few simple graphical tests will reveal the relative lengths of the several candidate translations.

The following outline is suggested for a systematic graphical search for the shortest translations. Let the original (presumably unreduced) cell have edges a, b, and c (in order of increasing length) and interaxial angles α, β, and γ. Now imagine a sphere of radius equal to the longest axis c drawn about some lattice point as origin (Fig. 10A). Only lattice points which fall within this sphere can be candidates for terminations of translations shorter than the original set a, b, and c.

It is convenient to draw the lattice projected on the plane of the two largest translations b and c, as shown in Fig. 10A. Let this plane be the equatorial plane of the sphere. It is very simple to locate the lattice points that lie in this plane, since they are at the corners of a parallelogram of edges b and c, and angle α. The rest of the lattice points that are contained in the sphere occur in several equally spaced levels parallel

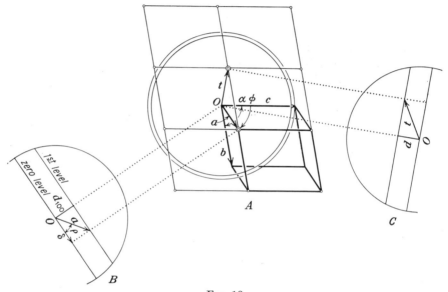

FIG. 10

to the equatorial plane. On each of these levels, the pattern of points is the same as on the equatorial plane, except that the pattern is shifted by an amount δ (Fig. 10B) in a direction φ (Fig. 10A). The determination of these parameters will now be discussed.

The levels of the lattice have spacing d_{100} (Fig. 10B). Since the reciprocal lattice has presumably already been found, this can be computed from the reciprocal-lattice edge a^*:

$$d_{100} = \frac{1}{a^*}. \tag{2}$$

If a^* is not already known, it can be determined in the standard way from

$$a^* = \frac{bc \sin \alpha}{V}. \tag{3}$$

The latitude angle ρ of the a axis is shown in Fig. 10B. It can evidently be computed from

$$\sin \rho = \frac{d_{100}}{a}. \tag{4}$$

Figure 10B shows that the magnitude of the displacement δ between adjacent levels of the lattice is related to d_{100} by

$$\cos \rho = \frac{\delta}{a}, \tag{5}$$

so that

$$\delta = a \cos \sin^{-1}\left(\frac{d_{100}}{a}\right). \tag{6}$$

In order to find the direction φ of the displacement of an adjacent level of the lattice, consider the spherical triangle (Fig. 11A) whose vertices

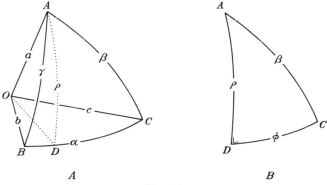

Fig. 11

A, B, and C are the directions of the axes a, b, and c, and whose sides are the interaxial angles α, β, and γ. Draw the arc of a great circle from A perpendicular to side BC. This arc segment is the angle ρ, computed in (4). Point D gives the angular position of the projection of a on the plane of b and c. The direction can be defined by azimuth angle φ. Figure 11B shows the right spherical triangle on the right side of Fig. 11A. The relation between ρ, φ, and C is

$$\tan C = \frac{\tan \rho}{\sin \varphi}. \tag{7}$$

Since $C = 180° - \gamma^*$, this can be converted to

$$- \tan \gamma^* = \frac{\tan \rho}{\sin \varphi} \tag{8}$$

from which φ can be determined.

Relations (6) and (8) permit one to shift the pattern of lattice points on the equatorial level to become the pattern of lattice points on the first level up. Since the same displacement occurs between each pair of levels, lattice points on each level can be plotted as they project on the equatorial level. Only those lattice points which fall within the sphere are of interest. These are the ones which are within the circle of that level, and indicated in Fig. 10*A* by the inner circle. Usually only one level above the equatorial level is necessary.

Each translation which terminates at a lattice point which falls within the circle of its level (such as translation **t**, Fig. 10*A*) can be quickly measured as shown in Fig. 10*C*. This shows an elevation of the sphere as seen normal to the line of projection of the translation. The true vertical component of the translation is d_n, where n is the level on which the lattice point occurs. Each candidate **t** can be sketched in true length very rapidly. The three shortest **t**'s which are noncoplanar are accepted as the edges of the reduced cell. If three **t**'s are coplanar, the plane of their terminal lattice points contains the origin.

Search for shortest translations by computation. The three translations of the reduced cell can be searched for by direct computation. Each candidate vector can be written as the vector sum of simple integral multiples of the edges of the unreduced cell, as given in (1). The absolute length of this vector can be found by forming the scalar product of the vector with itself:

$$\mathbf{t} \cdot \mathbf{t} = (u\mathbf{a} + v\mathbf{b} + w\mathbf{c}) \cdot (u\mathbf{a} + v\mathbf{b} + w\mathbf{c}), \tag{9}$$

$$\therefore t^2 = u^2a^2 + v^2b^2 + w^2c^2$$
$$+ 2uv\ ab\ \cos\gamma$$
$$+ 2vw\ bc\ \cos\alpha$$
$$+ 2wu\ ca\ \cos\beta. \tag{10}$$

The problem of finding the reduced cell by computation amounts to computing (10) for a comparatively few low values of the integers u, v, and w, and then selecting the three minimum values so found as edges of the reduced cell. This is not a very tedious computation, since $a, b, c, \alpha, \beta,$ and γ are constant for the problem. The computing in (10) can be reduced to the following form

$$t^2 = u^2k_1 + v^2k_2 + w^2k_3$$
$$+ uvk_4 + vwk_5 + wuk_6 \tag{11}$$

where the constants for the entire problem are:

$$k_1 = a^2, \quad k_4 = 2\ ab\ \cos\gamma,$$
$$k_2 = b^2, \quad k_5 = 2\ bc\ \cos\alpha,$$
$$k_3 = c^2, \quad k_6 = 2\ ca\ \cos\beta. \tag{12}$$

The integers u, v, and w are small ones, and usually limited to 1, 0, -1, and possibly 2 and -2. The following suggestions are offered for making the search for the edges of the reduced cell: The work of computation amounts only to supplying integral coefficients to the constants in (11). The results should be recorded systematically, for example, in matrix form, as follows:

$u = 1$						$u = 2$				
	$\bar{2}$	$\bar{1}$	0	1	2		$\bar{2}$	$\bar{1}$	0	1	2
2						2					
1						1					
0						0					
$\bar{1}$						$\bar{1}$					
$\bar{2}$						$\bar{2}$					

. . . etc.

(with w across the top and v down the side of each matrix)

The matrix for $u = 0$ is a two-dimensional problem in which

$$t^2 = v^2 k_2 + w^2 k_3 + vw k_5. \tag{13}$$

One proceeds from one matrix to the next by adding to the terms of the preceding matrix the appropriate quantity

$$\Delta t^2 = (u_2^2 - u_1^2)k_1 + (u_2 - u_1)vk_4 + (u_2 - u_1)wk_6. \tag{14}$$

It is best to compute the central region of a matrix first, and then work out since, if an increasing value is noted as the distance from the center increases, further values need not be computed.

When a sufficient number of these t^2's have been computed to show up three nonzero smallest values, these can be accepted as the lengths of the three edges of the reduced cell *provided they do not correspond to three coplanar translations.* If the three translations are coplanar, this can be recognized because the plane of the three t^2's contains the point $u = 0$, $v = 0$, $w = 0$ of the several matrices taken together and regarded as a three-dimensional matrix. When this occurs, the reduced cell is based upon the two minimum t's plus the next shortest t which is noncoplanar with these two shortest.

This analysis provides absolute lengths of the three primitive translations of the reduced cell:

$$\begin{aligned}
\mathbf{t}_1 &= u_1\mathbf{a} + v_1\mathbf{b} + w_1\mathbf{c} \\
\mathbf{t}_2 &= u_2\mathbf{a} + v_2\mathbf{b} + w_2\mathbf{c} \\
\mathbf{t}_3 &= u_3\mathbf{a} + v_3\mathbf{b} + w_3\mathbf{c}.
\end{aligned} \tag{15}$$

The interaxial angles of the reduced cell are $\mathbf{t}_2 \wedge \mathbf{t}_3$, $\mathbf{t}_3 \wedge \mathbf{t}_1$, and $\mathbf{t}_1 \wedge \mathbf{t}_2$ (which correspond to the new α, β, and γ, respectively). These angles can be found from the following analysis:

$$\mathbf{t}_1 \cdot \mathbf{t}_2 = |\mathbf{t}_1| \cdot |\mathbf{t}_2| \cos(\mathbf{t}_1 \wedge \mathbf{t}_2), \tag{16}$$

$$\therefore \cos(\mathbf{t}_1 \wedge \mathbf{t}_2) = \frac{\mathbf{t}_1 \cdot \mathbf{t}_2}{t_1 t_2} \tag{17}$$

$$= \frac{(u_1\mathbf{a} + v_1\mathbf{b} + w_1\mathbf{c}) \cdot (u_2\mathbf{a} + v_2\mathbf{b} + w_2\mathbf{c})}{t_1 t_2}$$

$$= \frac{1}{t_1 t_2} \{ u_1 u_2 a^2 + v_1 v_2 b^2 + w_1 w_2 c^2$$

$$+ u_1 v_2 \mathbf{a} \cdot \mathbf{b} + u_1 w_2 \mathbf{a} \cdot \mathbf{c}$$

$$+ v_1 u_2 \mathbf{b} \cdot \mathbf{a} + v_1 w_2 \mathbf{b} \cdot \mathbf{c}$$

$$+ w_1 u_2 \mathbf{c} \cdot \mathbf{a} + w_1 v_2 \mathbf{c} \cdot \mathbf{b} \}. \tag{18}$$

$$\therefore \cos(\mathbf{t}_1 \wedge \mathbf{t}_2) = \frac{1}{t_1 t_2} \{ u_1 u_2 a^2 + v_1 v_2 b^2 + w_1 w_2 c^2$$

$$+ (u_1 v_2 + u_2 v_1) ab \cos \gamma$$

$$+ (v_1 w_2 + v_2 w_1) bc \cos \alpha$$

$$+ (w_1 u_2 + w_2 u_1) ca \cos \beta \}. \tag{19}$$

Transformation to the reduced cell by an algorithm. There is a third method of finding the reduced cell, which consists of systematically transforming the unreduced cell toward the reduced cell by altering one translation at a time. To understand this method some preliminary theory will be presented first.

Since a reduced cell has the three shortest noncoplanar translations as edges, the following theorem holds:

Theorem 1: The three distinct faces of a reduced cell are reduced plane cells.

The converse is not true unless it is limited to primitive cells:

Theorem 2: If three distinct faces of a three-dimensional cell are reduced plane cells, the three-dimensional cell is reduced also, provided it is a primitive cell.

That this is true can be understood from the following reasoning: If the three-dimensional cell contains a translation shorter than its three plane-cell faces, this translation must terminate in a lattice point in the cell's interior. But this cannot be if the cell is primitive. (If it is *not* primitive, there may be a shorter translation terminating in a point of the cell's interior.)

Since the edges of the reduced plane cell are the shortest translations of the net, they are shorter than either of the two diagonals of the cell. For this reason the projection of either cell edge on the other cannot be greater than half the other. As a consequence of this it follows that, in a reduced cell

$$|a \cos \gamma| \leqq \frac{b}{2} \tag{20}$$

and
$$|b \cos \gamma| \leqq \frac{a}{2}. \tag{21}$$

If both sides of (20) are multiplied by b, and both sides of (21) by a, there result

$$|ab \cos \gamma| \leqq \frac{b^2}{2} \tag{22}$$

$$|ab \cos \gamma| \leqq \frac{a^2}{2}. \tag{23}$$

These can be written in vector notation as

$$|\mathbf{a} \cdot \mathbf{b}| \leqq \tfrac{1}{2}b^2, \tag{24}$$
$$|\mathbf{a} \cdot \mathbf{b}| \leqq \tfrac{1}{2}a^2. \tag{25}$$

By virtue of the theorems given above, these relations also apply to the reduced three-dimensional cell:

Theorem 3: *The necessary and sufficient conditions that a cell is reduced are that the absolute values of the scalar product of each pair of axial vectors is less than half the square of either factor in the product.*

Relations (24) and (25) are useful for finding the cell which is most nearly reduced in a "band." In the nomenclature of Bravais,[1] two neighboring parallel rows of lattice points delimit a *band*. Such a band is shown in Fig. 12. The direction of a band is that of some translation $\mathbf{a}$, which need not be the smallest translation of the net. Let the translations which connect lattice points on one border of the band with a lattice point on the other border be labeled $\mathbf{b}_1$, $\mathbf{b}_2$, . . . , $\mathbf{b}_n$. None of these transverse translations need be the smallest translation of the net. Let s be the width of the band and let $\mathbf{b}_1$ and $\mathbf{b}_2$ be the transverse translations nearest to s.

Now consider some properties of the transverse translations, as depicted in Fig. 12. Some representative translations are examined in Table 3. These translations are expressed in terms of the shortest translation $\mathbf{b}_1$. It is evident from the last two columns that, if one started with $\mathbf{b}_4$ and found its scalar product with $\mathbf{a}_1$, a comparatively large numerical value $\mathbf{b}_4 \cdot \mathbf{a}_1$ would result. The second line shows that, if a^2 is subtracted from this value, the numerical value is reduced, and this is equivalent to transforming the vector $\mathbf{b}_4$ to $\mathbf{b}_3$. If successive units of a^2 are subtracted from the scalar product, each step is equivalent to transforming to the next transverse translation of a sheaf in Fig. 12. In this procedure the scalar product becomes smaller and smaller, passes

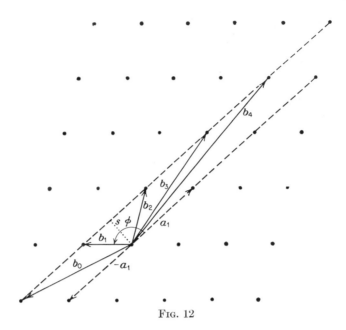

Fig. 12

through zero, and becomes negative. On one side of zero or the other it attains its smallest value (in the example, it attains the minimum value when it is negative, corresponding to $\mathbf{b}_1$). When this smallest value of $\mathbf{b} \cdot \mathbf{a}_1$ is attained, the shortest transverse translation $\mathbf{b}$ has been reached. This procedure provides a neat analytic way of starting with two conjugate translations of a net and transforming the pair so that, with one translation held constant, the other is *reduced* to its shortest value.

This procedure, applied successively to one translation and then the other, and repeated if necessary, can be used to find the two shortest translations of a net. The procedure can be diagramed as shown in Figs. 13 and 14. Let the initial primitive cell be defined by translations $\mathbf{a}_1$

Table 3
Some properties of the transverse translations of a band

Translation	Projection on $\mathbf{a}_1$	Product of projection with a_1	Product expressed in vector notation	
$\mathbf{b}_4 = \mathbf{b}_1 + 3\mathbf{a}_1$	$b_1 \cos \varphi + 3a_1$	$b_1 a_1 \cos \varphi + 3a_1^2$	$\mathbf{b}_1 \cdot \mathbf{a}_1 + 3a_1^2 = \mathbf{b}_4 \cdot \mathbf{a}_1$	
$\mathbf{b}_3 = \mathbf{b}_1 + 2\mathbf{a}_1$	$b_1 \cos \varphi + 2a_1$	$b_1 a_1 \cos \varphi + 2a_1^2$	$\mathbf{b}_1 \cdot \mathbf{a}_1 + 2a_1^2 = \mathbf{b}_3 \cdot \mathbf{a}_1 = \mathbf{b}_4 \cdot \mathbf{a}_1 -$	a_1^2
$\mathbf{b}_2 = \mathbf{b}_1 + \mathbf{a}_1$	$b_1 \cos \varphi + a_1$	$b_1 a_1 \cos \varphi + a_1^2$	$\mathbf{b}_1 \cdot \mathbf{a}_1 + a_1^2 = \mathbf{b}_2 \cdot \mathbf{a}_1 = \mathbf{b}_4 \cdot \mathbf{a}_1 - 2a_1^2$	
$\mathbf{b}_1 = \mathbf{b}_1$	$b_1 \cos \varphi$	$b_1 a_1 \cos \varphi$	$\mathbf{b}_1 \cdot \mathbf{a}_1 = \mathbf{b}_1 \cdot \mathbf{a}_1 = \mathbf{b}_4 \cdot \mathbf{a}_1 - 3a_1^2$	
$\mathbf{b}_0 = \mathbf{b}_1 - \mathbf{a}_1$	$b_1 \cos \varphi - a_1$	$b_1 a_1 \cos \varphi - a_1^2$	$\mathbf{b}_1 \cdot \mathbf{a}_1 - a_1^2 = \mathbf{b}_0 \cdot \mathbf{a}_1 = \mathbf{b}_4 \cdot \mathbf{a}_1 - 4a_1^2$	

and $\mathbf{b_1}$, where $\mathbf{a_1}$ is shorter than $\mathbf{b_1}$. If the interaxial angle is γ, the scalar product is $a_1 b_1 \cos \gamma = \mathbf{a_1} \cdot \mathbf{b_1}$. The initial state of the axes can be diagrammatically represented as at the left of Fig. 13. Since $|\mathbf{b}| > |\mathbf{a}|$, translation $\mathbf{b}$ is a good candidate for reduction. This is done by subtracting (or adding) n times vector $\mathbf{a_1}$ from vector $\mathbf{b_1}$, giving a new $\mathbf{b_1} - n\mathbf{a_1}$, to be labeled $\mathbf{b_2}$. At the same time, n times a_1^2 is subtracted from the scalar product, i.e., $\mathbf{a_1} \cdot \mathbf{b_1}$ transforms to $\mathbf{a_1} \cdot \mathbf{b_1} - na^2$, to be

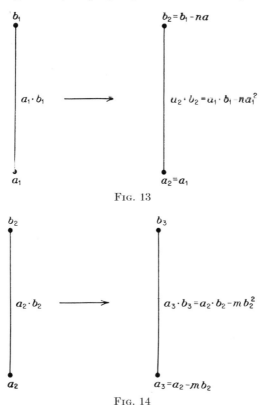

Fig. 13

Fig. 14

labeled $\mathbf{a_2} \cdot \mathbf{b_2}$. The test that this procedure *does* reduce the length of $\mathbf{b}$ is that $\mathbf{a_1} \cdot \mathbf{b_1} - na_1^2$ is numerically smaller than the original $\mathbf{a_1} \cdot \mathbf{b_1}$. The representation of the new state is shown at the right of Fig. 13.

When this stage is complete, the new scalar product $\mathbf{a_2} \cdot \mathbf{b_2}$ is less than $\frac{1}{2}a_2^2$, and according to (20), the most reduced cell for the fixed value of $\mathbf{a_1}$ has been found. To reduce $\mathbf{a_2}$, units of b_2^2 are successively subtracted from (or added to) the scalar product until a minimum is achieved, as shown in Fig. 14. At the end of this stage, the scalar product $\mathbf{a_3} \cdot \mathbf{b_3}$ is less than $\frac{1}{2}b_2^2$, and according to (24), the most reduced cell for the fixed value of $\mathbf{b_2}$ has been found. Next the new $\mathbf{b}$ is reduced again. This iterative process is discontinued when no further reduction occurs. The resulting pair of axes, and the angle implied by their scalar product, define the reduced cell of that net.

This procedure can be used to reduce a three-dimensional primitive cell. Let the axes be a, b, c in order of increasing length. Then the original condition is as diagramed at the left of Fig. 15. Since c is the longest axis, it is a good candidate for reduction (especially if the scalar $\mathbf{a} \cdot \mathbf{b}$ has a large value). It can be reduced by subtracting (or adding)

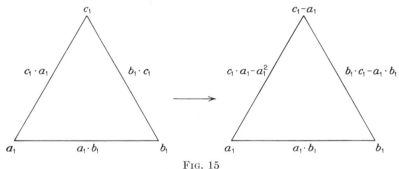

Fig. 15

either the vector $\mathbf{a}$ or $\mathbf{b}$. Suppose $\mathbf{a}$ is subtracted. Then the original cell and new cell are characterized as follows:

Original cell	*New cell*
$\mathbf{a}_1$	$\mathbf{a}_2 = \mathbf{a}_1$
$\mathbf{b}_1$	$\mathbf{b}_2 = \mathbf{b}_1$
$\mathbf{c}_1$	$\mathbf{c}_2 = \mathbf{c}_1 - \mathbf{a}_1$
$\mathbf{b}_1 \cdot \mathbf{c}_1$	$\mathbf{b}_2 \cdot \mathbf{c}_2 = \mathbf{b}_1 \cdot (\mathbf{c}_1 - \mathbf{a}_1) = \mathbf{b}_1 \cdot \mathbf{c}_1 - \mathbf{a}_1 \cdot \mathbf{b}_1$
$\mathbf{c}_1 \cdot \mathbf{a}_1$	$\mathbf{c}_2 \cdot \mathbf{a}_2 = (\mathbf{c}_1 - \mathbf{a}_1) \cdot \mathbf{a}_1 = \mathbf{c}_1 \cdot \mathbf{a}_1 - \mathbf{a}_1^2$
$\mathbf{a}_1 \cdot \mathbf{b}_1$	$\mathbf{a}_2 \cdot \mathbf{b}_2 = \mathbf{a}_1 \cdot \mathbf{b}_1$

The characteristics of the new cell are diagramed on the right of Fig. 15. The procedure consists of vectorially subtracting units of $\mathbf{a}_1$ from $\mathbf{c}_1$ until the absolute value of the new scalar product $\mathbf{c}_1 \cdot \mathbf{a}_1 - na^2$ begins to increase. The next previous value corresponds to the (temporary) shortest value of $\mathbf{c}$. This general process is continued until addition or subtraction of a^2, b^2, or c^2 from its neighboring scalar products increases its absolute value. The result is a diagram of the reduced cell. Let the three translations of the reduced cell be labeled $\mathbf{t}_1$, $\mathbf{t}_2$, and $\mathbf{t}_3$. They are composed of vector sums of the original axes, as given in (15). The absolute length of each axis is given by a relation like (10). For example, the absolute length of $\mathbf{t}_1$ is computed from

$$t_1^2 = u_1^2 a^2 + v_1^2 b^2 + w_1^2 b^2$$
$$+ 2u_1 v_1 \, ab \cos \gamma$$
$$+ 2v_1 w_1 \, bc \cos \alpha$$
$$+ 2w_1 u_1 \, ca \cos \beta. \tag{26}$$

The absolute lengths of t_1, t_2, and t_3 are computed by relations like this. There are then available the absolute lengths of each axis of the reduced cell, and the values of each of its scalar products.   From these quantities, the angles $t_1 \wedge t_3$, $t_3 \wedge t_1$, and $t_1 \wedge t_2$ (corresponding to the values of α, β, and γ for the reduced cell) can be found.   These are derived by relation (19).

The specific reduced-cell types

Cell representation.   A cell can be constructed or identified from a list of the three cell edges and the three interaxial angles.   Thus a cell can be represented by the six quantities a, b, c, α, β, γ (Fig. 16A).   The

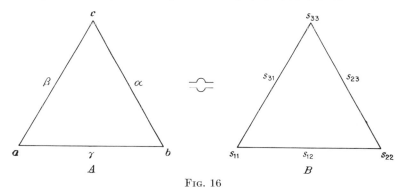

FIG. 16

discussion of the last section makes it evident that the six scalar products $a \cdot a$, $b \cdot b$, $c \cdot c$, $b \cdot c$, $c \cdot a$, $a \cdot b$ can also be taken as an exact representation of the cell, since the six quantities a, b, c, α, β, γ can be readily derived from them.   For purposes of identification, these six scalar products are set down in a particular order, specifically in the form of a rectangular matrix,

$$\begin{pmatrix} a \cdot a & b \cdot b & c \cdot c \\ b \cdot c & c \cdot a & a \cdot b \end{pmatrix}. \tag{27}$$

To emphasize the numerical aspect of these scalar products, the following equivalent designation is customarily used to represent a cell:

$$\begin{pmatrix} s_{11} & s_{22} & s_{33} \\ s_{23} & s_{31} & s_{12} \end{pmatrix}. \tag{28}$$

Here s_{12} represents the scalar product between the axes designated 1 and 2, etc.   Any particular cell is represented by setting down in the matrix the numerical values of these six quantities.   The cell may also be represented on a triangular diagram as shown in Fig. 16B.

Characterizing features of the scalar representation. When the reduced cell has been found by any of the methods discussed in the foregoing sections, its representation can be set down. This set of numerical values contains features which characterize both the symmetry of the lattice and the type of lattice centering. To grasp the nature of this, consider the representation of two monoclinic cells, one primitive and the other *C*-centered, whose shortest axes are shown in Figs. 17*A* and 18*A*.

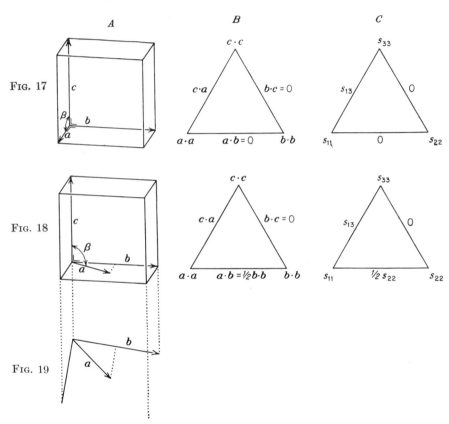

The primitive cell (Fig. 17*A*) contains two interaxial angles of 90°, and so its set of scalars contains two corresponding zeros. The *C*-centered cell contains only one 90° interaxial angle; so the set of scalars contains only one zero. The set has, however, another feature which reveals a centered cell. In Fig. 19, it can be seen that, since the projection of $\mathbf{a}$ on $\mathbf{b}$ is $\frac{1}{2}\mathbf{b}$, the scalar product $\mathbf{a} \cdot \mathbf{b} = \frac{1}{2}\mathbf{b} \cdot \mathbf{b} = \frac{1}{2}b^2$. The scalar product corresponding to $\mathbf{a} \cdot \mathbf{b}$, namely, s_{12}, is therefore numerically equal to $\frac{1}{2}b^2 = \frac{1}{2}s_{22}$. Thus the scalar in the position of s_{12} in Fig. 18*C* has half the numerical value of s_{22}.

There are fourteen space-lattice types. Their distribution among the six crystal systems is given in Table 4. It might be supposed that, corresponding to these fourteen types, there are fourteen reduced-cell representations. Actually, some lattice types have several different representations which depend upon the different possible dimensional relations

Table 4
The distribution of cell types among the crystal systems

	P	I	F	C	R
Isometric............	1	1	1		
Tetragonal...........	1	1			
Hexagonal....	1				1
Orthorhombic..........	1	1	1	1	
Monoclinic............	1	1		[1]	
Triclinic.............	1				

Table 5
Distribution of scalar representations among the crystal systems
(The number of scalar representations having different form is given in brackets)

	P		I		F		C		R		Totals	
Isometric	1	[1]	1	[1]	1	[1]					3	[3]
Tetragonal	2	[1]	3	[3]							5	[4]
Hexagonal	2	[1]							4	[3]	6	[4]
Orthorhombic	1	[1]	3	[3]	2	[2]	5	[2]			11	[8]
Monoclinic	3	[1]	3	[3]			8	[3]			14	[7]
Triclinic	2	[1]									2	[1]
Totals	11	[6]	10	[10]	3	[3]	13	[5]	4	[3]	41	[27]

among the three axes of the unit cells. The total number of representations is forty-one, and their distribution among the crystal systems and lattice types is shown in Table 5. In some cases there are several representations which differ in some way which does not affect the symmetry of the scalars. For example, they may differ by interchanges of short, intermediate, and long axes. The number of representations, omitting such similarity duplications, is given in brackets in Table 5.

The reduced-cell types and their representations have been discussed in detail by Niggli.[3] Their matrix representations are listed in Table 6. Those representations which are related by similarity are connected together.

If one has determined the scalar matrix for a particular crystal, his first problem is to identify it with one of the standard forms of Table 6. The general form of the reduced-cell representation matrix is

$$\begin{pmatrix} s_{11} & s_{22} & s_{33} \\ s_{23} & s_{31} & s_{12} \end{pmatrix}.$$

Let the upper row of scalars in this matrix be called *symmetrical scalars* and the lower row be called *unsymmetrical scalars*. It is possible to identify the particular reduced-cell type by classifying the matrices of Table 6 according to the relations among the symmetrical and unsymmetrical scalars. A way of doing this was presented by Niggli,[3] whose classification is presented in abridged form in Table 8. To use the table one first examines the symmetrical scalars of his particular crystal, and as a result, places his data in one of four categories; namely

$$s_{11} = s_{22} = s_{33}$$
$$s_{11} = s_{22} \neq s_{33}$$
$$s_{11} \neq s_{22} = s_{33}$$
$$s_{11} \neq s_{22} \neq s_{33}.$$

Next the unsymmetrical scalars are examined and the data placed in one of two subcategories, depending upon whether the unsymmetrical scalars are all-positive or all-negative (which corresponds to an all-acute or all-obtuse reduced cell). Finally, the unsymmetrical scalars are examined for specializations which include one or more being zero, one or more being related to each other, or one or more being related to the symmetrical scalars.

Transformation from reduced cell to unit cell. The axes of the reduced cell are identical with the axes of the unit cell only if the unit cell is primitive. In all other cases, the axes $\mathbf{A}$, $\mathbf{B}$, $\mathbf{C}$ of the unit cell must be found from the axes of the reduced cell $\mathbf{t}_1$, $\mathbf{t}_2$, $\mathbf{t}_3$ with the aid of a transformation. In the general case, this transformation has the form

$$\begin{aligned} \mathbf{A} &= u_1\mathbf{t}_1 + v_1\mathbf{t}_2 + w_1\mathbf{t}_3 \\ \mathbf{B} &= u_2\mathbf{t}_1 + v_2\mathbf{t}_2 + w_2\mathbf{t}_3 \\ \mathbf{C} &= u_3\mathbf{t}_1 + v_3\mathbf{t}_2 + w_3\mathbf{t}_3. \end{aligned} \qquad (29)$$

Since only the coefficients of these equations differ from case to case, the transformation is usually represented by a transformation matrix. For transformation (29) the transformation matrix is

$$\begin{pmatrix} u_1 & v_1 & w_1 \\ u_2 & v_2 & w_2 \\ u_3 & v_3 & w_3 \end{pmatrix}. \tag{30}$$

When the crystal has symmetry, one or more of the terms of the matrix are specialized. In Table 7, there are listed the transformation matrices to be used with the reduced-cell representation in corresponding positions of Table 6.

No transformation of interaxial angles is required in the triclinic case, since the reduced cell is primitive. The only interaxial angle required in the other cases is β for the monoclinic crystals. This can be computed from (19).

Application of reduced-cell theory

In this chapter, three different procedures have been described for finding the reduced cell of a given lattice. It is natural to ask at this point, therefore, which of these procedures is preferable in an actual application. This is a matter of personal preference since the reduced cell can be found by either procedure. Experience has shown that the algorithm described last is accurate and rapid, particularly if the initial cell is not one of the partly reduced cells already. On the other hand, the graphical method clearly shows the geometry of the transformations and may be preferred for that reason. In order to help the reader make his own selection, both these methods are illustrated below.

Graphical determination of the reduced cell in a primitive lattice. A primitive cell of $MgWO_2$, determined by Ito's method in Chapter 10, is used to illustrate this reduction procedure. It is first necessary to transform the reciprocal unit cell to the correct direct cell. The transformation of this particular cell was computed at the end of Chapter 9. The direct-cell dimensions, renamed in order of increasing magnitude, were found to be

$$\begin{aligned} a &= 4.693 \text{ Å}, & \alpha &= 131°00', \\ b &= 4.933 \text{ Å}, & \beta &= 89°34', \\ c &= 7.525 \text{ Å}, & \gamma &= 90°40'. \end{aligned} \tag{31}$$

Figure 20 shows the graphical representation of the lattice defined by the unit cell in (31). The lattice is drawn projected on the plane of the two largest translations **b** and **c**, the important part being inscribed in a sphere of radius equal to the magnitude of the longest axis **c**. In practice, one first draws the **bc** net to scale (solid lines in Fig. 20). The intersection of this equatorial net with the sphere is denoted by the circle of radius c (also drawn in a solid line). The rest of the lattice can be

Table 6
The Niggli matrix representation of the reduced cells
(Numbers are Niggli's figure numbers)

	P	I	F	C	R
Isometric	$\begin{pmatrix} s_{11} & s_{11} & s_{11} \\ 0 & 0 & 0 \end{pmatrix}$ (44A)	$\begin{pmatrix} s_{11} & s_{11} & s_{11} \\ \frac{1}{3}s_{11} & \frac{1}{3}s_{11} & \frac{1}{3}s_{11} \end{pmatrix}$ (44B)	$\begin{pmatrix} s_{11} & s_{11} & s_{11} \\ \frac{1}{2}s_{11} & \frac{1}{2}s_{11} & \frac{1}{2}s_{11} \end{pmatrix}$ (44C)		
Tetragonal	$\begin{pmatrix} s_{11} & s_{11} & s_{33} \\ 0 & 0 & 0 \end{pmatrix}$ (45A) $\begin{pmatrix} s_{11} & s_{22} & s_{22} \\ 0 & 0 & 0 \end{pmatrix}$ (45B)	$\begin{pmatrix} s_{11} & s_{11} & s_{33} \\ \frac{1}{2}s_{11} & \frac{1}{2}s_{11} & 0 \end{pmatrix}$ (45C) $\begin{pmatrix} s_{11} & s_{11} & s_{11} \\ \frac{1}{2}(\bar{s}_{11} - \bar{s}_{12}) & \frac{1}{2}(\bar{s}_{11} - \bar{s}_{12}) & \bar{s}_{12} \end{pmatrix}$ (45D) $\begin{pmatrix} s_{11} & s_{22} & s_{22} \\ \frac{1}{2}s_{11} & \frac{1}{2}s_{11} & \frac{1}{2}s_{11} \end{pmatrix}$ (45E)			
Hexagonal	$\begin{pmatrix} s_{11} & s_{11} & s_{33} \\ 0 & 0 & \frac{1}{2}s_{11} \end{pmatrix}$ (48A) $\begin{pmatrix} s_{11} & s_{22} & s_{22} \\ \frac{1}{2}s_{22} & 0 & 0 \end{pmatrix}$ (48B)				$\begin{pmatrix} s_{11} & s_{11} & s_{33} \\ \frac{1}{2}s_{11} & \frac{1}{2}s_{11} & \frac{1}{2}s_{11} \end{pmatrix}$ (49B) $\begin{pmatrix} s_{11} & s_{11} & s_{11} \\ s_{23} & s_{23} & s_{23} \end{pmatrix}$ (49C) $\begin{pmatrix} s_{11} & s_{11} & s_{11} \\ \bar{s}_{23} & s_{23} & s_{23} \end{pmatrix}$ (49D) $\begin{pmatrix} s_{11} & s_{22} & s_{22} \\ \frac{1}{2}(\bar{s}_{22} - \frac{1}{3}s_{11}) & \frac{1}{3}s_{11} & \frac{1}{3}s_{11} \end{pmatrix}$ (49E)

Table 6 (*Continued*)

	P	I	F	C	R
Orthorhombic	$\begin{pmatrix} s_{11} & s_{22} & s_{33} \\ 0 & \bar{s}_{31} & 0 \end{pmatrix}$ (50C)	$\begin{pmatrix} s_{11} & s_{11} & s_{11} \\ \bar{s}_{23} & (\bar{s}_{11} - \bar{s}_{23} & -\bar{s}_{31}) \end{pmatrix}$ (52A) $\begin{pmatrix} s_{11} & s_{22} & s_{22} \\ s_{23} & \frac{1}{2}s_{11} & \frac{1}{2}s_{11} \end{pmatrix}$ (52B) $\begin{pmatrix} s_{11} & s_{22} & s_{33} \\ \frac{1}{2}s_{22} & \frac{1}{2}s_{11} & 0 \end{pmatrix}$ (52C)	$\begin{pmatrix} s_{11} & s_{11} & s_{33} \\ \frac{1}{2}s_{22} & (\bar{s}_{11} & -2\bar{s}_{23}) \end{pmatrix}$ (51A) $\begin{pmatrix} s_{11} & s_{22} & s_{33} \\ s_{23} & s_{11} & s_{11} \end{pmatrix}$ (51B)	$\begin{pmatrix} s_{11} & s_{22} & s_{33} \\ 0 & \frac{1}{2}s_{11} & 0 \end{pmatrix}$ (50A) $\begin{pmatrix} s_{11} & s_{22} & s_{33} \\ 0 & 0 & \frac{1}{2}s_{11} \end{pmatrix}$ (50B) $\begin{pmatrix} s_{11} & s_{22} & s_{33} \\ \frac{1}{2}s_{22} & 0 & 0 \end{pmatrix}$ (50C) $\begin{pmatrix} s_{11} & s_{11} & s_{33} \\ 0 & & \frac{1}{2}s_{12} \end{pmatrix}$ (50D) $\begin{pmatrix} s_{11} & s_{22} & s_{22} \\ \frac{1}{2}s_{23} & 0 & 0 \end{pmatrix}$ (50E)	
Monoclinic	$\begin{pmatrix} s_1 & s_{22} & s_{33} \\ 0 & \bar{s}_{31} & 0 \end{pmatrix}$ (53A) $\begin{pmatrix} s_1 & s_{22} & s_{33} \\ \frac{1}{2}s_{13} & 0 & 0 \end{pmatrix}$ (53B) $\begin{pmatrix} s_1 & s_{22} & s_{33} \\ 0 & & s_{12} \end{pmatrix}$ (53C)	$\begin{pmatrix} s_{11} & s_{22} & s_{33} \\ \frac{1}{2}s_{22} & \frac{1}{2}s_{11} & \frac{1}{2}s_{12} \end{pmatrix}$ (57A) $\begin{pmatrix} s_{11} & s_{22} & s_{33} \\ s_{23} & \bar{s}_{31} & (\bar{s}_{11} -\bar{s}_{23} & -\bar{s}_{31}) \end{pmatrix}$ (57B) $\begin{pmatrix} s_{11} & s_{22} & s_{33} \\ s_{23} & \frac{1}{2}s_{11} & s_{11} \end{pmatrix}$ (57C)	$\begin{pmatrix} s_{11} & s_{22} & s_{33} \\ \frac{1}{2}s_{23} & 0 & \frac{1}{2}s_{11} \end{pmatrix}$ (54A) $\begin{pmatrix} s_{11} & s_{22} & s_{33} \\ \frac{1}{2}s_{22} & \frac{1}{2}s_{31} & 0 \end{pmatrix}$ (54B) $\begin{pmatrix} s_{11} & s_{22} & s_{33} \\ s_{23} & \frac{1}{2}s_{11} & 0 \end{pmatrix}$ (54C)	$\begin{pmatrix} s_{11} & s_{22} & s_{33} \\ s_{23} & \frac{1}{2}s_{23} & \frac{1}{2}s_{12} \end{pmatrix}$ (55A) $\begin{pmatrix} s_{11} & s_{22} & s_{22} \\ s_{23} & \bar{s}_{21} & \bar{s}_{21} \end{pmatrix}$ (55B)	$\begin{pmatrix} s_{11} & s_{11} & s_{33} \\ \frac{1}{2}s_{12} & \frac{1}{2}s_{11} & s_{12} \end{pmatrix}$ (56A) $\begin{pmatrix} s_{11} & s_{22} & s_{33} \\ \frac{1}{2}s_{22} & s_{12} & s_{12} \end{pmatrix}$ (56B) $\begin{pmatrix} s_{11} & s_{22} & s_{33} \\ \frac{1}{2}s_{31} & s_{31} & \frac{1}{2}s_{11} \end{pmatrix}$ (56C)
Triclinic	$\begin{pmatrix} s_{11} & s_{22} & s_{33} \\ s_{23} & s_{31} & s_{12} \end{pmatrix}$ (58A) $\begin{pmatrix} -s_{11} & s_{22} & s_{33} \\ \bar{s}_{23} & \bar{s}_{31} & \bar{s}_{12} \end{pmatrix}$ (58B)				

Table 7

Transformation matrices from reduced cells to unit cells

(Numbers are Niggli's figure numbers)

	P	I	F	C	R
Isometric	$\begin{pmatrix}1&0&0\\0&1&0\\0&0&1\end{pmatrix}$ (44A)	$\begin{pmatrix}1&0&1\\1&1&0\\0&1&1\end{pmatrix}$ (44B)	$\begin{pmatrix}1&\bar1&1\\1&1&\bar1\\\bar1&1&1\end{pmatrix}$ (44C)		
Tetragonal	$\begin{pmatrix}1&0&0\\0&1&0\\0&0&1\end{pmatrix}$ (45A) $\begin{pmatrix}0&0&1\\0&1&0\\1&0&0\end{pmatrix}$ (45B)	$\begin{pmatrix}1&0&0\\0&1&0\\1&1&2\end{pmatrix}$ (45C) $\begin{pmatrix}1&0&1\\1&1&0\\0&\bar1&\bar1\end{pmatrix}$ (45D) $\begin{pmatrix}0&\bar1&\bar1\\1&1&\bar1\\1&0&0\end{pmatrix}$ (45E)			
Hexagonal	$\begin{pmatrix}1&0&0\\0&1&0\\0&0&1\end{pmatrix}$ (48A) $\begin{pmatrix}0&1&0\\0&0&1\\1&0&0\end{pmatrix}$ (48B)				$\begin{pmatrix}1&0&0\\\bar1&1&0\\\bar1&\bar1&3\end{pmatrix}$ (49B) $\begin{pmatrix}\bar1&1&0\\1&\bar1&0\\1&1&\bar1\end{pmatrix}$ (49C) $\begin{pmatrix}\bar1&0&1\\\bar1&\bar1&0\\1&1&\bar1\end{pmatrix}$ (49D) $\begin{pmatrix}\bar1&2&1\\\bar1&\bar1&0\\1&0&0\end{pmatrix}$ (49E)

Table 7 (Continued)

	P	I	F	C	R
Orthorhombic	$\begin{pmatrix}1&0&0\\ \bar1&1&0\\ 0&0&1\end{pmatrix}$ (50C)	$\begin{pmatrix}1&1&0\\ 1&0&1\\ 0&\bar1&\bar1\end{pmatrix}$ (52A) $\begin{pmatrix}\bar1&1&0\\ 1&0&\bar1\\ 0&1&\bar1\end{pmatrix}$ (52B) $\begin{pmatrix}1&1&0\\ 0&1&0\\ \bar1&\bar1&\bar2\end{pmatrix}$ (52C)	$\begin{pmatrix}1&\bar1&0\\ 1&1&\bar2\\ \bar1&\bar1&0\end{pmatrix}$ (51A) $\begin{pmatrix}\bar1&2&0\\ 1&0&2\\ 1&0&0\end{pmatrix}$ (51B)	$\begin{pmatrix}\bar1&0&0\\ 0&0&\bar2\\ \bar1&0&0\end{pmatrix}$ (50A) $\begin{pmatrix}\bar1&0&0\\ \bar1&0&2\\ 0&1&0\end{pmatrix}$ (50B) $\begin{pmatrix}0&0&\bar1\\ \bar1&0&\bar2\\ 1&\bar1&0\end{pmatrix}$ (50F) $\begin{pmatrix}\bar1&0&0\\ \bar1&1&0\\ 0&1&\bar1\end{pmatrix}$ (50D) $\begin{pmatrix}0&\bar1&\bar1\\ \bar1&0&0\\ 0&0&1\end{pmatrix}$ (50E)	
Monoclinic	$\begin{pmatrix}1&0&0\\ 0&1&0\\ \bar1&0&\bar1\end{pmatrix}$ (53A) $\begin{pmatrix}1&0&0\\ 0&\bar1&0\\ \bar1&0&\bar1\end{pmatrix}$ (53B) $\begin{pmatrix}1&0&0\\ 0&0&\bar1\\ \bar1&\bar1&0\end{pmatrix}$ (53C)	$\begin{pmatrix}1&0&0\\ 0&\bar1&\bar2\\ \bar1&\bar1&0\end{pmatrix}$ (57A) $\begin{pmatrix}1&0&\bar1\\ \bar1&0&0\\ 1&\bar1&0\end{pmatrix}$ (57B) $\begin{pmatrix}\bar1&0&0\\ \bar1&\bar1&0\\ 0&0&1\end{pmatrix}$ (57C)	$\begin{pmatrix}1&2&0\\ 1&0&0\\ 0&\bar1&0\end{pmatrix}$ (54A) $\begin{pmatrix}\bar1&2&0\\ \bar1&0&0\\ 0&1&0\end{pmatrix}$ (54B) $\begin{pmatrix}1&2&0\\ 1&0&0\\ 1&0&0\end{pmatrix}$ (54C)	$\begin{pmatrix}\bar1&1&0\\ \bar1&\bar1&0\\ 0&0&1\end{pmatrix}$ (55A) $\begin{pmatrix}0&\bar1&\bar1\\ 0&\bar1&1\\ 1&0&0\end{pmatrix}$ (55B)	$\begin{pmatrix}\bar1&0&0\\ 0&2&0\\ \bar1&\bar1&0\end{pmatrix}$ (56A) $\begin{pmatrix}0&\bar1&2\\ 0&\bar1&0\\ \bar1&0&0\end{pmatrix}$ (56B) $\begin{pmatrix}0&0&0\\ 2&0&1\\ \bar1&\bar1&0\end{pmatrix}$ (56C)
Triclinic	$\begin{pmatrix}1&0&0\\ 0&1&0\\ 0&0&1\end{pmatrix}$ (58A) $\begin{pmatrix}0&0&1\\ 0&1&0\\ \bar1&0&0\end{pmatrix}$ (58B)				

Table 8
Determination of reduced-cell type by classification of symmetrical and unsymmetrical scalars

Symmetrical scalars $s_{11} = s_{22} = s_{33}$

Unsymmetrical scalars +	$\frac{1}{2}s_{11}$ $\frac{1}{2}s_{11}$ $\frac{1}{2}s_{11}$ s_{23} s_{23} s_{23}	Isometric Rhombohedral
Unsymmetrical scalars −	0 0 0 s_{23} s_{23} s_{23}	Isometric Rhombohedral

	Sum = s_{11} All equal
	          Two equal
	          Three unequal
		Isometric
		 Tetragonal (2 cases)
		 Orthorhombic

Symmetrical scalars $s_{11} = s_{22} \neq s_{33}$

Unsymmetrical scalars +	$\frac{1}{2}s_{11}$ $\frac{1}{2}s_{11}$ $\frac{1}{2}s_{11}$ s_{23} s_{23} s_{12}	Rhombohedral Monoclinic
Unsymmetrical scalars −	0 0 0 0 0 $\frac{1}{2}s_{11}$ 0 0 s_{12} s_{23} s_{23} s_{12}	Tetragonal Hexagonal Orthorhombic Monoclinic
	Sum = s_{11} $\frac{1}{2}s_{11}$ $\frac{1}{2}s_{11}$ 0 s_{23} s_{23} $(s_{11} - 2s_{23})$ s_{23} s_{13} $(s_{11} - s_{23} - s_{13})$	Tetragonal Orthorhombic Monoclinic

Symmetrical scalars $s_{11} \neq s_{22} = s_{23}$

Unsymmetrical scalars +	$\frac{1}{4}s_{11}$ $\frac{1}{2}s_{11}$ $\frac{1}{2}s_{11}$ s_{23} $\frac{1}{2}s_{11}$ $\frac{1}{2}s_{11}$ s_{23} s_{31} s_{31}	Tetragonal Orthorhombic Monoclinic
Unsymmetrical scalars −	0 0 0 $\frac{1}{2}s_{22}$ 0 0 s_{23} 0 0 $\frac{1}{2}(s_{22} - \frac{1}{3}s_{11})$ $\frac{1}{3}s_{11}$ $\frac{1}{3}s_{11}$ s_{23} s_{31} s_{31}	Tetragonal Hexagonal Orthorhombic Rhombohedral Monoclinic

Symmetrical scalars $s_{11} \neq s_{22} \neq s_{33}$

Unsymmetrical scalars +	$\frac{1}{4}s_{11}$ $\frac{1}{2}s_{11}$ $\frac{1}{2}s_{11}$ s_{23} $\frac{1}{2}s_{11}$ $\frac{1}{2}s_{11}$ $\frac{1}{2}s_{12}$ $\frac{1}{2}s_{11}$ s_{12} $\frac{1}{2}s_{31}$ s_{31} $\frac{1}{2}s_{11}$ $\frac{1}{2}s_{22}$ $\frac{1}{2}s_{12}$ s_{12}	Orthorhombic Monoclinic Monoclinic Monoclinic Monoclinic
	None of above conditions	Triclinic
Unsymmetrical scalars	0 0 0 0 s_{31} 0 0 0 s_{12} s_{23} 0 0	Orthorhombic Monoclinic Monoclinic Monoclinic
	$s_{31} = \frac{1}{2}s_{11}$ 0 $\frac{1}{2}s_{11}$ 0 s_{23} $\frac{1}{2}s_{11}$ 0	Orthorhombic Monoclinic
	$s_{12} = \frac{1}{2}s_{11}$ 0 0 $\frac{1}{2}s_{11}$ s_{23} 0 $\frac{1}{2}s_{11}$	Orthorhombic Monoclinic
	$s_{23} = \frac{1}{2}s_{22}$ $\frac{1}{2}s_{22}$ 0 0 $\frac{1}{2}s_{22}$ s_{31} 0	Orthorhombic Monoclinic
	Sum $= \frac{1}{2}(s_{11} + s_{22})$ $\frac{1}{2}s_{22}$ $\frac{1}{2}s_{11}$ 0 None zero	Orthorhombic Monoclinic
	None of these conditions	Triclinic

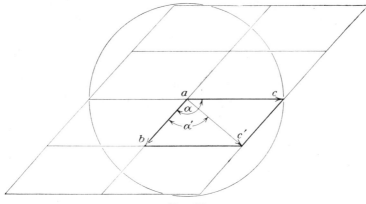

FIG. 20

represented by parallel nets **bc** whose origins are successively shifted by an amount δ in a direction φ. The magnitude of δ can be determined with the aid of (6). Since $d_{100} = 1/a^* = 1/.2131$,

$$\delta = a \cos \sin^{-1}\left(\frac{d_{100}}{a}\right)$$
$$= 4.693 \cos \sin^{-1}\left(\frac{4.692}{4.693}\right)$$
$$= 4.693 \cos \sin^{-1} .99978$$
$$= 4.693 \times .02094$$
$$= 0.098 \ \mathring{A}. \tag{32}$$

The azimuth angle φ can be determined with the aid of (8).

$$\varphi = \sin^{-1}\left(-\frac{\tan \rho}{\tan \gamma^*}\right)$$
$$= \sin^{-1}\left(-\frac{\tan 88°48'}{\tan 89°30'}\right)$$
$$= \sin^{-1}\left(-\frac{47.74}{114.59}\right)$$
$$= \sin^{-1}(-.4166)$$
$$= -24°37'. \tag{33}$$

One upper net only is drawn in Fig. 20. Since δ is so small, the upper net in this drawing superimposes on the original net drawn.

An examination of Fig. 20 immediately shows that $\mathbf{c}' = \mathbf{b} + \mathbf{c}$ is shorter than $\mathbf{c}$. Also, it is evident that the lattice point at the end of $\mathbf{a}$ is the point in the first level that is nearest to the origin. Hence, the

three shortest noncoplanar axes in Fig. 20 are **a**, **b**, and **c**′. The interaxial angles are all very close to 90°, too close, in fact, to be accurately
determined from a graphical construction. This example, therefore,
points out one of the difficulties of the graphical method, namely, its
limited accuracy in determining the interaxial angles.

In order to demonstrate the application of the graphical method more
clearly, another unit cell of the direct lattice of $MgWO_4$ is used to make

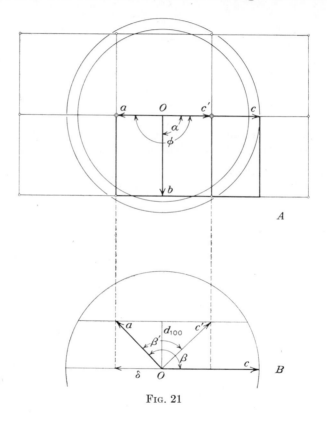

Fig. 21

the graphical construction shown in Fig. 21*A*. In this case, the lattice
is based on the unit cell derived by the Delaunay method in Chapter 12.

$$
\begin{aligned}
a &= 4.69 \text{ A,} & \alpha &= \ \ 90°00', \\
b &= 5.68 \text{ Å,} & \beta &= 133°15', \\
c &= 6.77 \text{ Å,} & \gamma &= \ \ 90°00'.
\end{aligned}
\tag{34}
$$

The two longest axes **c** and **b** are used to construct the equatorial net.
In order to project the next level, the displacement δ and the azimuth
angle φ are computed as before.

$$\delta = a \cos \sin^{-1}\left(\frac{d_{100}}{a}\right)$$
$$= 4.69 \cos \sin^{-1}\left(\frac{3.42}{4.69}\right)$$
$$= 4.69(.6843)$$
$$= 3.21 \text{ Å}, \tag{35}$$

and

$$\varphi = \sin^{-1}\left(-\frac{\tan \rho}{\tan \gamma^*}\right)$$
$$= \sin^{-1}\left(-\frac{1.066}{\infty}\right)$$
$$= \sin^{-1}(0)$$
$$= 180°. \tag{36}$$

Since the azimuth angle is zero, the **c** axis lying in the equator net, and in the next level above, overlap.

An examination of Fig. 21*A* shows that the lattice point at the end of the vector **c**′ is nearer to the origin than the lattice point at the end of the vector **c**. In order to help determine the three shortest noncoplanar translations of this lattice, an orthogonal projection along **b** is shown in Fig. 21*B*. It is evident that the two shortest axes in this plane are **c**′ and **a**. Their lengths and the angle β' can be readily determined from this construction. The reduced cell of $MgWO_4$ thus determined has

$$\begin{array}{ll} a' = 4.69 \text{ Å}, & \alpha' = 90°00', \\ b' = 5.68 \text{ Å}, & \beta' = 90°30', \\ c' = 4.95 \text{ Å}, & \gamma' = 90°00'. \end{array} \tag{37}$$

Analytical determination of the reduced cell in a primitive lattice. The algorithm for reduction in a band, described in a previous section, permits the determination of the reduced cell more rapidly than the graphical method. To illustrate this procedure, the direct cell of $MgWO_4$, derived by the Ito method in Chapter 10 and used in the first illustration above, is used as the initial cell.

From the initial-cell dimensions given in (31), one first computes

$$\begin{array}{ll} \mathbf{a}_1 \cdot \mathbf{a}_1 = a_1^2 = 22.02 & \cos \alpha_1 = -.6561 \\ \mathbf{b}_1 \cdot \mathbf{b}_1 = b_1^2 = 24.33 & \cos \beta_1 = +.0076 \\ \mathbf{c}_1 \cdot \mathbf{c}_1 = c_1^2 = 56.63 & \cos \gamma_1 = -.0116 \end{array} \tag{38}$$

$$\begin{array}{lll} \mathbf{b} \cdot \mathbf{c} = bc \cos \alpha = & (4.933)(7.525) \ (-.6561) = & -24.35 \\ \mathbf{c} \cdot \mathbf{a} = ca \cos \beta = & (7.525)(4.693) \ (+.0076) = & +0.27 \\ \mathbf{a} \cdot \mathbf{b} = ab \cos \gamma = & (4.693)(4.933) \ (-.0116) = & -0.27. \end{array}$$

These values are diagramed in Fig. 22*A*.

Since c_1 is greater than either a_1 or b_1, it is reduced first. Noting that

$|\mathbf{b}_1 \cdot \mathbf{c}_1| > \frac{1}{2}b_1^2$, this reduction is best carried out by successively adding vector $\mathbf{b}_1$ to vector $\mathbf{c}_1$ until the scalar product of $\mathbf{c}_2 = \mathbf{c}_1 + n\mathbf{b}_1$ with $\mathbf{b}_2 = \mathbf{b}_1$ is equal to or less than $\frac{1}{2}b_1^2$. This can be expressed analytically by

$$\mathbf{b}_2 \cdot \mathbf{c}_2 = \mathbf{b}_1 \cdot \mathbf{c}_1 + nb_1^2$$
$$= (-24.35) + n(24.33). \tag{39}$$

It is possible, by means of relation (39), to determine in one step how many translations $\mathbf{b}_1$ must be added vectorially to $\mathbf{c}_1$ to reduce c in this

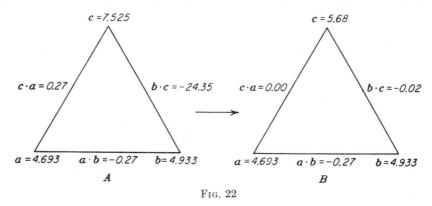

Fig. 22

band. For this case, adding $\mathbf{b}_1$ once reduces the value of the scalar to zero. The new axes, therefore, are

$$\mathbf{a}_2 = \mathbf{a}_1,$$
$$\mathbf{b}_2 = \mathbf{b}_1, \tag{40}$$
$$\mathbf{c}_2 = \mathbf{c}_1 + \mathbf{b}_1.$$

Accordingly, the new unsymmetrical scalar products are

$$\mathbf{a}_2 \cdot \mathbf{b}_2 = \mathbf{a}_1 \cdot \mathbf{b}_1$$
$$= -0.27.$$
$$\mathbf{b}_2 \cdot \mathbf{c}_2 = \mathbf{b}_1 \cdot \mathbf{c}_1 + b_1^2$$
$$= -24.35 + 24.33 \tag{41}$$
$$= -0.02$$
$$\mathbf{c}_2 \cdot \mathbf{a}_2 = \mathbf{c}_1 \cdot \mathbf{a}_1 + \mathbf{a}_1 \cdot \mathbf{b}_1$$
$$= 0.27 - 0.27$$
$$= 0.00.$$

These values are represented in the triangular diagram in Fig. 22*B*. It is evident in this figure that any further attempts at reduction would increase the magnitudes of the scalars and, hence, the lengths of the translations indicated at the corners of the triangle. The reduction is, therefore, complete.

The lengths of **a** and **b** in the reduced cell are unchanged. The length of **c** can be determined from (40) and (26).

$$\mathbf{c} = \mathbf{c}_1 + \mathbf{b}_1$$
$$\therefore c^2 = c_1^2 + 2\mathbf{c}_1 \cdot \mathbf{b}_1 + b_1^2$$
$$= (56.63) + (-48.70) + (24.33)$$
$$= 32.26, \tag{42}$$
$$c = \sqrt{c^2}$$
$$= \sqrt{32.26}$$
$$= 5.68 \text{ Å}.$$

Since the scalars $\mathbf{b} \cdot \mathbf{c}$ and $\mathbf{c} \cdot \mathbf{a}$ are zero, the cosines of the included angles must be zero, and $\alpha = \beta = 90°$. The value of γ can be determined from (17).

$$\cos \gamma = \frac{\mathbf{a} \cdot \mathbf{b}}{ab}$$
$$= \frac{-0.27}{23.16} \tag{43}$$
$$= -0.0112$$
$$\therefore \gamma = 90°39'.$$

The squares of the new axes and the data of (41) can be assembled in the form of a Niggli matrix of the reduced cell, as follows:

$$\begin{pmatrix} s_{11} & s_{22} & s_{33} \\ s_{23} & s_{31} & s_{12} \end{pmatrix} = \begin{pmatrix} 22.02 & 24.33 & 32.26 \\ -0.02 & 0.00 & -0.27 \end{pmatrix}. \tag{44}$$

The crystal system can be readily determined from the form of this matrix. Since $s_{11} \neq s_{22} \neq s_{33}$, the data fall into the last subdivision of Table 8. Since zero is regarded as negative (that is, since 90° is regarded as an obtuse angle for purposes of classification), the data fall into the subcategory "Unsymmetrical scalars $-$." In this subdivision, the crystal is immediately identified as monoclinic if the value -0.27 is accepted as significantly different from zero, while -0.02 is accepted as an approximation to zero. Reference to Table 6 shows that, in the monoclinic division, (44) represents a primitive monoclinic cell. Table 7 shows that the edges of the unit cell are the same as the edges of the reduced cell.

Analytical determination of the reduced cell in a centered lattice. It can be seen from the above examples that the band reduction method provides a rapid and accurate algorithm for finding the reduced cell in a lattice. It has the further advantage that the reduced-cell edges and their scalar products are determined directly, facilitating the comparison of the resulting matrix with the forty-one possible matrices in Table 6.

As a further illustration of this procedure, a primitive unit cell from the body-centered orthorhombic lattice of $Zn(NH_3)_2Br_2$ is reduced below. The initial unit-cell constants are

$$a = 8.41 \text{ Å}, \qquad \alpha = 14°11',$$
$$b = 11.68 \text{ Å}, \qquad \beta = 45°44', \qquad (45)$$
$$c = 18.97 \text{ Å}, \qquad \gamma = 43°57'.$$

From these one finds

$$\mathbf{a}_1 \cdot \mathbf{a}_1 = a_1^2 = 70.73, \qquad \cos \alpha_1 = +.9695,$$
$$\mathbf{b}_1 \cdot \mathbf{b}_1 = b_1^2 = 136.42, \qquad \cos \beta_1 = +.6980,$$
$$\mathbf{c}_1 \cdot \mathbf{c}_1 = c_1^2 = 326.52, \qquad \cos \gamma_1 = +.7200.$$

$$(46)$$

$$\mathbf{b}_1 \cdot \mathbf{c}_1 = bc \cos \alpha_1 = (11.68)(18.07)(.9695) = +204.63$$
$$\mathbf{c}_1 \cdot \mathbf{a}_1 = ca \cos \beta_1 = (18.07)(8.41)(.6980) = 106.08$$
$$\mathbf{a}_1 \cdot \mathbf{b}_1 = ab \cos \gamma_1 = (8.41)(11.68)(.7200) = +70.72.$$

These values are used to record the scalars on the triangular representation of Fig. 23A.

Let the largest translation $\mathbf{c}_1$ be the first candidate for reduction. Consider, therefore, the relation

$$\mathbf{b}_2 \cdot \mathbf{c}_2 = \mathbf{b}_1 \cdot \mathbf{c}_1 - nb^2$$
$$= (204.63) - n(136.42), \qquad (47)$$

from which it is evident that $\mathbf{c}_1$ can be reduced to the same length by vectorially subtracting either† $1\mathbf{b}_1$ or $2\mathbf{b}_1$. This first step in the reduction is illustrated in Fig. 23B. Note that the reduction of $\mathbf{b}_1$ reduces

† When (47) is examined, it becomes clear that n can equal 1 or 2 without changing the absolute magnitude of the scalar. Thus

$$\mathbf{b}_2 \cdot \mathbf{c}_2 = \mathbf{b}_1 \cdot \mathbf{c}_1 - 1b^2$$
$$= (204.63) - 1(136.42)$$
$$= 68.21,$$

or
$$\mathbf{b}_2 \cdot \mathbf{c}_2 = \mathbf{b}_1 \cdot \mathbf{c}_1 - 2b^2$$
$$= (204.63) - 2(136.42)$$
$$= -68.21.$$

Whenever a choice of sign exists, an all-obtuse cell is customarily chosen for the matrix representation in Table 6. Therefore, subsequent comparisons are facilitated by selecting an obtuse angle during the reduction procedure, provided that the translation being reduced is not lengthened thereby. In the above case, it is easy to show that the length of $\mathbf{c}_2$ is not affected. Let $\mathbf{c}_2 = \mathbf{c}_1 - \mathbf{b}_1$; then

$$c_2^2 = c_1^2 - 2\mathbf{b}_1 \cdot \mathbf{c}_1 + b_1^2$$
$$= (326.52) - (409.26) + (136.42)$$
$$= 53.68. \qquad (c = 7.33 \text{ Å})$$

Alternatively, let $\mathbf{c}_2 = \mathbf{c}_1 - 2\mathbf{b}_1$; then

$$c_2^2 = c_1^2 - 4\mathbf{b}_1 \cdot \mathbf{c}_1 + 4b_1^2$$
$$= (326.52) - (818.52) + (545.68)$$
$$= 53.68. \qquad (c = 7.33 \text{ Å})$$

not only the scalar product $\mathbf{b}_1 \cdot \mathbf{c}_1$ but also the scalar product $\mathbf{c}_1 \cdot \mathbf{a}_1$.

The longest remaining translation is $\mathbf{b}_2$. This translation can be reduced either by subtracting a_2 (since $a_2 \cdot b_2 > \frac{1}{2}b_2^2$) or by adding $\mathbf{c}_2$ (since $\mathbf{b}_2 \cdot \mathbf{c}_2 > c_2^2$). Since $\mathbf{c}_2$ already has been reduced, it is added to $\mathbf{b}_2$. The reduction proceeds as shown in Fig. 23C. The longest remaining translation $\mathbf{a}_3$ is next reduced by subtracting $\mathbf{b}_3$ from it. (Note that

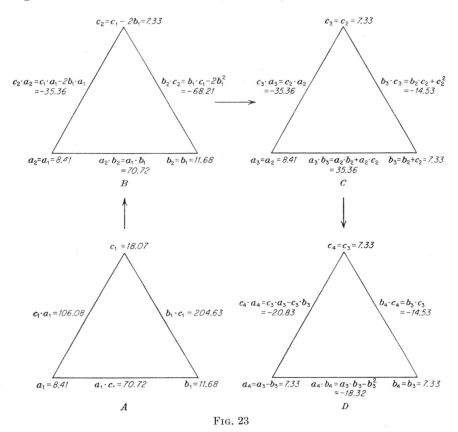

Fɪɢ. 23

alternatively $\mathbf{c}_3$ can be added to $\mathbf{a}_3$. This has the effect, however, of increasing the magnitude of the scalar $\mathbf{a}_3 \cdot \mathbf{b}_3$, necessitating one further step in the reduction.) This reduction is illustrated in Fig. 23D. As can be seen from this figure, no further reduction is possible.

The matrix representation of the scalars of the reduced cell in Fig. 23D is

$$\begin{pmatrix} s_{11} & s_{22} & s_{33} \\ s_{23} & s_{31} & s_{12} \end{pmatrix} = \begin{pmatrix} 53.68 & 53.68 & 53.68 \\ -14.53 & -20.83 & -18.32 \end{pmatrix}. \tag{48}$$

The crystal system corresponding to this matrix can be determined with the aid of Table 8.   One first notes that $s_{11} = s_{22} = s_{33}$, so that the matrix occurs in the first division of Table 8.   Since the unsymmetrical scalars are negative, the matrix belongs to the second subdivision.   Since the scalars are neither all zero, nor equal, the possibility is tested that their sum equals s_{11}:

$$14.53 + 20.83 + 18.32 = 53.68.$$

Since this proves to be the case, and since no two are equal, one is readily led to the conclusion that the crystal belongs to the orthorhombic system. Referring to the orthorhombic section of Table 6, it is evident that, since $s_{11} = s_{22} = s_{33}$, the crystal must have an I lattice, and correspond to the first case listed.

According to Table 7, the matrix of the transformation of this reduced cell to the unit cell is

$$\begin{pmatrix} 1 & 0 & 1 \\ 1 & 1 & 0 \\ 0 & 1 & 1 \end{pmatrix}. \tag{49}$$

Accordingly,

$$\begin{aligned}
\mathbf{A} &= \mathbf{a}_4 + \mathbf{c}_4 \\
\therefore A^2 &= a_4^2 + 2\mathbf{a}_4 \cdot \mathbf{c}_4 + c_4^2 \\
&= (53.68) + (-41.66) + (53.68) \\
&= 65.70 \qquad \therefore A = 8.01 \text{ Å}, \\
\mathbf{B} &= \mathbf{a}_4 + \mathbf{b}_4 \\
\therefore B^2 &= a_4^2 + 2\mathbf{a}_4 \cdot \mathbf{b}_4 + b_4^2 \\
&= (53.68) + (-36.44) + (53.68) \\
&= 70.92 \qquad \therefore B = 8.32 \text{ Å}, \\
\mathbf{C} &= \mathbf{b}_4 + \mathbf{c}_4 \\
\therefore C^2 &= b_4^2 + 2\mathbf{b}_4 \cdot \mathbf{c}_4 + c_4^2 \\
&= (53.68) + (-29.06) + (53.68) \\
&= 78.30 \qquad \therefore C = 8.85 \text{ Å}.
\end{aligned} \tag{50}$$

Since the crystal is orthorhombic, the three interaxial angles α, β, and γ are each right angles.

Literature

Crystallographic treatments of reduced cells

[1] M. A. Bravais, *Mémoire sur les systèmes formés par des points distribués régulièrement sur un plan ou dans l'espace*, J. école polytech. Cahier 33, Tome XIX (1850) 1–128.   Translated by Amos J. Shaler, Memoir 1, Crystallographic Society of America, 1949, especially 19–22, 48–50.

[2] A. Johnsen, *Kristallstruktur*, Fortschr. Mineral. 5 (1916) 17–130, especially 36–40.

[3] P. Niggli, *Kristallographische und strukturtheoretische Grundbegriffe*, Handbuch der Experimentalphysik **7**, Teil 1 (Akademische Verlagsgesellschaft, Leipzig, 1928) 108–176.

[4] M. J. Buerger, *X-ray crystallography* (John Wiley & Sons, Inc., New York, 1942) 364–365.

[5] M. J. Buerger, *Elementary crystallography* (John Wiley & Sons, Inc., New York, 1956) 107–111.

[6] V. Bolashov, *The choice of the unit cell in the triclinic system*, Acta Cryst. **9** (1956) 319–320.

[7] M. J. Buerger, *Reduced cells*, Z. Krist. **109** (1957) 42-60.

Mathematical treatment of reduced ternary quadratic forms

[8] L. A. Seeber, *Untersuchungen über die Eigenschaften der positiven ternären quadratischen Formen* (Freiburg, 1831).

[9] C. F. Gauss, *Recension der "Untersuchung über die Eigenschaften der positiven etrnären quadratischen Formen" von Ludwig August Seeber* (a review of Seeber's work), J. reine angew. Math. **20** (1840) 312–320.

[10] G. Legeune Dirichlet, *Uber die Reduction der positiven quadratischen Formen mit drei unbestimmten ganzen Zahlen*, J. reine angew. Math. **40** (1850) 209–227.

[11] G. Eisenstein, *Tabelle der reducierten positiven ternären quadratischen Formen, nebst den Resultaten neuer Forschungen über diese Formen, inbesondere Rücksicht auf ihre tabellarische Berechnung*, J. reine angew. Math. **41** (1851) 141–190, 227–242.

12

Homogeneous axes and the Delaunay reduction

When Ito devised his method of indexing powder photographs[3] he required a procedure for transforming the arbitrary cell, on the basis of which the photograph was first indexed, to a standard cell. He chose for this purpose a procedure suggested by Delaunay.[2] The background, theory, and practice of this procedure are given in this chapter. Actually, three separate and independent features are involved. These include the use of homogeneous coordinates, a particular program of transformations of the cell by steps, and the definition of the end point of the sequence of transformations. In this chapter these features are discussed separately.

Homogeneous axes

Two dimensions. A set of parameters which add to zero can be written as a homogeneous equation. For example, if the coordinates of a point P in a plane are x, y (Fig. 1), these values may be regarded as a vector of length x parallel to the X axis and a vector of length y parallel to the Y axis. Together they determine the location of point P. Let the vector from P to O be called $\mathbf{w}$. Then the three vectors $\mathbf{x}$, $\mathbf{y}$, and $\mathbf{w}$ form a closed circuit, so that

$$\mathbf{x} + \mathbf{y} + \mathbf{w} = 0. \tag{1}$$

The three vectors $\mathbf{x}$, $\mathbf{y}$, and $\mathbf{w}$ are called the homogeneous coordinates of point P.

Homogeneous coordinates can be applied to the geometrical properties of a lattice. In particular, if the axes of a plane unit cell are $\mathbf{a}$ and $\mathbf{b}$, the geometry of the cell can be referred to *homogeneous axes* $\mathbf{a}$, $\mathbf{b}$, and $\mathbf{d}$, which are related by

$$\mathbf{a} + \mathbf{b} + \mathbf{d} = 0. \tag{2}$$

This requires that the extra axis $\mathbf{d}$ has the value

$$\mathbf{d} = -(\mathbf{a} + \mathbf{b}). \tag{3}$$

This shows that the extra axis $\mathbf{d}$ is the negative of the sum of the other two. It is therefore the negative of the cell diagonal (Fig. 2).

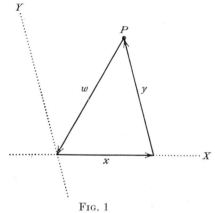

FIG. 1

Relation (2) shows that each of the three homogeneous axes ranks equally with the other two. Therefore, not only do the homogeneous axes $\mathbf{a}$, $\mathbf{b}$, and $\mathbf{d}$ define the cell based upon $\mathbf{a}$ and $\mathbf{b}$, but they also define the two additional cells based upon $\mathbf{a}$, $\mathbf{d}$ and upon $\mathbf{b}$, $\mathbf{d}$ (Fig. 3). In other words, homogeneous axes in a plane define a set of three closely related cells. In each of these cells, the cell diagonal is the reversed third homogeneous axis. The three cells can be called *fraternal*, and the set of three cells called a *fraternal set*.

A particular symmetrical variety of homogeneous axes is in common use by crystallographers for the hexagonal system. If one chooses axes

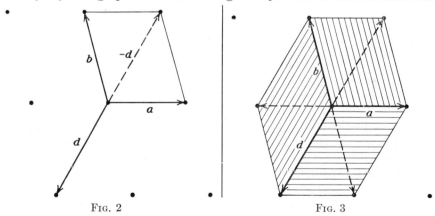

FIG. 2   FIG. 3

$\mathbf{a}_1$ and $\mathbf{a}_2$ at 120° to one another (Fig. 4) then symmetry requires an equivalent axis $\mathbf{a}_3$. Many of the relations which obtain for the three symmetrical homogeneous axes of the hexagonal system also exist for any two-dimensional set of homogeneous axes, although most of these properties are not specifically required here.

Let the homogeneous axes of a fraternal set of two-dimensional cells be **a**, **b**, and **d**.   Let the angles between these be given by

	a	b	d
a		γ	λ
b			μ
d			.

Then the scalar products of the pairs of axes in the fraternal set are

$$\mathbf{a} \cdot \mathbf{b} = ab \cos \gamma,$$
$$\mathbf{b} \cdot \mathbf{d} = bd \cos \mu, \qquad (4)$$
$$\mathbf{d} \cdot \mathbf{a} = da \cos \lambda.$$

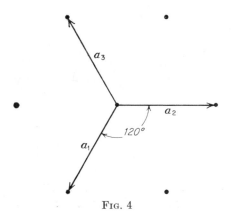

Fig. 4

Homogeneous axes have a computational advantage in performing transformations between cells.   In the last chapter it was seen that these transformations involve transformations between scalar products. The advantage comes about as follows: Since homogeneous axes are defined by (2), this can be solved for **a**, **b**, or **d**:

$$\mathbf{a} = -(\mathbf{b} + \mathbf{d}), \qquad (5)$$
$$\mathbf{b} = -(\mathbf{d} + \mathbf{a}), \qquad (6)$$
$$\mathbf{d} = -(\mathbf{a} + \mathbf{b}). \qquad (7)$$

If one forms the scalar product of **a** with (5), **b** with (6), and **d** with (7), there results

$$\mathbf{a} \cdot \mathbf{a} = a^2 = -(\mathbf{a} \cdot \mathbf{b} + \mathbf{a} \cdot \mathbf{d}), \qquad (8)$$
$$\mathbf{b} \cdot \mathbf{b} = b^2 = -(\mathbf{b} \cdot \mathbf{d} + \mathbf{b} \cdot \mathbf{a}), \qquad (9)$$
$$\mathbf{d} \cdot \mathbf{d} = d^2 = -(\mathbf{d} \cdot \mathbf{a} + \mathbf{d} \cdot \mathbf{b}). \qquad (10)$$

This shows that the square of the length of an axis can be computed from the scalar products of that axis with the other axes.   Specifically, the square of the length of an axis is the negative of the sum of its scalar products with the other axes.

Three dimensions.   For three-dimensional cells, the body diagonal of a cell is the vector sum of the cell edges.   Let the body diagonal be represented by $\bar{\mathbf{d}}$.   Then

$$\bar{\mathbf{d}} = \mathbf{a} + \mathbf{b} + \mathbf{c}. \qquad (11)$$

Let the vector which is the reverse of the body diagonal be defined as

$$\mathbf{d} = -\bar{\mathbf{d}}. \qquad (12)$$

Then, from (11), it follows that

$$\mathbf{a} + \mathbf{b} + \mathbf{c} + \mathbf{d} = 0. \qquad (13)$$

Since these four vectors sum to zero, they constitute a set of homogeneous axes in three dimensions (Fig. 5).

The four homogeneous axes can be taken three at a time in four ways:

$$
\begin{array}{cccc}
\mathbf{a,} & \mathbf{b,} & \mathbf{c} & \\
\mathbf{a,} & \mathbf{b,} & & \mathbf{d} \\
\mathbf{a,} & & \mathbf{c,} & \mathbf{d} \\
& \mathbf{b,} & \mathbf{c,} & \mathbf{d}
\end{array}
$$

The four cells defined by these sets of axes can be regarded as a set of fraternal cells in three dimensions.

The angles between the pairs of axes of the homogeneous set are

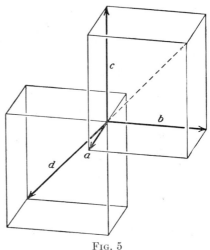

	a	b	c	d
a		γ	β	λ
b			α	μ
c				ν
d				

and the six scalar products are

$$\mathbf{a \cdot b} = ab \cos \gamma,$$
$$\mathbf{a \cdot c} = ac \cos \beta,$$
$$\mathbf{a \cdot d} = ad \cos \lambda,$$
$$\mathbf{b \cdot c} = bc \cos \alpha,$$
$$\mathbf{b \cdot d} = bd \cos \mu,$$
$$\mathbf{c \cdot d} = cd \cos \nu.$$

Fig. 5

Homogeneous axes in three dimensions offer the same computational advantage as homogeneous axes in two dimensions, namely, that the square of the length of any axis is equal to the negative of the sum of its scalar products with the other axes. The proof is similar to the proof for two dimensions. From (13), the several axes can be expressed in terms of the others, thus:

$$\mathbf{a} = -(\mathbf{b + c + d}), \tag{14}$$
$$\mathbf{b} = -(\mathbf{c + d + a}), \tag{15}$$
$$\mathbf{c} = -(\mathbf{d + a + b}), \tag{10}$$
$$\mathbf{d} = -(\mathbf{a + b + c}). \tag{17}$$

If the scalar product of each quantity on the left with itself is formed there results

$$\mathbf{a \cdot a} = a^2 = -(\mathbf{a \cdot b + a \cdot c + a \cdot d}), \tag{18}$$
$$\mathbf{b \cdot b} = b^2 = -(\mathbf{b \cdot c + b \cdot d + b \cdot a}), \tag{19}$$
$$\mathbf{c \cdot c} = c^2 = -(\mathbf{c \cdot d + c \cdot a + c \cdot b}), \tag{20}$$
$$\mathbf{d \cdot d} = d^2 = -(\mathbf{d \cdot a + d \cdot b + d \cdot c}). \tag{21}$$

Representations of a lattice

It is customary to represent a particular lattice by the geometry of one of its cells. In particular, if a cell has edges of lengths a, b, and c, between pairs of which the angles are α, β, and γ, then the set of six values a, b, c, α, β, and γ constitute a representation of the lattice.

There are, however, other, less obvious representations which can be derived from this one. These are based upon scalar products of pairs of axes. If the **a** axis is thought of as axis 1 and the **b** axis as axis 2, then the scalar product s_{12} is defined as

$$s_{12} = \mathbf{a} \cdot \mathbf{b} = ab \cos \gamma. \tag{22}$$

The total set of scalar products can be arranged in a matrix, as follows:

$$
\begin{array}{c|ccc}
 & \mathbf{a} & \mathbf{b} & \mathbf{c} \\
\hline
\mathbf{a} & s_{11} & s_{12} & s_{13} \\
\mathbf{b} & s_{21} & s_{22} & s_{23} \\
\mathbf{c} & s_{31} & s_{32} & s_{33}
\end{array}
\tag{23}
$$

Since $s_{12} = s_{21}$, etc., there are only six distinct numerical values in (23). They are customarily set down as a representation of the lattice as follows:

$$\begin{pmatrix} s_{11} & s_{22} & s_{33} \\ s_{23} & s_{31} & s_{12} \end{pmatrix}. \tag{24}$$

This constitutes a perfectly valid representation of the lattice, and the fundamental geometrical features of the cell can be derived from it, as follows:

$$
\begin{aligned}
a &= \sqrt{a^2} = \sqrt{\mathbf{a} \cdot \mathbf{a}} = \sqrt{s_{11}} \\
b &= \sqrt{b^2} = \sqrt{\mathbf{b} \cdot \mathbf{b}} = \sqrt{s_{22}} \\
c &= \sqrt{c^2} = \sqrt{\mathbf{c} \cdot \mathbf{c}} = \sqrt{s_{33}} \\
\cos \alpha &= \frac{bc \cos \alpha}{bc} = \frac{\mathbf{b} \cdot \mathbf{c}}{bc} = \frac{s_{23}}{\sqrt{s_{22} s_{33}}} \\
\cos \beta &= \frac{ca \cos \beta}{ca} = \frac{\mathbf{c} \cdot \mathbf{a}}{ca} = \frac{s_{31}}{\sqrt{s_{33} s_{11}}} \\
\cos \gamma &= \frac{ab \cos \gamma}{ab} = \frac{\mathbf{a} \cdot \mathbf{b}}{ab} = \frac{s_{12}}{\sqrt{s_{11} s_{22}}}.
\end{aligned}
\tag{25}
$$

If homogeneous axes are used, then there are also several ways of representing the lattice. For example, all four homogeneous axes and their interaxial angles could be given as a, b, c, d, α, β, γ, λ, μ, ν. Obviously, this set contains four parameters more than are necessary to characterize the six variables of the cell.

For four homogeneous axes there are $4^2 = 16$ scalar products:

	a	b	c	d
a	s_{11}	s_{12}	s_{13}	s_{14}
b	s_{21}	s_{22}	s_{23}	s_{24}
c	s_{31}	s_{32}	s_{33}	s_{34}
d	s_{41}	s_{42}	s_{43}	s_{44}

$$(26)$$

Since $s_{12} = s_{21}$, etc., there are ten distinct scalar products. But not all of these are independent. The sum of the terms in any row or column is zero. For example, since

$$s_{11} = \mathbf{a} \cdot \mathbf{a}$$
$$s_{12} = \mathbf{a} \cdot \mathbf{b}$$
$$s_{13} = \mathbf{a} \cdot \mathbf{c}$$
$$s_{14} = \mathbf{a} \cdot \mathbf{d},$$

$$(27)$$

if these are added, there results

$$s_{11} + s_{12} + s_{13} + s_{14} = \mathbf{a} \cdot \mathbf{a} + \mathbf{a} \cdot \mathbf{b} + \mathbf{a} \cdot \mathbf{c} + \mathbf{a} \cdot \mathbf{d}$$
$$= \mathbf{a} \cdot (\mathbf{a} + \mathbf{b} + \mathbf{c} + \mathbf{d}).$$

$$(28)$$

But, according to the definition of homogeneous axes and relation (13), the term in parentheses is zero. In a similar way, the sum of the terms in any row or column is $\mathbf{a} \cdot 0$, $\mathbf{b} \cdot 0$, $\mathbf{c} \cdot 0$, or $\mathbf{d} \cdot 0$.

This means that a limited number of scalars in (26) are independent. A particular choice of these scalars is interesting. Since, according to (28), any scalar of a row can be omitted without losing the characteristics of the cell, let the terms of the main diagonal be omitted. The remaining scalars are

—	s_{12}	s_{13}	s_{14}
s_{21}	—	s_{23}	s_{24}
s_{31}	s_{32}	—	s_{34}
s_{41}	s_{42}	s_{43}	—

$$(29)$$

Now the scalars symmetrical in the main diagonal are equal in pairs, so that only the following are discrete:

—	s_{12}	s_{13}	s_{14}
—	—	s_{23}	s_{24}
—	—	—	s_{34}
—	—	—	—

$$(30)$$

Thus, if homogeneous axes are employed, a lattice can be represented by the six scalar products of the form s_{ij} (thus omitting scalars of the form s_{ii}). The number of these parameters corresponds exactly to the number of parameters in the representation a, b, c, α, β, γ.

The six parameters of (30) were first used by Selling[1] in connection with ternary quadratic forms. Selling preferred to set these six quantities down in a rectangular matrix, as follows:

$$\begin{pmatrix} s_{23} & s_{13} & s_{12} \\ s_{14} & s_{24} & s_{34} \end{pmatrix} \tag{31}$$

which is shorthand notation for

$$\begin{pmatrix} bc \cos \alpha & ac \cos \beta & ab \cos \gamma \\ ad \cos \lambda & bd \cos \mu & cd \cos \nu \end{pmatrix}. \tag{32}$$

It will be seen that, whereas the parameters a, b, c, α, β, γ consist of a mixture of lengths and angles, the terms in (32) are all peculiar half-breeds of lengths and angles, and all have the same form, that is, they

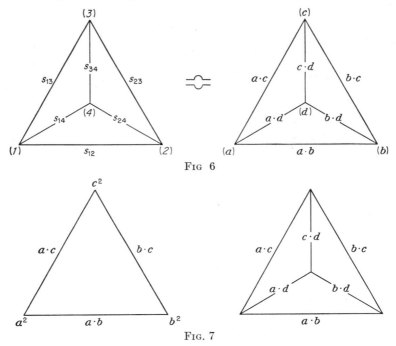

Fig 6

Fig. 7

are homogeneous. In fact, Selling called them *homogeneous coefficients* of the ternary quadratic form. They are more generally known as *Selling's parameters.*

Delaunay,[2] who made extensive use of Selling's parameters, called attention to the inhomogeneity of the form of (31) and (32). Since these six quantities are interrelated in threes, a rectangular matrix is poorly suited to display this. Delaunay introduced a tetrahedron, which he carefully called a "vierseit" (doubtless to avoid confusion with the Bravais tetrahedron, discussed in Chapter 11, here translated "*four-sider*." Delaunay's foursider is shown in Fig. 6. In these diagrams the

labels in parentheses at the corners are not part of the representation but
are added here merely to draw at-
tention to the fact that neighboring
each corner are three scalars each
involving the label in parentheses
at that corner.   While Delaunay's
foursider has the appearance of the
projection of a tetrahedron on its
base, it should be understood that
it is merely a four-cornered matrix.
One of Delaunay's claims for the
superiority of Fig. 6 over Selling's
original matrix (31) is that the foursider is "topologically homogeneous."   This means that it bears the
same close resemblance to the geometry it portrays that a good analogy
bears to the situation it illustrates.

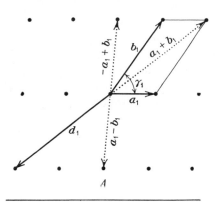

Note that each of the four sides
of Delaunay's foursider is the same
as the triangle representation of the
last chapter, except that, in the
Delaunay representation, there are
no scalars at the points (corresponding to axes) but only scalars along
the six lines.   These two representations of the same cell are shown
in Fig. 7.

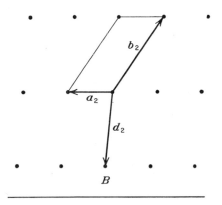

Delaunay's reduction

Two dimensions.   Delaunay[2]
discovered a transformation which,
when used with Selling's parameters, gives rise to an easily remembered transformation of scalar products.   The transformation trends
toward cells with short axes for a
reason which can be appreciated by
first studying the transformation in
two dimensions.

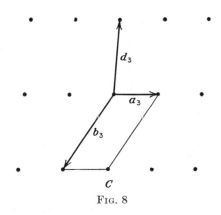

Fig. 8

In Fig. 8A, there is shown an oblique cell with edges $\mathbf{a}_1$, $\mathbf{b}_1$ and acute
angle γ_1.   Since γ is acute, the diagonal $\bar{\mathbf{d}} = \mathbf{a}_1 + \mathbf{b}_1$, which determines

the length of the third homogeneous axis $\mathbf{d}$, is longer than the other diagonal $\mathbf{a}_1 - \mathbf{b}_1 = -(-\mathbf{a}_1 + \mathbf{b}_1)$. If the direction of the $\mathbf{a}$ axis were reversed, thus making γ obtuse, the homogeneous set of axes would have the same a and b lengths, but a shorter d length. The result of this transformation is illustrated in Fig. 8B. Exactly the same transformation occurs if the direction of the $\mathbf{b}$ axis is reversed (Fig. 8C).

The transformations of axes and scalars involved in this cell transformation are as follows:

$$\begin{aligned}
\mathbf{a}_2 &= -\mathbf{a}_1 \\
\mathbf{b}_2 &= \mathbf{b}_1 \\
\mathbf{d}_2 &= \mathbf{d}_1 + 2\mathbf{a}_1, \\
\mathbf{a}_2 \cdot \mathbf{b}_2 &= (-\mathbf{a}_1) \cdot \mathbf{b}_1 = -\mathbf{a}_1 \cdot \mathbf{b}_1 \\
\mathbf{b}_2 \cdot \mathbf{d}_2 &= \mathbf{b}_1 \cdot (\mathbf{d}_1 + 2\mathbf{a}_1) = \mathbf{b}_1 \cdot \mathbf{d}_1 + 2\mathbf{a}_1 \cdot \mathbf{b}_1 \\
\mathbf{d}_2 \cdot \mathbf{a}_2 &= (\mathbf{d}_1 + 2\mathbf{a}_1) \cdot (-\mathbf{a}_1) = -\mathbf{d}_1 \cdot \mathbf{a}_1 - 2\mathbf{a}_1 \cdot \mathbf{a}_1 \\
&= -\mathbf{d}_1 \cdot \mathbf{a}_1 + 2\mathbf{a}_1(\mathbf{b}_1 + \mathbf{d}_1) \\
&= -\mathbf{d}_1 \cdot \mathbf{a}_1 + 2\mathbf{a}_1 \cdot \mathbf{b}_1 + 2\mathbf{a}_1 \cdot \mathbf{d}_1 \\
&= \mathbf{d}_1 \cdot \mathbf{a}_1 + 2\mathbf{a}_1 \cdot \mathbf{b}_1.
\end{aligned} \tag{33}$$

The two-dimensional analogue of Delaunay's foursider is a "threesider." It transforms as shown in Fig. 9.

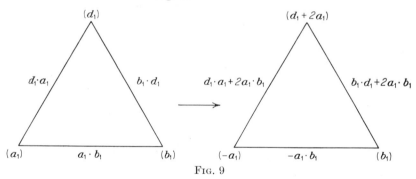

Fig. 9

This transformation can be described as follows: *If one scalar product (in this case, $\mathbf{a} \cdot \mathbf{b}$) is found to be positive (this is because γ is acute), the transformation consists of setting it down as negative and adding twice its value to the other two scalar products.*

This procedure reduces the numerical value of $a^2 + b^2 + d^2$. That this is so can be proved as follows:

$$\begin{aligned}
\mathbf{a}_2^2 + \mathbf{b}_2^2 + \mathbf{d}_2^2 &= (-\mathbf{a}_1)^2 + (\mathbf{b}_1)^2 + (\mathbf{d}_1 + 2\mathbf{a}_1)^2 \\
&= \mathbf{a}_1^2 + \mathbf{b}_1^2 + \mathbf{d}_1^2 + 4\mathbf{d}_1 \cdot \mathbf{a}_1 + 4\mathbf{a}_1^2 \\
&= \mathbf{a}_1^2 + \mathbf{b}_1^2 + \mathbf{d}_1^2 + 4\mathbf{a}_1(\mathbf{d}_1 + \mathbf{a}_1) \\
&= \mathbf{a}_1^2 + \mathbf{b}_1^2 + \mathbf{d}_1^2 + 4\mathbf{a}_1(\mathbf{d}_1 - \mathbf{b}_1 - \mathbf{d}_1) \\
&= \mathbf{a}_1^2 + \mathbf{b}_1^2 + \mathbf{d}_1^2 - 4\mathbf{a}_1 \cdot \mathbf{b}_1.
\end{aligned} \tag{34}$$

Thus the new sum of the squares of the homogeneous axes is less than the old sum by the quantity $4\mathbf{a}_1 \cdot \mathbf{b}_1$. It is from this reduction of the sum of the squares of the lengths of the axes of the homogeneous set that the transformation gets the name *Delaunay reduction*.

If one step of this transformation does not make all scalars negative (or just zero) it should be repeated until all three scalars are negative (or zero). In Delaunay's nomenclature a threesider would be called a *reduced threesider* when this condition is achieved.

Each threesider defines a set of axes to which correspond a set of fraternal cells. When a reduced threesider has been found, it implies a fraternal set which consists of a reduced cell, plus a half-reduced cell, plus a quarter-reduced cell.† It should be observed that the

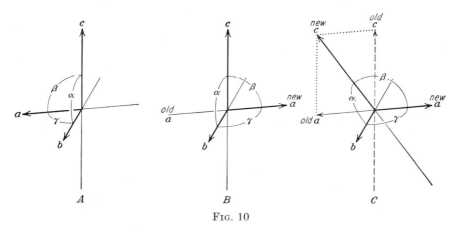

Fɪɢ. 10

word "reduced" is used here in two different senses. The reduced three-sider is the end product of the Delaunay reduction in two dimensions. The reduced cell is the cell which has for its edges the two shortest translations of the lattice. Fortunately, in two dimensions the reduced threesider includes the reduced cell among its three cells. In two dimensions, therefore, finding the reduced threesider is tantamount to finding the reduced cell. In three dimensions, on the other hand, this is not necessarily true, as discussed in the next section.

Three dimensions. That the Delaunay reduction reduces the magnitude of $a^2 + b^2 + d^2$ depends entirely upon changing a long diagonal for a shorter one. When an attempt is made to do this in three dimensions, complications result. Suppose that the original cell has two obtuse angles α and β but the third one, γ, is acute (Fig. 10*A*). The acute angle can be transformed into an obtuse angle by reversing the direction of either one of its legs, say $\mathbf{a}$ (Fig. 10*B*). But, as pointed out in Chapter

† See Chapter 11, page 126.

11, this procedure not only reverses the character of $\mathbf{a} \wedge \mathbf{b} = \gamma$ and renders it obtuse, but also reverses the character of $\mathbf{a} \wedge \mathbf{c} = \beta$. Since this last was obtuse it becomes acute.

There are several ways of handling this difficulty. Delaunay dealt with it specifically by changing the direction of the $\mathbf{c}$ axis so that the new $\mathbf{c}$ becomes old $\mathbf{c} + \mathbf{a}$ (Fig. 10*C*). The result is a cell corner with obtuse angles. This assures that the diagonal of the cell is a short diagonal. This is the basic background of the Delaunay reduction scheme.

This procedure can be expressed analytically by considering the relation between the four homogeneous axes

$$\mathbf{a} + \mathbf{b} + \mathbf{c} + \mathbf{d} = 0. \tag{35}$$

The acute angle γ is rendered obtuse by reversing the direction of $\mathbf{a}$. In order to preserve the homogeneity of (35), add and subtract $\mathbf{a}$ in (35) to give

$$\mathbf{a} + \mathbf{b} + \mathbf{c} + \mathbf{d} + \mathbf{a} - \mathbf{a} = 0. \tag{36}$$

The terms in (36) can be regrouped to give

$$-\mathbf{a} + \mathbf{b} + (\mathbf{c} + \mathbf{a}) + (\mathbf{d} + \mathbf{a}) = 0. \tag{37}$$

This can be described as a transformation of axes in which $\mathbf{a}$ is replaced by $-\mathbf{a}$, $\mathbf{b}$ remains unchanged, $\mathbf{c}$ is replaced by $\mathbf{c} + \mathbf{a}$, and $\mathbf{d}$ by $\mathbf{d} + \mathbf{a}$.

More specific aspects of the Delaunay reduction are as follows: The original cell is represented by the scalar products of its homogeneous axes arranged on the Delaunay foursider (Fig. 6). Delaunay defined a *reduced foursider*† as one for which all six scalars are nonpositive (that is, negative, but with the possibility that one or more could be just zero). If one or more are positive, the foursider is not reduced, but it can be reduced by the following transformation: Suppose that the scalar $\mathbf{a} \cdot \mathbf{b}$ is positive. It can be rendered negative by replacing $\mathbf{a} \cdot \mathbf{b}$ by $-\mathbf{a} \cdot \mathbf{b}$, and this can be caused by replacing $\mathbf{a}$ by $-\mathbf{a}$ or $\mathbf{b}$ by $-\mathbf{b}$. Suppose $\mathbf{a}$ is replaced by $-\mathbf{a}$; then, as discussed above, $\mathbf{c}$ should be replaced by $\mathbf{c} + \mathbf{a}$. The transformation of all four homogeneous axes then becomes

$$
\begin{aligned}
\mathbf{a}_2 &= -\mathbf{a}_1 \\
\mathbf{b}_2 &= \mathbf{b}_1 \\
\mathbf{c}_2 &= \mathbf{c}_1 + \mathbf{a}_1 \\
\mathbf{d}_2 &= -\mathbf{a}_2 - \mathbf{b}_2 - \mathbf{c}_2 \\
&= \mathbf{a}_1 - \mathbf{b}_1 - \mathbf{c}_1 - \mathbf{a}_1 \\
&= \mathbf{d}_1 + \mathbf{a}_1 \\
&= -\mathbf{b}_1 - \mathbf{c}_1.
\end{aligned}
\tag{38}
$$

This transformation was diagramed in Fig. 10.

† This does not necessarily imply that any of the cells of the foursider is reduced.

For this transformation, the scalars transform as follows:

$$
\begin{aligned}
\mathbf{a}_2 \cdot \mathbf{b}_2 &= -\mathbf{a}_1 \cdot \mathbf{b}_1, \\
\mathbf{b}_2 \cdot \mathbf{c}_2 &= \mathbf{b}_1 \cdot (\mathbf{c}_1 + \mathbf{a}_1) \\
&= \mathbf{b}_1 \cdot \mathbf{c}_1 + \mathbf{b}_1 \cdot \mathbf{a}_1, \\
\mathbf{c}_2 \cdot \mathbf{a}_2 &= (\mathbf{c}_1 + \mathbf{a}_1) \cdot (-\mathbf{a}_1) \\
&= -\mathbf{c}_1 \cdot \mathbf{a}_1 - \mathbf{a}_1 \cdot \mathbf{a}_1 \\
&= -\mathbf{c}_1 \cdot \mathbf{a}_1 + (\mathbf{b}_1 + \mathbf{c}_1 + \mathbf{d}_1) \cdot \mathbf{a}_1 \\
&= -\mathbf{c}_1 \cdot \mathbf{a}_1 + \mathbf{b}_1 \cdot \mathbf{a}_1 + \mathbf{c}_1 \cdot \mathbf{a}_1 + \mathbf{d}_1 \cdot \mathbf{a}_1 \\
&= \mathbf{b}_1 \cdot \mathbf{a}_1 + \mathbf{d}_1 \cdot \mathbf{a}_1, \\
\mathbf{a}_2 \cdot \mathbf{d}_2 &= -\mathbf{a}_1 \cdot (-\mathbf{b}_1 - \mathbf{c}_1) \\
&= \mathbf{a}_1 \cdot \mathbf{b}_1 + \mathbf{a}_1 \cdot \mathbf{c}_1, \\
\mathbf{b}_2 \cdot \mathbf{d}_2 &= \mathbf{b}_1 \cdot (-\mathbf{b}_1 - \mathbf{c}_1) \\
&= -\mathbf{b}_1 \cdot \mathbf{b}_1 - \mathbf{b}_1 \cdot \mathbf{c}_1 \\
&= (\mathbf{a}_1 + \mathbf{c}_1 + \mathbf{d}_1) \cdot \mathbf{b}_1 - \mathbf{b}_1 \cdot \mathbf{c}_1 \\
&= \mathbf{a}_1 \cdot \mathbf{b}_1 + \mathbf{c}_1 \cdot \mathbf{b}_1 + \mathbf{d}_1 \cdot \mathbf{b}_1 - \mathbf{b}_1 \cdot \mathbf{c}_1 \\
&= \mathbf{a}_1 \cdot \mathbf{b}_1 + \mathbf{d}_1 \cdot \mathbf{b}_1, \\
\mathbf{c}_2 \cdot \mathbf{d}_2 &= (\mathbf{c}_1 + \mathbf{a}_1) \cdot (-\mathbf{b}_1 - \mathbf{c}_1) \\
&= -\mathbf{c}_1 \cdot \mathbf{b}_1 - \mathbf{c}_1 \cdot \mathbf{c}_1 - \mathbf{a}_1 \cdot \mathbf{b}_1 - \mathbf{a}_1 \cdot \mathbf{c}_1 \\
&= -\mathbf{c}_1 \cdot \mathbf{b}_1 + (\mathbf{a}_1 + \mathbf{b}_1 + \mathbf{d}_1) \cdot \mathbf{c}_1 - \mathbf{a}_1 \cdot \mathbf{b}_1 - \mathbf{a}_1 \cdot \mathbf{c}_1 \\
&= -\mathbf{c}_1 \cdot \mathbf{b}_1 + \mathbf{a}_1 \cdot \mathbf{c}_1 + \mathbf{b}_1 \cdot \mathbf{c}_1 + \mathbf{d}_1 \cdot \mathbf{c}_1 - \mathbf{a}_1 \cdot \mathbf{b}_1 - \mathbf{a}_1 \cdot \mathbf{c}_1 \\
&= \mathbf{d}_1 \cdot \mathbf{c}_1 - \mathbf{a}_1 \cdot \mathbf{b}_1.
\end{aligned}
\tag{39}
$$

This transformation can be expressed in Selling's parameters as follows:

$$
\begin{aligned}
s_{12} &\longrightarrow -s_{12} \\
s_{23} &\longrightarrow s_{23} + s_{12} \\
s_{13} &\longrightarrow s_{14} + s_{12} \Big| \\
s_{14} &\longrightarrow s_{13} + s_{12} \Big/ \\
s_{24} &\longrightarrow s_{24} + s_{12} \\
s_{34} &\longrightarrow s_{34} - s_{12}.
\end{aligned}
\tag{40}
$$

The feature which gives this particular transformation of six scalars its popularity is the simplicity of the way these scalars transform. One notes that transformation of scalars involves a manipulation of the scalar s_{12} which was changed from positive to negative. This scalar is either added to or subtracted from each scalar on the foursider, except that the results are interchanged for the scalar s_{13} (which connects the reversed axis and the changed axis) and the scalar s_{14} (which is the third scalar attached to the reversed axes). This interchanged pair is bracketed in (40). The transformation is diagramed in Fig. 11. The way to manipulate the diagrams during a reduction will be discussed in the next section.

Each step in such a reduction reduces the quantity $a^2 + b^2 + c^2 + d^2$, which is the sum of the quantities in the main diagonal of a matrix composed of all scalar products. That this is so is proved as follows:

$$\mathbf{a}_2^2 + \mathbf{b}_2^2 + \mathbf{c}_2^2 + \mathbf{d}_2^2 = (-\mathbf{a}_1)^2 + \mathbf{b}_1^2 + (\mathbf{c}_1 + \mathbf{a}_1)^2 + (\mathbf{d}_1 + \mathbf{a}_1)^2$$

$$= \mathbf{a}_1^2 + \mathbf{b}_1^2 + \mathbf{c}_1^2 + 2\mathbf{a}_1 \cdot \mathbf{c}_1 + \mathbf{a}_1^2$$
$$+ \mathbf{d}_1^2 + 2\mathbf{d}_1 \cdot \mathbf{a}_1 + \mathbf{a}_1^2$$
$$= \mathbf{a}_1^2 + \mathbf{b}_1^2 + \mathbf{c}_1^2 + \mathbf{d}_1^2 + 2\mathbf{a}_1(\mathbf{a}_1 + \mathbf{c}_1 + \mathbf{d}_1)$$
$$= \mathbf{a}_1^2 + \mathbf{b}_1^2 + \mathbf{c}_1^2 + \mathbf{d}_1^2 - 2\mathbf{a}_1 \cdot \mathbf{b}_1. \tag{41}$$

That is, each transformation which changes a positive scalar $\mathbf{a} \cdot \mathbf{b}$ to a negative scalar $-\mathbf{a} \cdot \mathbf{b}$ reduces the quantity $a^2 + b^2 + c^2 + d^2$ by the

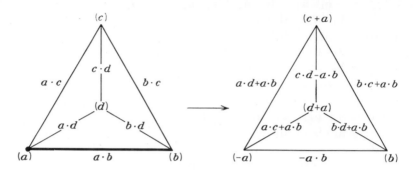

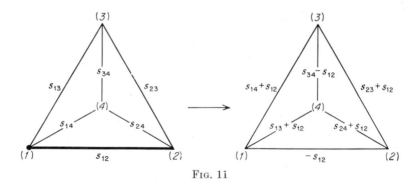

F IG. 1i

value of $2\mathbf{a} \cdot \mathbf{b}$.   If the new set of scalar products still has a nonnegative member, the transformation may be repeated, with a consequent new reduction of the quantity $a^2 + b^2 + c^2 + d^2$.   After a limited number of such transformations the reduced foursider is obtained.

The twenty-four special cases of reduced foursiders

To each of the fourteen space-lattice types there corresponds one or more special types of Delaunay foursiders.   This has been investigated and tabulated by Delaunay,[2] and the twenty-four possible foursiders are

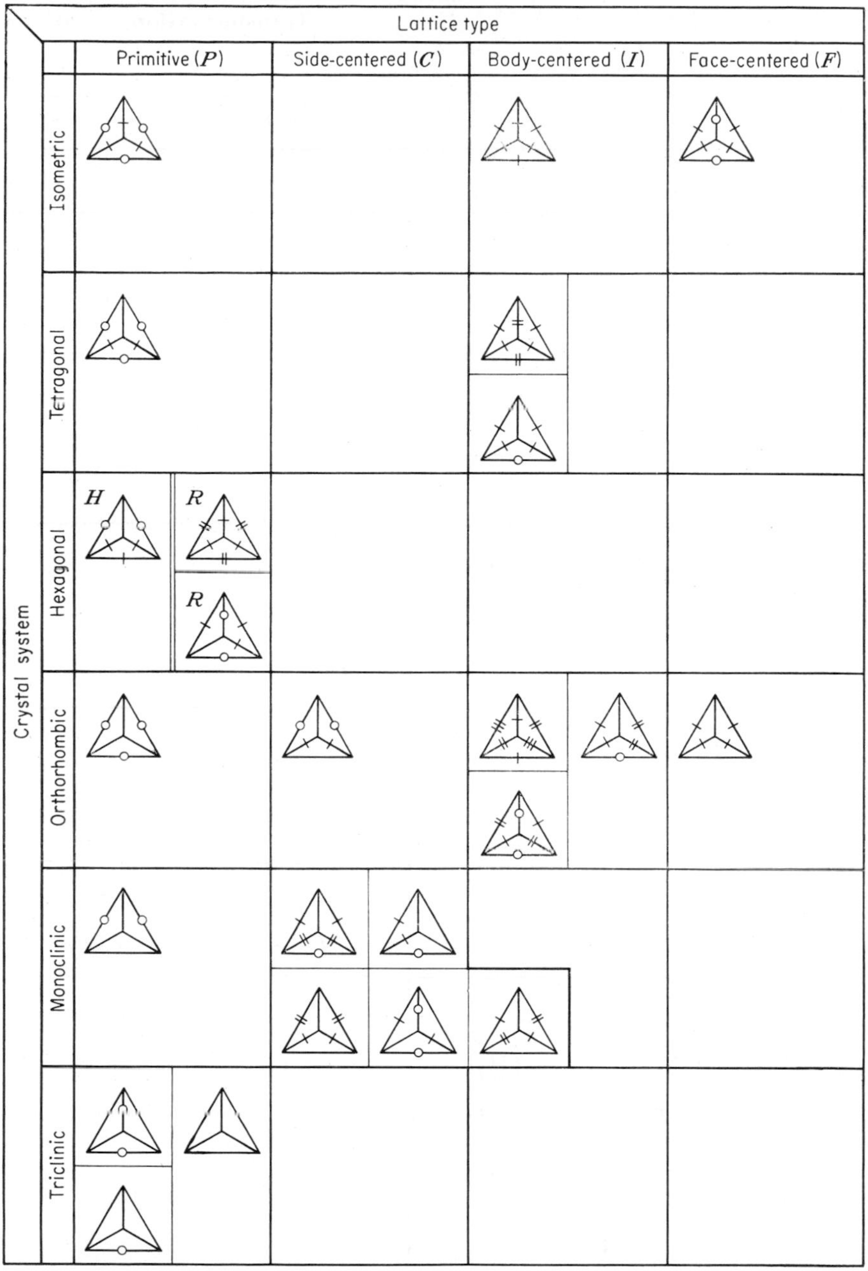

Lattice type
Primitive (*P*)
Side-centered (*C*)
Body-centered (*I*)
Face-centered (*F*)
Crystal system
Isometric
Tetragonal
Hexagonal
Orthorhombic
Monoclinic
Triclinic
H
R
R

Fig. 12

Table 1

Transformation equations

Crystal system	Lattice type		
	Primitive (P)		Side-centered (C)
Isometric	$a = b = c = \sqrt{-s_{14}}$ $\alpha = \beta = \gamma = 90°$		
Tetragonal	$a = b = \sqrt{-s_{14}}$ $c = \sqrt{-s_{34}}$ $\alpha = \beta = \gamma = 90°$		
Hexagonal	$a = b = \sqrt{-2s_{14}}$ $c = \sqrt{-s_{34}}$ $\alpha = \beta = 90°$ $\gamma = 120°$	$a = \sqrt{-s_{14} - 2s_{12}}$ $\alpha = \cos^{-1}\left(\dfrac{s_{12}}{-s_{14} - 2s_{12}}\right)$ $a = \sqrt{-s_{14} - s_{31}}$ $\alpha = \cos^{-1}\left(\dfrac{-s_{14}}{-s_{14} - s_{31}}\right)$	
Orthorhombic	$a = \sqrt{-s_{14}}$ $b = \sqrt{-s_{24}}$ $c = \sqrt{-s_{34}}$ $\alpha = \beta = \gamma = 90°$		$a = \sqrt{-2s_{14} - 4s_{12}}$ $b = \sqrt{-2s_{14}}$ $c = \sqrt{-s_{34}}$ $\alpha = \beta = \gamma = 90°$
Monoclinic	$a = \sqrt{-s_{14} - s_{12}}$ $b = \sqrt{-s_{12} - s_{24}}$ $c = \sqrt{-s_{34}}$ $\alpha = \beta = 90°$ $\gamma = \cos^{-1}\left(\dfrac{s_{12}}{ab}\right)$		$a = \sqrt{2s_{31} - 2s_{14}}$ $b = \sqrt{-2s_{31} - s_{34}}$ $c = \sqrt{-2s_{31} - 2s_{14}}$ $\alpha = \beta = 90°$ $\gamma = \cos^{-1}\left(\dfrac{2s_{31}}{ab}\right)$ $a = \sqrt{-4s_{34} - 4s_{23} - 2s_{14}}$ $b = \sqrt{-s_{24} - s_{23}}$ $c = \sqrt{-2s_{14}}$ $\alpha = \beta = 90°$ $\gamma = \cos^{-1}\left(\dfrac{2s_{23}}{ab}\right)$ $a = \sqrt{-2s_{31} - 2s_{14}}$ $b = \sqrt{-2s_{31} - s_{34}}$ $c = \sqrt{-2s_{31} - 2s_{14} - 4s_{12}}$ $\alpha = \beta = 90°$ $\gamma = \cos^{-1}\left(\dfrac{2s_{31}}{ab}\right)$ $a = \sqrt{-4s_{23} - 2s_{31}}$ $b = \sqrt{-s_{23} - s_{14}}$ $c = \sqrt{-2s_{31}}$ $\alpha = \beta = 90°$ $\gamma = \cos^{-1}\left(\dfrac{2s_{23}}{ab}\right)$
Triclinic	$a = \sqrt{-s_{14} - s_{31}}$ $b = \sqrt{-s_{24} - s_{23}}$ $c = \sqrt{-s_{23} - s_{31}}$ $\alpha = \cos^{-1}\left(\dfrac{s_{23}}{bc}\right)$ $\beta = \cos^{-1}\left(\dfrac{s_{31}}{ca}\right)$ $\gamma = 90°$ $a = \sqrt{-s_{14} - s_{31}}$ $b = \sqrt{-s_{24} - s_{23}}$ $c = \sqrt{-s_{23} - s_{34} - s_{31}}$ $\alpha = \cos^{-1}\left(\dfrac{s_{23}}{bc}\right)$ $\beta = \cos^{-1}\left(\dfrac{s_{31}}{ca}\right)$ $\gamma = 90°$	$a = \sqrt{-s_{12} - s_{14} - s_{31}}$ $b = \sqrt{-s_{12} - s_{24} - s_{23}}$ $c = \sqrt{-s_{23} - s_{34} - s_{31}}$ $\alpha = \cos^{-1}\left(\dfrac{s_{23}}{bc}\right)$ $\beta = \cos^{-1}\left(\dfrac{s_{31}}{ca}\right)$ $\gamma = \cos^{-1}\left(\dfrac{s_{12}}{ab}\right)$	

for Delaunay's foursiders in Fig. 12

	Lattice type		
	Body-centered (I)		**Face-centered (F)**
	$a = b = c = 2\sqrt{-s_{14}}$ $\alpha = \beta = \gamma = 90°$		$a = b = c = 2\sqrt{-s_{14}}$ $\alpha = \beta = \gamma = 90°$
	$a = b = \sqrt{-2s_{12} - 2s_{14}}$ $c = 2\sqrt{-s_{14}}$ $\alpha = \beta = \gamma = 90°$ $a = b = \sqrt{-2s_{14}}$ $c = 2\sqrt{-s_{14} - s_{34}}$ $\alpha = \beta = \gamma = 90°$		
	$a = \sqrt{-2s_{12} - 2s_{14}}$ $b = \sqrt{-2s_{14} - 2s_{31}}$ $c = \sqrt{-2s_{31} - 2s_{12}}$ $\alpha = \beta = \gamma = 90°$ $a = \sqrt{-2s_{31}}$ $b = \sqrt{-2s_{14}}$ $c = \sqrt{-2s_{31} - 2s_{14}}$ $\alpha = \beta = \gamma = 90°$	$a = \sqrt{-2s_{24}}$ $b = \sqrt{-2s_{14}}$ $c = \sqrt{-2s_{14} - 2s_{24} - 4s_{34}}$ $\alpha = \beta = \gamma = 90°$	$a = \sqrt{-4s_{14} - 4s_{12}}$ $b = \sqrt{-4s_{11} - 4s_{34}}$ $c = \sqrt{-4s_{14}}$ $\alpha = \beta = \gamma = 90°$
	$a = \sqrt{-2s_{31} - 2s_{14} - 4s_{34}}$ $b = \sqrt{-2s_{31} - s_{12} - s_{34}}$ $c = \sqrt{-2s_{31} - 2s_{14}}$ $\alpha = \beta = 90°$ $\gamma = \cos^{-1}\left(\dfrac{2s_{31} + 2s_{34}}{ab}\right)$		

shown in Fig. 12. If the space-lattice type is primitive, one of the four cells of the Delaunay foursider can be chosen directly as the unit cell. If the lattice type is not primitive, however, it is necessary to transform the Delaunay cell to the appropriate centered unit cell. To facilitate doing this, the necessary transformation equations based on the four-sider representation in Fig. 6 are tabulated in Table 1.

The finally reduced foursider can be compared with the twenty-four possible cases in Fig. 12. In making this comparison, two points should be kept in mind. The first concerns the presence of zeros in the four-sider diagram. When a zero occurs, an ambiguity arises as to the choice of direction for the two axes whose scalar product is zero. This is because the reversal of one axis changes the interaxial angle from 90° to 270° but does not change the value of the cosine, which remains zero. This ambiguity is further discussed below.

The second point concerns the symmetry of the twenty-four cases in Fig. 12. The symmetry tabulated by Delaunay and reproduced in Fig. 12 is minimal in the sense that additional equalities between scalars may occur fortuitously. When this additional symmetry is not specifically required by one of the other cases in Fig. 12, these extra equalities do not mean that a new type of reduced cell has been obtained, but rather that the cell has equalities not required by the symmetry of the lattice.

After the reduction is completed, the unit-cell parameters can be determined from the final set of s_{ij}'s. It should be realized that, because of experimental errors in the measured d values, the s_{ij}'s may not be exactly equal to zero or to each other. On the other hand, the differences may be real, in which case the lattice has only the pseudosymmetry indicated by the appropriate foursider in Fig. 12. When such ambiguity arises, it can usually be resolved by refining the unit-cell parameters as described in Chapter 15.

Application of the Delaunay reduction procedure

In principle, the Delaunay reduction procedure can be carried out equally well in reciprocal space or in crystal space. In actual practice, it is not advisable to carry out the reduction in reciprocal space since additional reduction procedures may be required after transforming the reciprocal cell to the direct cell. This is because an all-obtuse reciprocal cell does not transform to an all-obtuse direct cell. Thus if the reduction is carried out in reciprocal space, the reduced all-obtuse reciprocal cell transforms to an acute direct cell if the lattice is triclinic. If the lattice is monoclinic, the transformed cell has one acute angle which is the supplement to the correct obtuse angle. This problem does not arise if the lattice is orthogonal.

In the last chapter, a primitive cell for $MgWO_4$ found by Ito's method from powder data was reduced. The manipulation of this same cell by the Delaunay reduction procedure is given below. The initial data, from (31) of Chapter 9, but with axes relabeled so that $a < b < c$, are

$$\begin{aligned}
a &= 4.693 \text{ Å} & \alpha &= 131°00' \\
b &= 4.936 & \beta &= 89°34' \\
c &= 7.524 & \gamma &= 90°40'.
\end{aligned} \tag{42}$$

One needs, in addition, the following cosine values

$$\begin{aligned}
\cos \alpha &= -.6561 \\
\cos \beta &= +.0076 \\
\cos \gamma &= -.0116
\end{aligned} \tag{43}$$

The s_{ij}'s are first computed according to (31) and (32):

$$\begin{aligned}
s_{11} &= a^2 = (4.693)^2 = 22.02 \\
s_{22} &= b^2 = (4.936)^2 = 24.36 \\
s_{33} &= c^2 = (7.524)^2 = 56.61 \\
s_{12} &= ab \cos \gamma = (4.693)(4.936)(-.0116) = - \quad .27 \\
s_{23} &= bc \cos \alpha = (4.936)(7.524)(-.6561) = -24.37 \\
s_{31} &= ca \cos \beta = (7.524)(4.693)(.0076) \quad = + \quad .27
\end{aligned} \tag{44}$$

and, according to (28),

$$\begin{aligned}
s_{14} &= -s_{11} - s_{12} - s_{13} \\
&= -22.02 + 0.27 - 0.27 \\
&= -22.02 \\
s_{24} &= -s_{21} - s_{22} - s_{23} \\
&= +0.27 - 24.36 + 24.37 \\
&= +.28 \\
s_{34} &= -s_{31} - s_{32} - s_{33} \\
&= -0.27 + 24.37 - 56.61 \\
&= -32.51.
\end{aligned} \tag{45}$$

The values of the scalars computed in (44) and (45) are substituted in the appropriate places in the Delaunay foursider shown in Fig. 6. The actual reduction is carried out by means of the foursider diagrams illustrated in Fig. 13.

When the reduced foursider is compared with the twenty-four cases tabulated in Fig. 12, it appears at first glance that a match cannot be found. Actually, the foursider for the monoclinic primitive case has the same symmetry as the final foursider in Fig. 13, namely, two adjacent edges have zero scalars. Remembering that the Delaunay foursider represents a three-dimensional matrix drawn in projection on one of its sides, the foursider can be rolled over onto any other side as well. If the final four-sider in Fig. 13 is rolled over as shown in Fig. 14, the identity of the two diagrams is obvious.

Once the reduced foursider is obtained, it remains only to select the unit cell characteristic of the lattice type. Since the lattice is primitive

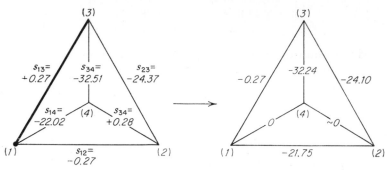

Fɪɢ. 13

monoclinic, the unique axis is the one that makes a right angle with the other two axes. In Fig. 13, this is the axis **d** which forms a right angle with **a** and **b** as evidenced by the zero scalars s_{14} and s_{24} lying along the edges of the four-sider joining **d** with **a** and **d** with **b**, respectively. According to (26), (27), and (28), the magnitudes of new **a**, **d**, and **b** are

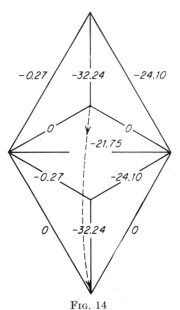

Fɪɢ. 14

$$a = \sqrt{s_{11}}$$
$$= \sqrt{-s_{12} - s_{13} - s_{14}}$$
$$= \sqrt{+21.75 + .27 - 0}$$
$$= 4.69 \text{ Å},$$
$$d = \sqrt{s_{44}}$$
$$= \sqrt{-s_{41} - s_{42} - s_{43}} \qquad (46)$$
$$= \sqrt{0 - 0 + 32.24}$$
$$= 5.68 \text{ Å},$$
$$b = \sqrt{s_{22}}$$
$$= \sqrt{-s_{21} - s_{23} - s_{24}}$$
$$= \sqrt{+21.75 + 24.10 - 0}$$
$$= 6.77 \text{ Å}.$$

The monoclinic angle is the obtuse angle formed by the two edges a and b as given by

$$s_{12} = ab \cos a \wedge b. \qquad (47)$$

The angle $(a \wedge b)$ can be determined from (47) as follows:

$$(a \wedge b) = \cos^{-1}\left\{\frac{s_{12}}{ab}\right\}$$

$$= \cos^{-1}\left\{\frac{-21.75}{(4.69)(6.77)}\right\}$$

$$= \cos^{-1}(-.6850)$$

$$= 133°15'. \tag{48}$$

The three cell edges and the monoclinic axis can now be renamed according to either the c unique or b unique convention.

The above discussion gives an example of the procedure to be used in determining the unit cell from the scalars of the reduced foursider. If the lattice should turn out to be a centered lattice, then the transformation equations in Table 1 can be used to obtain the appropriate centered cell directly.

The monoclinic ambiguity[4]

It has already been stated in a previous section that, whenever one scalar is zero, two different cells can be chosen, and whenever two scalars are zero, three different cells can be chosen. The latter case is typical of the primitive monoclinic lattice and is described in detail below.

Assuming the b unique convention, the cell obtained in the previous section according to (47) and (48) is

$$a = 4.69 \text{ Å},$$
$$b = 5.68 \text{ A},$$
$$c = 6.77 \text{ Å}, \tag{49}$$
$$\beta = 133°15'.$$

The Delaunay foursider corresponding to this cell is shown in the center of Fig. 15A. The two alternative cells are obtained by reversing the direction of either **a** or **c** and are also shown in the foursider representation in Fig. 15A. At first glance, it is not obvious which of the three foursiders in this figure contains the *reduced cell*. To answer this question, the three respective cells can be determined directly from the new scalar products as in the previous section, and the cell with the three shortest axes selected.

In the monoclinic system, it is very easy to do this graphically by preparing a scale drawing of the lattice. It is a simple matter to select the reduced cell with the aid of such a drawing, since the choice lies in two dimensions. Figure 15B shows the axes of the three cells corresponding to the three cases discussed above drawn to scale. The unit cell in the center of this figure is the unit cell given at the beginning of this section.

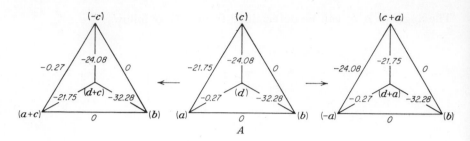

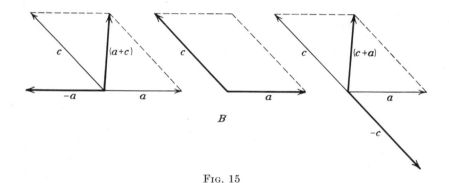

FIG. 15

It is obvious that the face diagonal **a** + **c** is shorter than **c**, and that the reduced cell is the one shown on the left, i.e., the cell that has $\mathbf{c}' = \mathbf{c} + \mathbf{a}$ and $\mathbf{a}' = -\mathbf{a}$ as the clino axes. Therefore, the reduced cell for $MgWO_4$ is

$$
\begin{aligned}
a &= 4.69 \text{ Å} \\
b &= 5.68 \\
c &= 4.93 \\
\beta &= 90°20'.
\end{aligned}
\tag{50}
$$

It should be clear from the above that the unit cell obtained from the reduced Delaunay foursider is not necessarily the reduced cell of the lattice. It is for this reason that the reduction procedures described in Chapter 11 are to be preferred.

Literature

[1] Eduard Selling, *Ueber die binäre und ternäre quadratischen Formen*, J. reine angew. Math. **77** (1874) 143–229, especially 164.

[2] B. Delaunay, *Neue Darstellung der geometrischen Kristallographie*, Z. Krist. **84** (1933) 109–149.

[3] T. Ito, *X-ray studies on polymorphism* (Maruzen Co. Ltd., Tokyo, 1950) 187–228.

[4] A. L. Patterson and Warner E. Love, *Remarks on the Delaunay reduction*, Acta Cryst. **10** (1957) 111–116.

13

Identification of substances
by the powder method

The identification of the constituents in a given substance with the aid
of the powder photograph of that substance has become one of the widest
applications of the powder method. Since the x-ray diffraction diagram
of a crystalline material is characteristic of the atomic arrangement in
that material, it is similar to a fingerprint in that no two substances give
rise to exactly identical diagrams. Conversely, if two materials do give
rise to identical powder photographs, the two materials must be identical. It follows, therefore, that powder photographs of known materials
can be used to identify the composition of an unknown material.

As early as 1919, Hull[†] used the powder method to show that a supposedly chemically pure sodium fluoride contained some acid-fluoride con-
taminant. This was done by comparing standard photographs of $NaHF_2$
and NaF with a photograph of the chemical investigated. Whenever
such standard films are available, they provide a convenient way for
identifying the constituents of an unknown. To become a practical
method, however, the direct comparison of films would require the existence of a vast library of films of known substances. Not only is such a
library difficult to obtain, but its utilization would entail a tedious comparison between the test photograph and successive standard films.
Although such a procedure, therefore, is obviously not suitable for routine identifications, it can be used to great advantage when a group of
structurally related compounds is being studied. For example, if one is
studying complex sulfosalts, it is possible to prepare a small library of
films which can then be used to identify an unknown sulfosalt. In such
a group, the common structural features often give rise to lines which are

[†] A. W. Hull, *A new method of chemical analysis*, J. Am. Chem. Soc. **41** (1919)
1168–1175.

common to the whole group. The lines not common to the group, however, are the ones most helpful in distinguishing between the various members of the groups.

If one deals with an isostructural series, such a library of standard films can be most helpful in the identification of any member in that series. Such a procedure has been suggested by Frevel[1] for the identification of powder photographs of unknown materials. Frevel's procedure assumes that the unknown has a structure similar to that of one of a limited number of known structure types. The interplanar spacings of the unknown are plotted to a logarithmic scale on a strip of paper. The strip of paper is then moved about on a plot of log d values of the known structure types, in a manner similar to the graphical indexing procedures described in Chapter 7, until all the lines marked on the paper strip correspond to lines on the plot. Frevel prepared several such plots[2–5] for groups of crystals belonging to the isometric and tetragonal systems. By using several isomorphous compounds of each structure type, a range of cell dimensions for each structure type is included in the chart. Although such charts can be useful in special cases, they cannot be used as a general procedure for identifying unknowns because this would necessitate making plots for the very large number of structure types known.

For a given experimental wavelength, a powder photograph has two unique characteristics: the position of the lines on the photograph and the relative blackness of these lines. The blackness of a line depends on the intensity of the diffracted beam which, in turn, is a function of the arrangement of atoms in the cell. On the other hand, the positions of the lines on a film depend solely on the unit-cell dimensions. It is possible, therefore, to use either one, or both of these characteristics to help identify an unknown. Thus, the methods of Chapters 7, 8, or 10 can be used to determine the unit cell from a powder photograph. These values can then be used to identify the unknown by looking up similar values listed in one of several compendia.[6–10] Such a procedure is, by far, the most satisfactory method for identifying an unknown because every line in the photograph is accounted for. It can, however, be very time-consuming if the unknown crystal has low symmetry. Moreover, if the unknown is a mixture of more than one kind of crystal, use of this method alone is futile, although it is a valuable adjunct to the procedure described below.

It is possible to use both characteristics of a powder photograph simultaneously. When a test film is compared with a standard film in the procedure described above, this is actually done. As an alternative to preparing a library of standard films, the pertinent characteristics of each film can be recorded on suitable cards, and a catalogue of such cards, schematically arranged, can then be used to identify an unknown. Such catalogues were independently suggested in 1938 by Hanawalt, Rinn, and Frevel[11] in the United States, and Boldyrev, Mikheev, and Kovalev[12] in Russia. Principally because of the efforts of Davey, the American Society for Testing Materials (ASTM), in conjunction with the National Research Council, agreed to publish the original Hanawalt catalogue printed on cards. The so-called *ASTM Index* was later supplemented by similar data gathered by the British Institute of Physics, which amalgamated the Russian data with data published by Harcourt[8] for ore materials, as well as other unpublished British data. Subsequently, the Joint Committee on Chemical Analysis by Powder Diffraction Methods, comprising representatives from ASTM, American Crystallographic Association, Institute of Physics, and National Association of Corrosion Engineers, was formed in order to assume responsibility for the *X-ray powder data file*. At the time of this writing, the *X-ray powder data file* comprises over five thousand entries printed on plain cards, Keysort cards, and IBM cards. The use of these cards is described in a subsequent section.

It should be remembered that the powder method is only one of several analytical procedures that can be used. Before 1938, the two principal methods of identifying an unknown were chemical analysis and optical crystallographic procedures using the polarizing microscope. Chemical methods are capable of disclosing the atomic composition but not the nature of the atomic aggregation. The polarizing microscope can be used in a manner similar to the powder method to disclose both the composition and state of aggregation of the atoms. This is because the indices of refraction for visible or infrared light are characteristic of a crystalline substance, as is its powder photograph. Optical methods cannot be used, however, if the particles are too small or if the substance is opaque to visible or infrared light. There are other analytical methods, such as optical spectroscopy and x-ray fluorescence analysis, that can also be used for analysis. Although all these methods have individual limitations not present in the powder method, their usefulness should not be overlooked; on the contrary, they should be used in conjunction with diffraction methods to simplify the problem of identification.

Identification procedures

In their original publication, Hanawalt, Rinn, and Frevel[11] tabulated the interplanar spacings and relative intensities of one thousand compounds. They also suggested a procedure for preparing a card file of these data so that the card describing a given compound could be readily located. The classification scheme proposed lists the *d* values of the three most intense (blackest) lines first. The relative intensities are expressed as a percentage of the intensity of the most intense line present, that is, the blackest line is assigned an intensity of 100, and the intensities of the other lines are scaled accordingly. To identify an unknown compound, therefore, it is necessary to arrange the measured interplanar spacings in the order of decreasing blackness. The d values of the three most intense lines can then be used to find a match in the card file. This classification scheme has been adopted, with only minor modifications, by the *X-ray powder data file*, and is in use today. Since intensity measurement is a part of this procedure, it will be described before the details of the identification procedure are discussed.

Elementary theory of intensity measurement. When electromagnetic radiation in the form of x-rays strikes a photographic film it causes a photochemical reaction to take place in the emulsion. After the film is processed the activated silver grains separate from the silver halide emulsion to form the opaque portions of the developed negative. The optical density of the freed silver is therefore a measure of the amount of radiation received by that spot on the film. Since the processed negative is viewed with visible light, it is convenient to define a density D or blackening B in terms of this light. This is defined as

$$D = B = \log_{10} \frac{I_0}{I},$$
(1)

where I_0 and I are, respectively, the intensities of the light incident upon and transmitted through the film. Thus, a blackening of unity means that one-tenth of the incident light is transmitted. To the average eye, a blackening of 2 $(I/I_0 = \frac{1}{100})$ appears very dark and constitutes an approximate upper limit. That is, the eye cannot distinguish between densities much greater than 2.

The photographic film normally used in x-ray diffraction studies is double-coated (that is, it has emulsions on both sides of the cellulose support); it is capable of recording blackening up to $B = 3.0$. Furthermore, provided that the film has been processed according to the manufacturer's instructions, the blackening is linearly related to the exposure on any one film. When a spot on the film reaches a blackening of 3, all the silver grains in that spot have been activated and any additional radiation falling on that spot cannot increase the value of B. To prevent overexposure of the film, it is necessary, therefore, to limit the exposures by some means. This can be done either by decreasing the time of exposure or by decreasing the intensity of the x-ray beam, or both. According to the *reciprocity law* of optics,

$$\text{Intensity} \times \text{time} = \text{constant blackening.}$$
(2)

The meaning of (2) is that the blackening of the lines in a powder photograph remains constant if the x-ray beam intensity is increased by the same factor that the exposure time is decreased. Furthermore, decreasing the exposure time, say, by a factor of 2, while keeping the x-ray beam intensity constant, decreases the blackening of the lines in the photograph by the same factor, 2.

Photometry. The eye's inability to perceive blackening greater than 2 necessitates the use of more sensitive instruments whenever greater accuracy is required. An instrument for doing this is called a densitometer, or, more correctly, a photometer. The basic components of the photometer are schematically illustrated in Fig. 1. The photometer consists of a light source, condenser, film holder, and detector. The beam of light is focused on the film, and the transmitted light is refocused on an aperture placed in front of a photoelectric cell or other type of detector. Although such a simple photometer is easily constructed and therefore

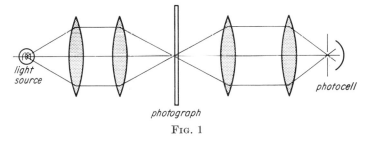

photograph

F_{IG}. 1

quite useful for routine intensity measurements, it is not very accurate. These inaccuracies are due to changes in the incident light intensity caused by line-voltage fluctuations, and the aging, or sagging, of the filament in the bulb. This shortcoming can be overcome by using a double photometer (Fig. 2) in which light from the same source is transmitted through the film and through a reference plate consisting of a gradation of densities. The so-called neutral wedge or reference plate is moved past the light beam until the density of the wedge is the same as that of the film. The blackening of the film is thus determined by the relative position of the wedge. This null method is independent of fluctuations in the light intensity and has the further advantage of determining the blackening, or density, directly. Any possible error caused by differences in sensitivity of the photocells can be eliminated by interchanging the cells and repeating the readings. Design details of several photometers are given in the references listed at the end of this chapter.

Practical considerations. Whether photometric or visual methods are used to measure the intensities, it is advisable to prepare a scale with which the observed line densities can be compared. Such a scale can be prepared in the laboratory by exposing different parts of a film to a uniform x-ray beam for predetermined lengths of time. According to the reciprocity law (2), the blackening of each part of the film is directly proportional to the length of the exposure. Such a scale can be prepared either by placing the film behind a metal shield in which a suitable opening has been made, and irradiating the opening with a direct x-ray beam coming from a tube at least 1 meter away (to ensure uniformity over the entire opening), or by allowing a beam diffracted by an actual powder sample to fall on a suitably shielded film.

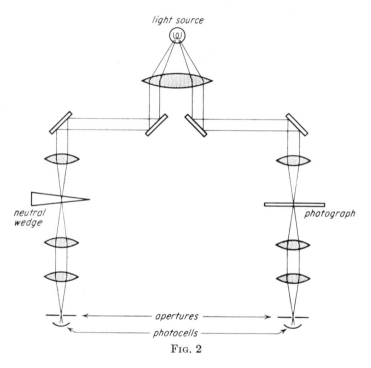

Fig. 2

A much more satisfactory procedure for preparing such a scale is to place a film in a light-tight envelope and position it behind a circular plate perforated by annular arcs of varying lengths (Fig. 3A). The entire assembly is placed several feet from an x-ray tube, and the circular plate is rotated at a constant speed. The angles of the arcs determine the total exposure times of different parts of the film and, therefore, their relative blackening. Instead of cutting arcs in a plate, it is simpler to construct a stepped or spiral shutter such as the ones shown in Fig. 3B and C. The spiral shutter produces a continuously changing gradation in place of the discrete steps produced by a stepped shutter. Since blackening depends on the processing conditions, greater accuracy can be achieved by placing the scale directly on the same film on which the diffraction lines are recorded.   Such a procedure, of course, requires that a part of the film be shielded from x-rays during the diffraction experiment.

It often happens that some of the lines are much blacker than others. It is not possible in such cases to record the weaker reflections without overexposing the stronger ones.   According to the reciprocity law, however, it is possible to expose different films for different lengths of time, thus making possible the measurement of strong intensities on one film and weak intensities on another.   An alternative procedure is to place several films behind each other into a camera and to expose all the films simultaneously.   Each thickness of film cuts down the intensity of the x-rays by a readily determined factor (to about one-fourth for $CuK\alpha$). If the design of the camera does not permit the insertion of more than one film at a time, a metal foil of known thickness and absorption can be

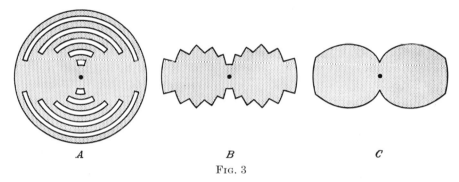

A   B   C

Fig. 3

used to cover one-half of the film.   Whenever these so-called multiple-film techniques are used, care should be exercised to process all the films in an identical manner.   The intensities of the different films can be placed on the same scale by selecting as internal calibration standards certain lines that appear on at least two films.

The X-ray powder data file.   A typical card from the current *X-ray powder data file* is reproduced in Fig 4.   On the first line, the interplanar spacings of the three most intense lines are arranged in order of decreasing intensity; the fourth d value is the largest observed spacing.   The relative intensities of these reflections, based on 100 for the strongest intensity observed, are listed on the next line directly below the corresponding d values.   The space to the right of these values is reserved for the chemical formula and the name of the material, as well as other descriptive information such as the mineral name or structural formula of an organic compound.   The next lines, which are self-explanatory, list the experimental data, crystallographic data, and optical data.   The last line provides miscellaneous information pertaining to any unusual

7-210

d	3.10	4.76	3.07	4.76	CaWO₄
I/I_1	100	53	31	53	

CALCIUM TUNGSTATE　　　　　　　　　SCHEELITE

Rad. CuKα₁ λ 1.5405　Filter Ni　Dia.
Cut off　　I/I₁ Spectrometer
Ref. NBS Circular 539, Vol 6 (1956)

Sys. Tetragonal　　S.G. $C^6_{4H} - I4_1/A$
a₀ 5.242　b₀　c₀ 11.372　A　　C
α　β　γ　Z 4　Dx 6.12
Ref. Ibid.

εα 1.918　nωβ 1.935　εγ　Sign +
2V　D　mp　Color Colorless
Ref. Ibid.

Sample from Kernville, California; spect. analysis
shows < 1.0% Mo; < 0.1% Na; < 0.01% Al, Si, Sr; < 0.001%
Ag, Cr, Cu, Mg, Mn.
Pattern made at 25°C.

Replaces 1-0806

d Å	I/I_1	hkl	d Å	I/I_1	hkl
4.76	53	101	1.3358	3	217
3.10	100	112	1.3106	3	400
3.072	31	103	1.2638	2	411
2.844	14	004	1.2488	13	316
2.622	23	200	1.2284	2	109
2.296	19	211	1.2074	5	332
2.256	3	114	1.2054	5	413
2.0864	5	105	1.1901	4	404,307
1.9951	13	213	1.1728	1	420
1.9278	28	204	1.1280	5	228
1.8538	12	220	1.1096	2	415
1.7278	5	301	1.0870	5	1.1.10,424
1.6882	16	116	1.0838	8	327,501
1.6332	10	215	1.0439	3	431,2.0.10
1.5921	30	312	1.0351	2	336
1.5532	14	224	1.0140	6	1.0.11
1.4427	6	321	1.0116	4	512
1.4219	2	008	0.9699	1	521,2.2.10
1.3859	3	305	.9636	4	408
1.3577	4	323	PLUS 26 LINES TO 0.7937		

Fɪɢ. 4

treatment of the sample, its source, or its chemical composition.   The complete list of observed interplanar spacings, the relative reflection intensities, and the reflection indices when known, are tabulated in the right-hand section of the card.

In its current form, *X-ray powder data file* and its accompanying *Index* volume consists of seven sets which may be used separately or in combination.   The cards in each set are arranged in order of decreasing magnitude of the *d* value of the most intense reflection.   Whenever more than one card lists the same spacing value first, the *d* value of the next most intense reflection determines the sequence.   For economy in printing, the latest sets are not thus arranged.   The user of the *Powder data*

Table 1
The Hanawalt grouping of file cards according to *d* values

Groups arranged according to d value of most intense reflection	Number of groups	
20.00 Å and larger	One	
19.99 to 18.00 Å	One	
17.99 to 16.00	One	
15.99 to 14.00	One	
13.99 to 12.00	One	
11.99 to 11.00	One	
10.99 to 10.00	One	
9.99 to 6.00	Eight	(in steps of 0.50 Å)
5.99 to 5.00	Four	(in steps of 0.25 Å)
4.99 to 3.50	Fifteen	(in steps of 0.10 Å)
3.49 to 1.00	Fifty	(in steps of 0.05 Å)
0.99 to 0.90	One	
0.89 to 0.80	One	
0.79 Å and smaller	One	

file will find it advantageous, however, to arrange the cards in this sequence as described below.   In the original *File*, three cards were used to represent each substance.   One card is similar to the one shown in Fig. 4, that is, it contains all the information about the substance that was known when the card was prepared.   The other two cards list only the three *d* values and the respective intensities.   The sequence of listing the *d*'s differs on these two cards: One lists the second most intense line first, and the other card lists the third most intense line first.   The reason for this triple listing is that independent observers may consider the most intense reflection to be a line different from the one selected by the author of the card.   Hanawalt suggested that the three cards of each substance be interleaved in the same index, in preference to having three separate files.   He further suggested that the cards be separated into groups by file markers, according to the scheme in Table 1.   Further suggestions regarding the arrangement of the cards can be found in the

Index to x-ray powder data file accompanying the card file. This index comes in two parts. The first part is a numerical index listing the three d values of each compound in the same three different sequences described above. The second part contains alphabetical listings of organic compounds, inorganic compounds, and minerals, and their principal d values. The first d values of each sequence in the numerical index are used to group the sets according to the scheme of Table 1. The succession of values tabulated in each group is then arranged in order of decreasing values of the second d listed.

Identification of unknown substance by means of the *Index*

General procedure. The interplanar spacings and the relative intensities of an unknown powder having been determined, a table is prepared listing all the observed reflections. (Visual estimates of intensity are usually sufficiently accurate for this purpose.) The spacings are arranged in order of decreasing d values, and the intensities are expressed as percentages of the intensities of the strongest reflections, that is, the most intense reflection is arbitrarily assigned the value 100 and the other intensities are scaled accordingly.

The large number of cards in the *File* makes it expedient to first consult the *Index* volume. The d value of the most intense reflection is used to determine the Hanawalt group in the *Index*. The d value of the next most intense line is then used to find a corresponding set of d's tabulated in this group. Several reasons may account for a failure to find an appropriate match. The most obvious is that the substance being investigated is not included in the *Index*. Excluding this possibility, it may be that slightly different d values were reported on the appropriate card in the *Index*. This necessitates a search for this card in the two adjacent Hanawalt groups. Alternatively, if the relative intensities of several lines are very nearly 100, different sequences of the three d values should be tried. Except when the unknown is a mixture of several components, these discrepancies are rarely troublesome. More often the reverse is true, that is, more than one set of values in the *Index* appears to agree with the observed set. In this case the respective cards in the *File* must be consulted. A comparison of the interplanar spacings and the relative intensities of the observed set with those listed in each card usually produces a correct identification. A given identification is confirmed when the values on the card match the values determined from the film. This procedure is illustrated by examples in the next two sections.

Unknowns containing only one component. The identification procedure outlined above is straightforward if the unknown powder is

Table 2

Powder data for an unknown, and comparison with the tabulated data for some known compounds

Unknown (CuKα)		BaF₂ (MoKα)		LiZn (CuKα)		LiI (radiation unknown)	
d	$\dfrac{I}{I_{max}} \times 100$	d	$\dfrac{I}{I_{max}} \times 100$	d	$\dfrac{I}{I_{max}} \times 100$	d	$\dfrac{I}{I_{max}} \times 100$
3.58 Å	100	3.59 Å	100	3.59 Å	90	3.50 Å	100
3.10	30	3.10	25				
2.20	80	2.19	100	2.20	100	2.14	60
1.87	50	1.86	80	1.88	80		
1.80	5	1.78	15	. . . . . .	. . .	1.80	60
1.55	5	1.55	15	1.56	60	1.52	10
1.423	15	1.423	32	1.43	60		
1.385	7	1.385	18	. . . . . .	. . .	1.37	30
1.266	17	1.265	32	1.27	80	1.23	20
1.194	5	1.192	20	1.20	60		
1.096	2	1.097	5	1.10	60		
1.048	7	1.047	15	1.05	60		
1.033	1	1.033	3				
0.980	2	0.980	6	0.984	70		
.946	1	.946	3	0.949	40		
.935	3	.935	2				
.895	1	. . . . . .	. . .	.898	40		
.868	3	.868	3	.871	60		
.860	4	.861	2				
.829	4	.829	5	.832	80		
.807	4						

composed of one kind of crystal only. It is worthwhile, therefore, to examine the powder under a microscope prior to the analysis. The observed spacing values and relative intensities of a homogeneous powder photographed with CuKα radiation are listed in the first column of Table 2. The three strongest reflections in Table 2 have the following d values:

$$
\begin{array}{c|c}
\dfrac{I}{I_{max}} \times 100 & d \\
\hline
100 & 3.58 \text{ Å} \\
80 & 2.20 \\
50 & 1.87 \\
\end{array}
\qquad (3)
$$

The d value of the strongest reflection is located in the *Index* in the Hanawalt group 3.59 to 3.50. The second d value, 2.20 Å, is then used to find a matching set of three d's in this group. The sets that appear to be most similar are listed in Table 3. As can be seen in this table, the relative intensities of BaF_2, LiZn, and LiI come closest to the observed values. The complete data on the cards of these three substances are listed in Table 2. A comparison with the data for the unknown indicates that the best agreement occurs between the data of BaF_2 and the data of the unknown. Whenever possible, such identifications should be independently checked by another method, particularly if a real choice exists between two possible compounds.

Table 3

Data for the three principal lines of an unknown, compared with several sets of data from the *Index* furnishing the best match

d			$\dfrac{I}{I_{max}} \times 100$			Substance
3.58 Å	2.20 Å	1.87 Å	100	80	50	Unknown
3.51	2.23	1.83	75	100	80	$5Pbs \cdot Sb_2S_3$
3.59	2.20	1.88	90	100	80	LiZn
3.58	2.19	1.86	100	100	80	BaF_2
3.53	2.16	1.84	100	50	33	SnI_4
3.52	2.15	1.84	100	100	90	Cu_2HgI_4
3.52	2.15	1.84	100	90	80	CuI
3.52	2.15	1.83	100	100	100	CuI
3.50	2.14	1.80	100	60	60	LiI

Unknowns containing a mixture. The procedure described in the preceding section is used in principle whether the powder contains one or several components. If more than one component is present, it is necessary to select properly the three most intense reflections for each component. A great deal of time and labor can be saved if the possible, or even likely, composition of the sample is known in advance. This information can often be inferred, in the absence of an actual analysis, from a knowledge of the sample's source, mode of preparation, or from its previous history. When such information is available, the alphabetical section of the *Index* can be used to determine the three d values of the suspected substances. Quite often, one or more components can be identified this way and their d's eliminated from further consideration.

Sometimes this cannot be done and more direct procedures become necessary. The procedure for identifying the components of a two-component powder photograph is illustrated by the data listed in the first column of Table 4. This information was obtained from a mixture of two inorganic compounds. The six most intense reflections listed in this table have the following d values:

$$
\begin{array}{c|c}
\dfrac{I_{max}}{I} \times 100 & d \\
\hline
100 & 2.97\ \text{Å} \\
90 & 2.45 \\
85 & 3.43 \\
54 & 2.10 \\
50 & 1.510 \\
39 & 1.284
\end{array}
\tag{4}
$$

Table 4

Powder data for an unknown containing a mixture of components, and comparison with the tabulated data of the identified constituents

Unknown (CuKα)		PbS (MoKα)		TeO$_2$ (MoKα)	
d	$\dfrac{I}{I_{max}} \times 100$	d	$\dfrac{I}{I_{max}} \times 100$	d	$\dfrac{I}{I_{max}} \times 100$
3.43 Å	85	3.43 Å	80	3.40 Å	80
3.01	2				
2.97	100	2.97	100	2.99	100
2.45	90	. . . . . .	. . .	2.41	16
2.12	28				
2.10	54	2.09	60		
1.80	29	1.79	32	1.869	56
1.70	15	1.71	16	1.699	8
				1.658	32
. . . . . .	. . .	. . . . . .	. . .		
1.510	50	. . . . . .	. . .	1.520	8
1.485	10	1.484	8	1.485	25
1.360	9	1.361	8		
1.327	15	1.327	16		
1.284	39	. . . . . .	. . .	1.260	14
1.229	4	. . . . . .	. . .	1.224	10
1.210	12	1.212	8	1.184	14
1.142	5	1.144	8	1.114	4
1.065	3	. . . . . .	. . .	1.091	6
1.050	2	1.050	4		
1.004	4	1.004	4		
0.989	6				
.979	6				
.956	2				
.936	4				
.905	1				
.871	2				
.831	3				
.823	2				
.821	1				

The d value of the most intense reflections lies in the Hanawalt group 2.99 to 2.95.   Several choices can be made in selecting the second most intense reflection of the same set.   Proceeding in a systematic manner, the next three most intense reflections are successively chosen as being the second most intense reflection of the same component, and for each choice the Hanawalt group is examined in a manner similar to the procedure of the previous section.   It should be realized that the third reflection completing the set may be one of the six listed in (4), or it may be one of the other reflections listed in Table 4.   Likely matches obtained by consulting the *Index volume* are listed in Table 5.   As can be seen in

Table 5
Possible combinations of strong lines of an unknown, compared with several sets of data from the *Index* furnishing the best match

	d			$\dfrac{I}{I_{max}} \times 100$			Substance
Choice 1	2.97 Å	2.45 Å	?	100	90	?	Unknown
	A "set" having similar relative intensities for the first two d's could not be found						
Choice 2	2.97	3.43	?	100	85	?	Unknown
	2.95	3.43	2.09 Å	90	100	60	CH_4
	2.97	3.43	2.09	100	80	60	PbS
	2.95	3.40	2.02	100	80	80	$Pb_2Bi_2S_5$
	2.99	3.40	1.87	100	80	56	TeO_2
	2.96	3.40	1.73	80	100	80	$Ca_2P_2O_7 \cdot H_2O$
Choice 3	2.97	2.10	?	100	54	?	Unknown
	2.99	2.12	3.47	100	100	67	C_3F
	2.99	2.11	1.72	100	32	22	$SrCeO_3$
	2.98	2.10	1.79	80	100	30	NaBr
	2.96	2.09	3.44	100	63	45	NaBr
	2.96	2.09	3.42	100	80	60	CaC_2
	2.96	2.08	1.79	100	90	90	PbS
Choice 4	2.97	1.51	?	100	50	?	Unknown
	A "set" having similar relative intensities for the first two d's could not be found						

this table, if $d = 2.45$ Å is chosen as the second most intense reflection of one component, it is not possible to find a set of three d's in the *Index* consistent with the experimentally observed data.   This indicates that $d = 2.45$ A is probably the most intense reflection of the second component.   Selecting next $d = 3.43$ Å as the second d value of a possible set, a satisfactory match is obtained for CH_4, PbS, and TeO_2.   Since it was known at the outset that the unknown was a mixture of inorganic compounds, CH_4 can be eliminated from further consideration. Another possibility for the second d value is $d = 2.10$ Å. For this choice, only $SrCeO_3$ appears to be likely. Again from previous knowledge, the presence of a rare earth oxide can be excluded as a likely constituent. (Note that PbS appears again in this listing. The relative intensities reported on this card, however, do not agree with the experimentally observed values; so this card need not be considered.) Finally, if $d = 1.51$ Å is chosen, it is not possible to obtain a match. From this it can be concluded that one component is probably either PbS or TeO_2. A comparison of the complete set of observed spacings with the d's listed on the *File* cards of PbS and TeO_2 (Table 4) shows a satisfactory agreement for PbS, and it may be concluded from this that one of the components in the unknown is lead sulfide.

Having identified one of the components, the reflections remaining in Table 4 which were not accounted for are used to identify the second

Table 6
Remainder of the powder data for an unknown listed in Table 4, and comparison with the tabulated data for the identified constituents

Unknown (CuKα)		Cu_2O† (CuKα)		Cu_2O† (MoKα)		Cu_2O† (MoKα)		Cu_2O† (CuKα)	
d	$\dfrac{I}{I_{max}} \times 100$	d	$\dfrac{I}{I_{max}} \times 100$	d	$\dfrac{I}{I_{max}} \times 100$	d	$\dfrac{I}{I_{max}} \times 100$	d	$\dfrac{I}{I_{max}} \times 100$
3.01 Å	2	3.04 Å	20	3.01 Å	40	3.00 Å	3	3.01 Å	3
2.45	100	2.46	100	2.46	100	2.45	100	2.45	100
2.12	31	2.14	70	2.13	50	2.12	31	2.12	31
		1.67	20	1.74	20				
1.510	55	1.51	90	1.506	60	1.51	44	1.51	44
		1.42	5						
		1.35	3						
1.284	43	1.28	80	1.283	50	1.283	31	1.286	31
1.229	4	1.23	50	1.230	30	1.228	5	1.231	5
		1.16	3						
1.065	3	1.07	4	1.064	50	1.065	3	1.067	3
0.989	7	1.00	5						
.979	7	0.978	60	0.978	40	0.977	5	0.979	5
.950	2	.955	50	.952	40	.953	3	.955	3
.936	4								
.905	1								
.895	4								
.871	2	.872	60	.870	40	.869	3	.871	3
.831	3	.846	3						
.823	2	.824	3	.822	40				
.821	1					.819	3	.821	3

† The presence in the *X-ray powder data file* of more than one card for a given substance, with apparently conflicting data, is one of the shortcomings of the *File* being corrected. See, for instance, *Natl. Bur. Standards* Circ. 539, from which the data in the last column were obtained.

component. It is advisable at this point to rescale the relative intensities of the remaining reflections by setting the strongest remaining intensity equal to 100. In the example in Table 4, this can be done by multiplying each intensity by $\frac{100}{90}$. A list of the remaining reflections, thus scaled, is shown in the first column of Table 6. The three most intense reflections of this component have the following d values:

$\dfrac{I}{I_{max}} \times 100$	d
100	2.45 Å
55	1.510
43	1.284

(5)

A match for these lines is sought in the Hanawalt groups 2.49 to 2.45 and 2.44 to 2.40. The possible identifications are determined with the aid of the second and third d value in (5), and are listed in Table 7. As

Table 7

Data for the three principal lines of the remainder of the powder data from Table 4, compared with several sets of data from the *Index*

d			$\dfrac{I}{I_{max}} \times 100$			Substance
2.45 Å	1.510 Å	1.284 Å	100	55	43	Unknown
2.46	1.51	1.28	100	90	80	Cu_2O

Note: Adjacent to the above values for Cu_2O, two other cards for Cu_2O are listed. The three principal reflections listed on these cards are:

2.46	1.51	2.13	100	60	50	Cu_2O
2.45	1.51	2.12	100	44	31	Cu_2O

can be seen from this table, only the reported values of Cu_2O appear to agree with the experimentally observed data. The interplanar spacings of Cu_2O reported in the card file are compared with the observed values in Table 6. The agreement appears to be satisfactory.

If the unknown powder is identified as a mixture of PbS and Cu_2O, the following lines in Table 6 remain unaccounted for:

d	$\dfrac{I}{I_{max}} \times 100$
0.989	7
.936	4
.905	1
.905	4
.831	3
.823	2

(6)

These extra lines can be accounted for in two ways. One possibility is that for some reason these reflections were not included in the card file data (see next section). The other possibility is that the unknown has not been properly identified.

The agreement obtained for most of the observed reflections appears to indicate that the identification is probably correct. In order to explain the additional lines observed, therefore, it is necessary to index these lines on the basis of the unit cell of either PbS or Cu_2O. When this is done, it turns out that $d = 0.989$ is the 600 reflection of PbS ($a = 5.935$ Å). Similarly, the other five reflections are, respectively, the 620, 523, 622, 711, and 640 reflections of PbS. This illustration clearly shows the advisability of indexing all possible reflections subsequent to the identification of each component in order to eliminate these reflections from further consideration.

If it turns out that extra lines cannot be accounted for in the above manner, it becomes necessary to reexamine the entire procedure for a possibly incorrect identification. Alternatively, it is possible that such lines can be caused by the presence of small amounts of an additional component. In either case, an analysis by some other analytical method would be very helpful. This does not mean that the powder method is inadequate for identification purposes, but rather that no one method is omnipotent.

Practical considerations. It can be seen even from the above two simple examples that the problem of identifying the constituents of a mixture can be quite complicated. Obviously, the more components present, the greater is the difficulty of identifying each individual component. Several factors may be responsible for increasing the attendant complexity. Superposition of certain lines common to two of the components can often be recognized when all the d values given on a card appear to agree with most of the observed d values, but the observed intensity of a particular line is far greater than the corresponding intensity reported in the card. Whenever this happens, it is safe to assume that more than one component contributes to the intensity of such a line. Another factor, often overlooked, is that most mixtures examined are rarely mixtures of pure components. That is, the mixture is ordinarily the result of some chemical process in which the various components can occur. In some cases solid-solution crystals occur. In other cases, partial replacement of some of the atoms in a component by other atoms not normally present may reduce the perfection of the crystals or limit the particle size of that component. Any one of these factors may affect the relative intensities and, possibly, the unit-cell dimensions, thus producing slight shifts in the observed d values. A simple example of this type of difficulty is found in the solid-solution series between cobalt monoxide and nickel monoxide. Since the x-ray scattering powers of nickel and cobalt are virtually the same, the intensities of the lines in a powder photograph do not change despite changes in composition from, say, (3 CoO):(1 NiO) to (1 CoO):(7 NiO). About the only way to determine the correct composition with x-rays is to determine accurately the length of the cell edge a (both oxides are isometric) which varies slightly with the (cobalt): (nickel) ratio. It is strongly recommended, therefore, that, whenever several components appear to be present in an unknown, a tentative identification of one of the components be followed by determination of its unit-cell dimensions (if not already known) after which all the possible d values of that component should be computed using the procedures of Chapter 7 or 8. By following this procedure it is often possible to account for certain observed d values not found listed on the file card, and to eliminate such lines from further consideration. This is particularly important for very large and very small d values, since these are often missing from the lists on the cards. In this connection, a great deal of time can be saved if powder photographs of single compounds are available for comparison purposes. Figure 5 shows a film comparator with several films placed side by side. If a small library of such "standard" films is available, therefore, an identification of a component in a mixture is followed by the use of a standard film to eliminate all lines belonging to this component from further consideration. The two films being compared, of course, must have been prepared with cameras of like size as well as with identical x-radiations.

Some of the *Powder data file* cards list more lines than appear to be present in the photograph being investigated. This may be due to the reasons mentioned above; that is, the intensities of certain reflections may have been altered by changes in composition of the constituents in the powder sample, or it may be due to errors in the data as listed on the card. Some of the cards list extraneous lines caused by contaminations on the x-ray tube target, that is, wavelengths other than the one supposedly used to compute the reported spacing values. Another cause of spurious lines may be the presence of small amounts of impurity in the "standard" sample. These and other errors in the *Powder data file* have been recognized for many years and steps have already been taken to remove them. In 1949 a research fellowship was established at the National Bureau of Standards (NBS) to review the data and to produce standard diagrams of higher quality. These diagrams have been published by NBS in circular form. File cards containing these data, as well as reliable data from other sources, are marked by a star on the card (note upper right corner of Fig. 4).

Another possible difficulty arises when the interplanar spacings tabulated on a card agree with those measured on the photograph but the relative intensities do not.   Barring the isomorphous-substitution effect discussed above, this discrepancy may be due to another cause.   The intensities of reflections appearing in a powder photograph are functions of the structure of the crystals in the powder, and of a geometrical factor, the Lorentz-polarization factor.   The latter is a function of the glancing angle θ and therefore of the wavelength of x-radiation used.   This effect is most pronounced for θ values close to zero and 90°.   If the x-radiation used in preparing the standard diagram differs from the radiation used in

FIG. 5

the laboratory, therefore, the relative intensities may be different.   The wavelength of x-radiation used in obtaining the file data is usually listed on the card.   If a discrepancy in the intensities arises, therefore, it can be resolved by repeating the photograph in the laboratory, using radiation of the same wavelength as the one used in the preparation of the card data.

Another possible cause of incorrect intensities on a given card is that the photograph measured was overexposed.   As described in an earlier section, the film and the human eye are both insensitive to differences in extremely black lines.   For this reason if a card lists a large number of very intense lines, it should be regarded with some suspicion. It obviously behooves the investigator to use equal care in selecting exposure conditions in the laboratory. Minor differences in the relative intensities may arise when methods of measuring the intensities in the laboratory are different from those used in preparing the card data. These discrepancies are rarely serious, however.

Although it has been stressed in the preceding section that all the spacing values appearing on a card should also be present in the list obtained from the photograph before a positive identification can be made, some of the factors discussed above may make such agreement impossible. A much more satisfactory agreement can be obtained by indexing the reflections in the photograph and comparing the indices. Unfortunately, only a small number of cards in the *Powder data file* list the indices of reported reflections. Another omission in many cards is the absence of information describing the method used in estimating relative intensities. However, these and possibly other criticisms of the *File* are being constantly studied and gradually corrected.

Constituents may be missed entirely if present in small amounts only. This is particularly true if the constituents have much lower absorption for the x-radiation used than the rest of the powder. The importance of relative absorption is discussed in detail in the next section where procedures for quantitative analysis of powder mixtures are treated. The effect of relative absorption can be observed, for instance, when glass capillaries are used in preparing powder mounts. When powders having different absorptions are used the diffraction halos due to the glass vary in intensity, despite identical conditions of exposure, etc. This leads to two conclusions: 1. Avoid the use of glass capillaries whenever possible. 2. Whenever accurate identification of *all* constituents in a powder is required, other analytical methods should be used to confirm the x-ray analysis.

Quantitative analysis

In the above description of identification procedures it was assumed that each component in the mixture was present in sufficient amount to give a characteristic diffraction diagram. In a careful analysis, therefore, it is necessary to know the smallest amount of a given component that can be detected by the powder method. Frequently it is necessary to determine the percentage composition of the major components as well. Unfortunately, the relations between the intensities of observed reflections and the percentage composition of a mixture are not simple and cannot be expressed by a usable mathematical expression. It is necessary, therefore, to use semi-empirical methods.

The applicability of the powder method to quantitative analysis has been recognized ever since the discovery of this method, but practically no attempts at such analysis were made until 1936, when Clark and Reynolds[14] published a description of a procedure for estimating the amount of silicon dioxide present in mine dust.   This pioneering work stimulated others to use the powder method for quantitative analysis, with the result that the literature[15-18] now abounds with descriptions of procedures and their applications.   These procedures compare the relative intensities of certain reflections of each component present with those observed on photographs of known mixtures.   Although it is beyond the scope of this book to discuss in detail the factors affecting the intensities of x-ray reflections, two important results of x-ray diffraction theory will be used, namely, the effect of the crystal structure and the effect of absorption on the intensity of diffracted radiation.

Effect of crystal structure.   The intensity of a diffracted beam depends on:

1. The intensity and wavelength of the incident beam
2. The crystal structure, that is, the arrangement of atoms within a unit cell
3. The volume of the diffracting crystals
4. The diffraction angle
5. The absorption of x-radiation by the crystals
6. The experimental arrangement used

Since the same direct beam is used for all reflections recorded on a film in the powder method, the relationship between the diffracted intensity and the above-named factors can be written

$$I_{hkl} = I_0 \frac{Cm}{\mu} F_{hkl}^2 VLp \qquad (7)$$

where I_0 = direct-beam intensity

C = an experimental constant, having the same value for all reflections recorded on one photograph

m = multiplicity of the reflecting planes, that is, the number of planes in a crystal having identical interplanar spacings

μ = linear absorption coefficient

F_{hkl} = structure factor, which depends on the atomic arrangement in a unit cell

V = total volume of the diffracting crystals

Lp = Lorentz-polarization factor, equal to

$$Lp = \frac{1 + \cos^2 2\theta}{\sin^2 \theta \cos \theta}. \qquad (8)$$

If variation in absorption is neglected, equation (7) can be abbreviated to

$$I_{hkl} = C'_{hkl}F^2_{hkl}V, \tag{9}$$

since I_0, C, and μ are constant for all reflections observed on any one photograph and mLp are known functions of hkl and θ. Equation (9) illustrates the fact that the intensity of a diffracted beam depends not only on the amount of sample being irradiated but also on the crystal structure because it depends on F^2_{hkl}. Thus, for a mixture of two different kinds of crystals A and B, it is possible to select one specific reflection of each component and, after evaluating the respective C'_{hkl}'s, to set up the ratio

$$\frac{{}^A(I/C')_{hkl}}{{}^B(I/C')_{hkl}} = \frac{{}^A(F^2_{hkl}V)}{{}^B(F^2_{hkl}V)}. \tag{10}$$

It is clearly seen in (10) that, since ${}^AF^2_{hkl} \neq {}^BF^2_{hkl}$, such a ratio is not proportional to the volumes of A and B alone.

It would appear from (7) and (10) that, if the crystal structure of a compound were known, the percentage volume of that compound present in a mixture could be determined directly from the observed diffracted intensities. For a simple two-component mixture, this is indeed possible. If the absorption coefficients of both components, as well as their structures, are known, a simple ratio between two expressions like (7) gives the respective volume ratios directly. In practice, however, it turns out that these two conditions are rarely satisfied. It is possible, though, to use a semi-empirical procedure, wherein it is not necessary to know the direct-beam intensity or, for that matter, the crystal structure.

For a given compound, equation (9) states in effect that, if F^2_{hkl} is held constant, that is, if only reflections from the same plane (hkl) are considered, the intensity is directly proportional to the crystal volume. The volume per cent v_A of a given component A can be expressed as a ratio between the volume of A and the total volume V_T:

$$v_A = \frac{V_A}{V_T}. \tag{11}$$

In the case of a mixture of substances having like absorption coefficients, for example, a polymorphous series, equation (9) can be used directly to determine the percentage volume of A present in the mixture since

$$I_{hkl} = V_T(C'F^2)_{hkl}v_A. \tag{12}$$

The meaning of (12) is that the intensity of a given reflection hkl is linearly dependent on the percentage volume of the crystals in the mixture, since doubling the volume doubles the intensity, etc. It is possible, therefore, to prepare mixtures containing known amounts of components and to use their photographs as reference standards. Since the relationship between intensity and percentage volume is linear, only two known mixtures are necessary to construct a graph of intensity vs. percentage volume. Such a graph can then be used to determine the percentage volume of any component from the measurement of the intensity of one line. Of course, it is necessary to maintain uniformity of sample preparation, film processing, etc., in this procedure. The details of this procedure are discussed in a later section.

Effect of absorption. If the absorption coefficients of the constituents of a mixture are different, it is necessary to use another form of equation (7) instead of (9). The intensity of a reflection from the (*hkl*) planes of component A in such a mixture can be expressed by the relation

$$I_A = K_A \frac{v_A}{\mu}, \tag{13}$$

where

$$K_A = I_0 C V_T (mF_{hkl}^2 Lp)_A, \tag{14}$$

and μ is the linear absorption coefficient of the mixture (*not* of component A).

It is interesting to observe what happens when a mixture contains two components A and B whose absorption coefficients are markedly different. According to (13), for a given percentage volume of A the intensity is inversely proportional to μ, which represents an "averaged" value of μ_A and μ_B. Assuming $\mu_A \gg \mu_B$, the averaged linear absorption coefficient is smaller than μ_A, and greater than μ_B. Thus I_A is relatively greater than would be the case if $\mu = \mu_A$. Conversely, the intensity of a reflection by component B appears less intense than would be the case if $\mu = \mu_B$. From this it can be concluded that the reflection intensity from a heavily absorbing substance is relatively exaggerated when the mixture contains less-absorbing components. Moreover, since I_A in (13) is a function of two variables v_A and μ the relationship between intensity and percentage volume is no longer linear. Nevertheless, it is possible to use standard mixtures to plot a curve relating the reflection intensity to the percentage volume in a manner similar to that in the case where the linear absorption coefficient does not change with composition. This procedure becomes complicated, however, when the number of components exceeds two. Since this is the case most frequently encountered in the laboratory, it is more convenient to use an alternative procedure suggested by Alexander and Klug, described below.

Internal-standard technique. Alexander and Klug's technique[19] consists of adding a known amount of a substance not already present in the mixture and determining the percentage volume of any component in terms of the volume of the added substance. The principal advantage of this method is that it is independent of the linear absorption coefficient; therefore the relationship between the relative intensities and relative percentage volumes is linear.

If it is desired to determine the fractional volume v_A, of component A, a known amount of the standard S is added to the mixture. If the fractional volumes of A and S in the new mixture are designated v'_A and v_S, their respective reflection intensities can be determined with the aid of (13):

$$I_A = K_A \frac{v'_A}{\mu}, \tag{15}$$

and
$$I_S = K_S \frac{v_S}{\mu}. \tag{16}$$

Dividing (15) by (16),

$$\frac{I_A}{I_S} = \frac{K_A v'_A}{K_S v_S}, \tag{17}$$

which can be solved for v'_A:

$$v'_A = \frac{K_S v_S}{K_A} \frac{I_A}{I_S}. \tag{18}$$

The relation between v_A, the fractional volume sought, and v'_A given by (18) can be determined as follows: Denoting by V_T the total volume of the new mixture, that is, the volume of the original mixture plus the volume V_S of the added substance,

$$v'_A = \frac{V_A}{V_T}, \tag{19}$$

and
$$v_A = \frac{V_A}{V_T - V_S}. \tag{20}$$

Dividing (19) by (20),

$$\frac{v'_A}{v_A} = \frac{V_A/V_T}{V_A/V_T - V_S}$$

$$= \frac{V_T - V_S}{V_T}$$

$$= 1 - \frac{V_S}{V_T}. \tag{21}$$

But,
$$\frac{V_S}{V_T} = v_S$$

by definition, and

$$\frac{v'_A}{v_A} = 1 - v_S,$$

or
$$\frac{v_A}{v'_A} = \frac{1}{1 - v_S}.$$ (22)

Finally,
$$v_A = \frac{v'_A}{1 - v_S}.$$ (23)

Substituting (18) for v'_A in (23):

$$v_A = \frac{K_S v_S}{K_A (1 - v_S)} \frac{I_A}{I_S}$$

$$= K \frac{I_A}{I_S},$$ (24)

where
$$K = \frac{K_S v_S}{K_A (1 - v_S)}.$$ (25)

Since, according to (14), K does not depend on μ, equation (24) is independent of μ also. It is possible, therefore, to prepare a graph of I_A/I_S as a function of v_A, for a fixed v_S, from known mixtures. Since (24) is a linear equation, the graph of v_A vs. I_A/I_S is a straight line whose slope is K. In preparing such a graph, it is important to keep v_S constant in each mixture prepared. The details for doing this are discussed in the next section.

Practical procedure. The procedure for quantitative analysis of a mixture containing different components having like absorption coefficients is illustrated by a mixture of calcite and aragonite, two polymorphous forms of $CaCO_3$. Since this mixture contains only two components, it is sufficient to determine the percentage volume of either component. In this example, the intensity of the 014 reflection of calcite ($d = 3.04$ Å) is used to determine the percentage volume of calcite present in the binary mixture.

Three mixtures are prepared, one containing, say 10 per cent calcite, the other 60 per cent calcite, and the third 100 per cent calcite. In practice, it is easier to measure the weight than the volume of a powder. Since

$$\text{Weight} = \text{volume} \times \text{density},$$

it is possible to use either quantity in the analysis. Next, the three mixtures are fashioned into similar rods and photographed according to procedures described in Chapter 5. It is important to use identical conditions for the exposure and the processing of the x-ray films. The intensities of the 014 reflection of calcite on all three photographs are measured, preferably with a photometer, and found to be 15, 90, and 150, respectively. The intensities of the 014 reflections of the two mixtures are expressed as fractions of the intensity of the sample containing 100 per cent calcite, namely, $\frac{15}{150} = .10$ and $\frac{90}{150} = .60$. These values can now be used to construct the graph shown in Fig. 6. As expected from (12), the straight line connecting these two points passes through the origin, that is, $I = 0$ when $v_A = 0$.

The unknown mixture is treated by the above procedure. By referring to Fig. 6, the intensity of the 014 reflection of calcite is then used to determine the percentage volume of calcite present. As can be seen in this figure, from the observed intensity ratio $(I_{unknown}/I_{100\%} = .45)$, the percentage volume of calcite is determined to be 45 per cent.

A similar procedure is used in analyzing a mixture containing several components whose absorption coefficients may be different. A separate graph is plotted for each component whose percentage volume is sought.

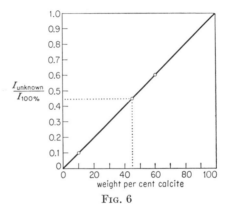

FIG. 6

In preparing such graphs, a constant amount of a standard powder is added to the same amounts of several mixtures containing known amounts of the components sought, in addition to a diluent which can be anything. As an example, a powder containing unknown amounts of $PbCl_2$ and KCl in addition to other components will be analyzed to determine the percentage composition of lead chloride and of potassium chloride.

As already mentioned, it is easier to determine the weight of a sample than it is to measure its volume. Accordingly, V_T in (14) is replaced by the weight of the powder W_T divided by the density of the powder. (The density can be included in K, which is not evaluated numerically anyway.) With this new value of K, equation (24) can be written for the weight per cent w_A of component A, as follows:

$$w_A = K \frac{I_A}{I_S}, \tag{26}$$

where I_A and I_S are the intensities of a selected reflection of the component A and the standard S, respectively. It is desirable that an intense reflection be chosen as I_A so that small amounts of the component can be detected. The selection of a reflection from the standard is based on the proximity on the film of this reflection to that of the component sought. It is important that other reflections of the standard should not overlap this reflection. Similar precautions are necessary in selecting a diluent for the known mixtures.

In this analysis, the 111 reflection of $PbCl_2$ ($d = 3.58$ Å) and the 200 reflection of KCl ($d_{200} = 3.14$ Å) were used. It is desirable that the internal standard have a fairly intense reflection lying between these two reflections. The 110 reflection of rutile ($d_{110} = 3.24$ Å) satisfies this requirement. Finally, NaCl was selected as the diluent since the d value

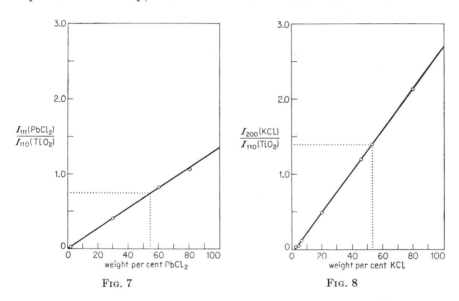

FIG. 7   FIG. 8

of the first strong reflection from NaCl is smaller than either of the above values and thus would not conflict with them. (The diluent was added merely to simulate a practical case in which other components may be present.)

Several mixtures containing different known amounts of lead chloride and potassium chloride mixed with NaCl were prepared. Eighty milli grams of each of these mixtures were then thoroughly mixed with twenty milligrams of rutile. The resulting powders were photographed and the intensities of the reflections mentioned above were measured with a photometer. The two graphs shown in Figs. 7 and 8 were then prepared by

plotting $\dfrac{I_{111}(PbCl_2)}{I_{110}(TiO_2)}$ vs. w_{PbCl_2} and $\dfrac{I_{200}(KCl)}{I_{110}(TiO_2)}$ vs. w_{KCl}, respectively.

Finally, 80 mg of the unknown mixture were thoroughly mixed with

20 mg of TiO_2 and the sample was photographed. The intensity ratios determined from the resulting film were then used to determine the weight per cent of both components (dotted lines in Figs. 7 and 8).

The methods described above are capable of detecting several per cent of a component with relatively high accuracy. The accuracy varies with the nature of the mixtures and also with the precautions used to ensure uniformity in each step of the procedure. If it is necessary to detect very small amounts of a component, the calibration curves must be prepared from data obtained from known mixtures containing comparably small amounts of the component. In addition to the accuracy that can be obtained with a careful experimental technique, this procedure has the advantage that the mixture can be analyzed for the component sought without regard for the other components present.

Literature

Tabulated diffraction data for isomorphous compounds

[1] Ludo K. Frevel, *Indexing powder diffraction patterns of isomorphous substances*, J. Appl. Phys. **13** (1942) 109–112.

[2] Ludo K. Frevel, *Tabulated diffraction data for cubic isomorphs*, Ind. Eng. Chem., Anal. Ed. **14** (1942) 687–693.

[3] Ludo K. Frevel, *Chemical analysis by powder diffraction*, Ind. Eng. Chem., Anal. Ed. **16** (1944) 209–218.

[4] L. K. Frevel, H. W. Rinn, and H. C. Anderson, *Tabulated diffraction data for tetragonal isomorphs*, Ind. Eng. Chem., Anal. Ed. **18** (1946) 83–93.

[5] Ludo K. Frevel and H. W. Rinn, *Tabulated diffraction data for hexagonal isomorphs*, Anal. Chem. **25** (1953) 1697–1718.

Tables for the identification of unknowns

[6] Victor Goldschmidt and Samuel G. Gordon, *Crystallographic tables for the determination of minerals* (Acad. Nat. Sci. Phila. Spec. Publ. 2, 1928).

[7] I. E. Knaggs and B. Karlik, *Tables of cubic crystal structures of elements and compounds* (Ada Hilger, London, 1932).

[8] G. A. Harcourt, *Tables for identification of ore minerals by x-ray diffraction powder patterns*, Am. Mineralogist **27** (1942) 63–113.

[9] M. W. Porter and R. C. Spiller, *The Barker index of crystals* (W. Heffer & Sons, Ltd., Cambridge, England, 1952).

[10] J. D. H. Donnay and Werner Nowacki, *Crystal data*, Geol. Soc. Amer. Mem. 60 (New York, 1954).

X-ray powder data file

[11] J. D. Hanawalt, H. W. Rinn, and L. K. Frevel, *Chemical analysis by x-ray diffraction. Classification and use of x-ray diffraction patterns*, Ind. Eng. Chem., Anal. Ed. **10** (1938) 457–512.

[12] A. K. Boldyrev, V. I. Mikheev, and G. A. Kovalev, *Index for identifying*

minerals from their powder diffraction diagrams, Zapiski Gornova Inst. **11** (1938) 1–157.

[13] A. J. C. Wilson, *The ASTM Index for identification by x-ray diffraction*, Norelco Reporter **1** (1954) 64–66.

Quantitative analysis

[14] G. L. Clark and D. H. Reynolds, *Quantitative analysis of mine dusts. An x-ray diffraction method*, Ind. Eng. Chem., Anal. Ed. **8** (1936) 36–40.

[15] V. Hicks, O. McElroy, and M. E. Warga, *Quartz in industrial dusts and deposits on human lung tissues: x-ray diffraction, chemical and spectrographic studies*, J. Ind. Hyg. Toxicol. **19** (1937) 177–182.

[16] J. W. Ballard, H. I. Oshry, and H. H. Schrenk, *U.S. Bur. Mines Rept. Invest.* 3520, June, 1940; 3638, April, 1940; 3888, June, 1946.

[17] H. J. Goldschmidt and G. T. Harris, *An examination of mechanical wear products*, J. Sci. Instr. **18** (1941) 94–97.

[18] C. E. Imhoff and L. A. Burkardt, *Crystalline compounds observed in water treatment*, Ind. Eng. Chem. **35** (1943) 873–882.

[19] L. Alexander and H. P. Klug, *The basic aspects of x-ray absorption in quantitative diffraction analysis of powder mixtures*, Anal. Chem. **20** (1948) 886–888.

Photometers

[20] A. H. Jay, *A simple photometer for the examination of x-ray films*, J. Sci. Instr. **18** (1941) 128–130.

[21] A. Taylor, *X-ray metallography* (Chapman & Hall, Ltd., London, 1945) 92–99.

[22] R. J. Brown, H. K. Moneypenny, and R. J. Wakelin, *A servo-controlled micro-densitometer for x-ray diffraction photographs*, J. Sci. Instr. **32** (1955) 55–59.

14

The sources of error in
measured spacings

The advantages deriving from an accurate knowledge of measured d values have been pointed out in the chapters concerned with indexing procedures for powder photographs. It is important, therefore, to understand what the possible sources of error are. In this chapter the nature of these errors is considered. In the next chapter the actual procedures for reducing the errors as far as possible are discussed.

Nature of the errors

Two types of errors are associated with any measurements: random errors of observation and experimental errors which are inherent in the arrangement. The principal observational error in the powder method lies in the location of the center of a line on the film. This reading error can be minimized by repeating the reading several times or by preparing and measuring several films. For the best accuracy, it is better to average the readings of several different investigators since some peculiarity on the film, such as a speck of dirt, may cause a particular investigator to make the same reading several times in succession. It has been shown that reading errors are not appreciably reduced by preparing a microphotometer tracing of the film, since the added detail present tends to obscure the position of the center of the peak.

In addition to reading difficulties, there are several situations which cause a shift in the center of an x-ray line from the position it ideally should occupy. These can be classified into physical causes and geometrical causes as follows:

210

Physical errors

1. Absorption of x-rays by the specimen
2. Refraction of x-rays by the specimen
3. Uneven distribution in the intensity of the background
4. Relative displacement of lines appearing on the two sides of a double-coated film

Geometrical errors

1. Eccentricity of the specimen with respect to the axis of the film cylinder
2. Lack of knowledge of the exact film radius
3. Divergence of the x-ray beam

All the above errors are often classified into two different groups according to whether the error is completely random or whether its magnitude depends in a systematic fashion on the position of the reflections on the film, that is, on the θ value of the reflection. These so-called systematic errors decrease as θ increases, vanishing completely when $\theta = 90°$. In the following sections these several categories of errors are discussed.

Physical errors

Absorption. The position of a reflection depends on the transparency of the specimen in the manner illustrated in Fig. 1. Assuming a parallel incident beam, Fig. 1*A* shows that the full sample diffracts radiation

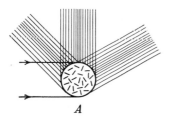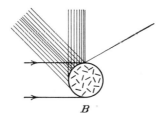

A   *B*

Fɪɢ. 1

equally well in all directions if the sample is nonabsorbing. If the crystal is so highly absorbing that it is substantially opaque to x radiation, then only the surface regions of the specimen can diffract x-rays, as shown in Fig. 1*B*. The suppression of part of the reflected beam has the effect of narrowing the line on the photograph and of displacing its center from the true center of the reflection. It should be observed that this effect decreases as θ increases, vanishing at $\theta = 90°$.

In laboratory arrangements normally used, the x-ray beam is not composed of parallel rays. Instead, they diverge from a finite portion of the x-ray tube target. This divergence greatly complicates the analysis of the angular dependence of the absorption error. Expressions for the error caused by absorption of a diverging x-ray beam have been presented by Bradley and Jay,[1] Buerger,[2] Taylor and Sinclair,[3] and Warren.[4] These expressions are tabulated in Table 1. Although the final expressions of these authors are different, they all agree that the error disappears when $\theta = 90°$. Since the procedures for eliminating errors discussed in the next chapter do not directly employ these expressions, they are not further discussed here.

Refraction. All electromagnetic radiation is refracted on passing from one medium to another having a different index of refraction. For x-rays normally used in the powder method, the index of refraction of virtually all substances lies between approximately 0.99997 and 1.00003. If maximum attainable accuracy is sought, it is necessary to correct for this refraction.[5]

Table 1
**Dependence of the error in the measured spacing on the glancing angle,
caused by absorption in the specimen**

Dependence of $\Delta d/d$ on θ	Author
$\dfrac{\cos^2 \theta}{\theta}$	Bradley and Jay[1]
$\cot \theta \cos^2 \theta$	Buerger[2]
$\dfrac{1}{2}\left(\dfrac{\cos^2 \theta}{\sin \theta} + \dfrac{\cos^2 \theta}{\theta}\right)$	Taylor and Sinclair[3]
$\dfrac{\cos^2 \theta}{\sin^2 \theta}$	Warren[4]

While this is evident if large single crystals are used, Barrett[6] (following a suggestion made by Zachariasen) has pointed out that such a correction should not be applied in the powder method. The effect of refraction is to increase the glancing angle if the reflection takes place at the surface of a crystal, but to decrease it if the reflected beam passes through the crystal. In a powder containing many individual crystallites, these two cases, as well as all intermediate possibilities, occur. The result is that, on the average, no deviation of the reflection develops. Straumanis[7] recently showed that this is not entirely correct. Because of absorption, the intensity of x-rays diffracted from central portions of the specimen is decreased relative to that of x-rays coming from crystallites at the surface. The surface crystallites diffract at angles that are affected by refraction; hence the observed maximum of the diffraction line on the film is shifted accordingly. The need for a refraction correction is borne out by actual tests in practice. Unless refraction corrections are made it is not possible to compare cell dimensions measured with x-rays of different wavelengths. If no refraction correction is made, therefore, the wavelength used for the determination of cell dimensions should be specifically stated.

Uneven background. The uneven background seen on a film in the form of a general blackening of the whole or parts of the film is due to x-radiation scattered in several ways. A few of the most important of these causes are listed below.

1. *Diffraction of x-rays having other than the desired Kα wavelength.* The use of a filter to monochromatize the x-ray beam effectively eliminates the $K\beta$ component but does not prevent all the rest of the continuous spectrum from passing through. X-radiation having each of the wavelengths passing through, therefore, is diffracted and recorded on the film. Since the range of wavelengths is continuous, the diffraction pattern is continuous also. It is not uniform over the entire film, however, and differs for different specimens as well as for different experimental conditions.

2. *Air scattering.* In addition to passing through the specimen, the direct beam passes through air which also scatters the x-rays. The amount of scattering depends on the wavelength of x-radiation employed. But since the scattering decreases rapidly with increasing glancing angle, it does not contribute very greatly to the background in the back-reflection region.

3. *Scattering by specimen support.* If the specimen is composed of material other than just the substance under investigation, this material also contributes to the background scattering.

4. *Other causes.* There are several other causes such as nonuniform film response to different wavelengths and geometrical factors. A somewhat more complete discussion of them is given in Chapter 16.

Nonuniformities in the background have the effect of shifting the peak position in the direction of increasing background intensity. This can be seen with the aid of Fig. 2. The observed intensity distribution on the film is the sum of the intensity of the background plus that of the reflection peak. The intensity of the nonuniform background is shown added to that of the peak in Fig. 2 by lines drawn above several points on the peak. The lengths of the lines represent the intensity of the background at those points. The resulting intensity curve is shown by the upper line. It is evident that the peak has been shifted slightly in the direction of increasing background intensity. Fortunately in most instances the amount of shift in the line position is not much greater than the normal error of reading, that is, about .03 mm. The actual amount of shift, however, differs from film to film so that, if maximum attainable accuracy is sought, it is necessary to avoid this effect. This can be done experimentally by the use of a crystal monochromator to ensure that only the desired wavelength radiation gets through, or by an analysis of the geometry of the peak, similar to Fig. 2.

Line displacement caused by double-coated film. When the film mounting used is not cylindrical, or when the specimen does not lie on the axis of the cylinder, the diffracted x-rays are inclined with respect to the film. This means that the diffracted ray strikes the first emulsion of a double-coated film at a slightly different place than the second emulsion, as shown in Fig. 3. When the film is viewed at right angles to its surface, the resulting line appears broader than it should be. Moreover,

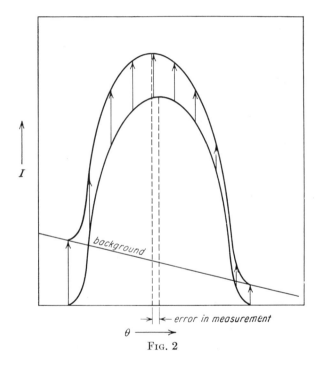

Fig. 2

the intensity of the beam is decreased by absorption in the first emulsion, and the intensity distribution across the widened reflection is not uniform.

The most direct procedure for eliminating this difficulty is to use single-emulsion film. Alternatively, it is possible to remove one of the emulsions by stripping it off the supporting film after the exposure is finished. This procedure is messy and difficult to execute. Parrish[9] has suggested an equivalent but much simpler procedure: After the film has been exposed, but before it is developed, the side of the film facing the sample is covered by masking tape. Usually, one-half of the film only is covered so that the complete record can be seen. The film is then developed in the usual way. Before the film is placed in the hypo, the masking tape

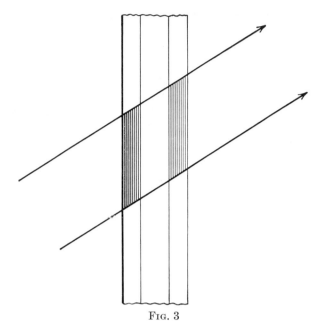

FIG. 3

is removed and the film is fixed. The tape prevents one emulsion from being developed; this side is rendered transparent by the hypo. In addition to preventing the line broadening described above, this procedure also decreases the background intensity since the radiation causing it is most absorbed by the first emulsion, which is not developed.

Geometrical errors

Specimen eccentricity. When the cylindrical cameras described in Chapter 4 are used, the specimen is automatically aligned at the center of the camera. If for any reason, such as poor camera design or mechanical damage to the camera, however, the sample axis is displaced from the cylinder axis during the rotation of the specimen, a displacement of the reflections occurs. The direction and amount of this displacement are illustrated in Fig. 4.

Suppose the specimen is displaced from the center M to a position N. The amount of displacement of the line position on the film can be seen best by resolving the displacement along a vertical and horizontal direction. Figure $4B$ shows the shift in the lines due to a displacement upward, that is, at right angles to the incident-beam direction. Although the pair of lines is displaced from its correct position, the arc length S between corresponding lines on the film has not been affected. Since θ is determined directly from the measurement of S, it is clear that the vertical component of the displacement does not introduce an error in θ and, hence, in d.   Figure $4C$ shows the shift in the lines due to a displacement along the incident-beam direction.   The effect of this displacement is to shift both lines toward (or away from) one another.   This shift does produce a change in the arc length S, and hence an error in d.   Bradley and Jay[1] showed that the fractional error in d caused by this displacement is proportional to $\cos^2 \theta$.

Radius error.   Standard commercial cameras are manufactured with fairly high precision.   Nevertheless, even careful measurement of the radius prior to starting an exposure is not sufficient to assure precision since it is the radius of the film rather than that of the camera that is important.   This uncertainty in the film radius is increased by unavoidable shrinkage of the film during the developing process.

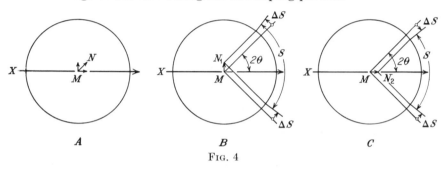

FIG. 4

The effect of the uncertainty of the radius on the line positions is shown in Fig. 5.   Suppose the arc length subtended by a pair of lines on the film is S.   After development of the film, however, this length has shrunk to S'.   If this new arc length is used with the assumed radius R, an obviously incorrect (in this case, smaller) value of the glancing angle θ is obtained.   The arc length S' would give the correct θ value if R' were substituted for R.   However, the length of R' is unknown.   Hence, this error is called the *radius error*.   If the positions of the reflections are measured from the position $\theta = 90°$, the fractional error in d caused by this error is proportional to $\left(\dfrac{\pi}{2} - \theta\right) \cot \theta$.

Divergence of the x-ray beam.   Since x-radiation, unlike some other types of electromagnetic radiation, cannot be controlled by means of lenses, it is virtually impossible to obtain a truly parallel beam. Although it is theoretically possible to take advantage of the rigorous Bragg reflection condition to obtain a parallel beam by successively reflecting the beam from nearly perfect single crystals, the extreme attenuation in the intensity of such a beam precludes the utilization of such an arrangement in practice.

The error introduced by using a divergent beam takes on two forms. One is to produce nonequivalent attenuation of the x-ray beam intensity by different parts of the sample. This effect has already been noted in the section discussing errors due to absorption. Another form of error introduced by the divergence of the x-ray beam is the so-called *umbrella effect*. The effect manifests itself in a distortion of the shape of the line as shown in Fig. 5, Chapter 3. Bradley and Jay[1] showed that the frac-

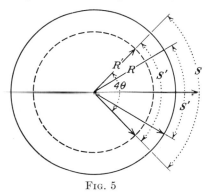

Fig. 5

tional error in d caused by the resulting displacement of the line is proportional to $\cot^2 \theta - 1$. Consequently, the lines are shifted to larger angles for $\theta < 45°$ and to smaller angles for $\theta > 45°$.† Fortunately the effect is very slight and can be virtually eliminated by using small circular pinholes to collimate the x-ray beam.

The ultimate limits of accuracy

Since x-ray wavelengths are known to about 1 part in 100,000, it would appear possible to determine the value of a cell edge to this accuracy. In practice, it has not been possible to obtain this chiefly because of the random subjective errors introduced in locating the true center of a reflection. Assuming that the systematic errors have all been reduced to the vanishing point, and that the position of a reflection at $\theta = 85°$ can be read with an error no greater than 0.03 mm, this reading error causes an error in the lattice paramotor of 2 to 4 parts in 100,000. The error is often increased further, because the position of the apex of the peak may differ from the center of gravity of the peak by approximately 0.003 per cent. A more realistic estimate of the highest accuracy attainable is thus approximately 3 parts in 100,000. For a given case, however, it is possible to estimate the maximum expected accuracy directly from the

† More recently, Lipson and Wilson[10] have shown that this error is proportional to the square of the length of the specimen irradiated and that it decreases as θ nears 90°. This has been confirmed independently by Eastabrook.[11]

photograph. This is accomplished by observing how well the $K\alpha$ doublets are resolved. If they are sharp and well resolved, the accuracy attainable is at least 0.005 per cent; if they are fuzzy, the limit is around 0.02 per cent. If the back-reflection lines are fuzzy and difficult to read, an accuracy no better than 1.0 per cent should be expected.

Literature

[A fairly complete list of the literature published prior to 1942 can be found at the end of Chapter 20 of *X-ray crystallography* (John Wiley & Sons, Inc., New York, 1942). For this reason, only those publications actually referred to are given below.]

Effect of absorption

[1] A. J. Bradley and A. H. Jay, *A method for deducing accurate values of the lattice spacing from x-ray powder photographs taken by the Debye-Scherrer method*, Proc. Phys. Soc. (London) **44** (1932) 563–579.

[2] M. J. Buerger, *X-ray crystallography* (John Wiley & Sons, Inc., New York, 1942) 402–407, 414.

[3] A. Taylor and H. Sinclair, *The influence of absorption on the shapes and positions of lines in Debye-Scherrer powder photographs*, Proc. Phys. Soc. (London) **57** (1945) 108–125.

[4] B. E. Warren, *The absorption displacement in x-ray diffraction by cylindrical samples*, J. Appl. Phys. **10** (1945) 614–620.

Effect of refraction

[5] A. J. C. Wilson, *On the correction of lattice spacings for refraction*, Proc. Cambridge Phil. Soc. **36** (1940) 485–489.

[6] Charles S. Barrett, *Structure of metals*, 2d ed. (McGraw-Hill Book Company, Inc., New York, 1952) 150–151.

[7] M. E. Straumanis, *The refraction correction for lattice constants calculated from powder or rotation crystal patterns*, Acta Cryst. **8** (1955) 654–656.

Effect of double-coated film

[8] David W. Levinson, *A method for improvement of contrast of x-ray diffraction photographs*, Rev. Sci. Instr. **24** (1953) 468–469.

[9] William Parrish, *Elimination of the second image in double-coated film*, Norelco Reporter **2** (1955) 67.

Effect of divergence of the x-ray beam

[10] H. Lipson and A. J. C. Wilson, *The derivation of lattice spacings from Debye-Scherrer photographs*, J. Sci. Instr. **18** (1941) 144–148.

[11] J. N. Eastabrook, *Effect of vertical divergence on the displacement and breadth of x-ray powder diffraction lines*, Brit. J. Appl. Phys. **3** (1952) 349–352. (See also first reference given above.)

[12] Leroy Alexander, *The effect of vertical divergence on x-ray powder diffraction lines*, Brit. J. Appl. Phys. **4** (1953) 92–93.

15

The practice of attaining accuracy

In the last chapter the sources of error in the measured d values were considered. In this chapter some practical methods for minimizing these errors, and thus improving the accuracy of the d values, are discussed in some detail. There are six methods for minimizing errors in spacing measurements in actual use. These are based upon

1. Utilization of spacings recorded in the back-reflection region
2. Practice of careful experimental technique
3. Use of comparison standards
4. Matching absorption errors
5. Graphical extrapolation
6. Analytical extrapolation combined with least-square fitting of extrapolated line to data

These methods are discussed in the following section.

Utilization of spacings recorded in the back-reflection region

As can be seen from the expressions relating the errors in the measured d values to θ, most errors vanish at $\theta = 90°$. Cell-edge determinations based upon measurements of reflections found in the back-reflection region, therefore, tend to be more accurate than those based upon measurements made in the front-reflection region, or averages of measurements made on reflections found in various parts of the film. There is another big advantage in using reflections having glancing angles close to 90°. The d values are controlled by the Bragg relation

$$n\lambda = 2d \sin \theta. \tag{1}$$

Equation (1) can be rewritten

$$\frac{n}{d} = \frac{1}{d_{hkl}} = \left(\frac{2}{\lambda}\right) \sin \theta_{hkl}. \qquad (2)$$

Figure 1 shows a plot of this relationship. It is evident that the same error in the measured value of θ produces a large error in d if θ is small, and a small error in d if θ is large.

The increasing accuracy of the measured d values as θ increases can be illustrated also analytically. Differentiating (1) and assuming λ is constant, there results

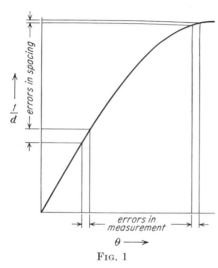

$$0 = 2\Delta d \sin \theta + 2d(\cos \theta)\Delta\theta. \qquad (3)$$

If equation (3) is divided by 2 and the terms are rearranged, one finds

$$\Delta d \sin \theta = -d\Delta\theta \cos \theta, \qquad (4)$$

so that

$$\frac{\Delta d}{d} = -\Delta\theta \cot \theta. \qquad (5)$$

Since $\cot \theta$ approaches zero as θ approaches 90°, any error $\Delta\theta$ in the measured glancing angle produces successively smaller errors Δd in the measured value of the interplanar spacing. For this reason, other things being equal, spacing values based upon measurements in the back-reflection region are comparatively more precise, the precision increasing as θ approaches 90°.

Fig. 1

Back-reflection cameras. Since there are advantages in using the back-reflection region for spacing measurements, several specialized cameras have been developed which record only reflections which lie in this region. Such cameras are called *back-reflection cameras*. (The ordinary cylindrical powder camera can be used as a back-reflection camera, as discussed in Chapter 4.)

A commonly used back-reflection camera consists of a flat cassette in the back-reflection arrangement, as shown in Fig. 3, Chapter 5. The simplicity of this apparatus makes it easy to manufacture and quite inexpensive. Unfortunately, a flat cassette can record such a limited range of reflections that this type of back-reflection camera has limited usefulness.

A much better piece of apparatus for recording the back-reflection field is the *symmetrical back-reflection focusing camera*, which is diagrammatically illustrated in Fig. 2. In this camera the x-ray beam enters at a pinhole or slit on the surface of the cylinder, from which it diverges until it reaches the specimen on the opposite surface. The specimen, commonly sprinkled on a pliable tacky support, is wrapped so as to conform with the cylindrical curvature of the surface. This geometry causes the radiation reflected from a particular plane to converge from all points in the sample to the same pair of lines on the film. In this way the specimen is said to *focus* the reflected radiation to the lines. The arrangement shown in Fig. 2 is symmetrical, so that each reflection is recorded twice on the same film, the two reflections being symmetrically disposed about the

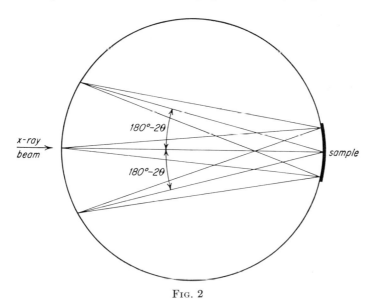

FIG. 2

direct-beam entrance. Halfway between two such lines is the exact location of $2\theta = 180°$. The possibility of locating this fiducial point accurately is the main advantage of the symmetrical arrangement over unsymmetrical arrangements.

Selection of wavelengths for back-reflection work. To gain the advantages inherent in back-reflection measurements it is necessary to assure that a sufficient number of reflections lie in this region. It is easy to see from the Bragg equation,

$$\lambda = 2d_{hkl} \sin \theta_{hkl}, \qquad (6)$$

that for a given reflection hkl the interplanar spacing d_{hkl} is constant, and the glancing angle θ_{hkl} can be varied according to the wavelength λ of the x-radiation selected.

Equation (2) can be rearranged to give

$$\frac{1}{d_{hkl}} = \frac{2}{\lambda} \sin \theta_{hkl}. \qquad (7)$$

Since $\sin \theta$ cannot exceed 1,

$$\frac{1}{d_{hkl}} \leq \frac{2}{\lambda}, \qquad (8A)$$

or in reciprocal-lattice notation,

$$\sigma_{hkl} \leq \frac{2}{\lambda}. \qquad (8B)$$

Remembering (Chapter 10) that

$$Q_{hkl} = \sigma_{hkl}^2 \qquad (9)$$

it also follows that

$$Q_{hkl} \leq \frac{4}{\lambda^2}. \qquad (10)$$

For isometric crystals, $Q_{hkl} = N_C a^{*2}$, and (10) affords a simple way for determining the indices of the last plane whose reflection can be recorded for any chosen x-radiation.

Straumanis' method for determining appropriate wavelengths. Straumanis[4] used this relation in his procedure for selecting x-radiation for precision lattice-constant determination. If a new quantity p is defined as

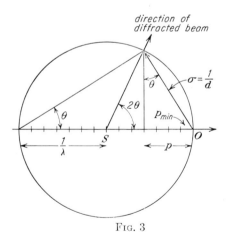

Fig. 3

$$p = \frac{\lambda}{2} Q_{hkl}, \qquad (11)$$

it follows directly from (10) that

$$p \leq \frac{2}{\lambda}. \qquad (12)$$

Defining next,

$$p_{\min} = \frac{\lambda}{2} Q_{100}$$
$$= \frac{\lambda}{2} (1a^{*2}) \qquad (13)$$

it follows from (11) that

$$p = N_C p_{\min}. \qquad (14)$$

It is possible, therefore, to plot the various values of p along a straight line, simply by laying off successive equal lengths representing $p_{\min}$. If, then, a circle of radius $1/\lambda$ is drawn on the same scale so that $p = 0$ lies on its circumference, then, according to (12), all permissible values of p are enclosed by this circle. Such a construction has several interesting properties which can be seen with the aid of Fig. 3. For instance, it is easy to see that p is the projection on the x-ray beam direction of the reciprocal-lattice vector σ since

$$p = \sigma \sin \theta. \tag{15}$$

Figure 3 shows that

$$\sin \theta = \frac{\sigma}{2/\lambda}, \tag{16}$$

or, since $\sigma = 1/d$, it is evident that

$$\sin \theta = \frac{\lambda}{2d}, \tag{17}$$

which is simply a rearrangement of the terms in (7). Thus Fig. 3 is simply the graphical representation of the Bragg equation already encountered in Chapter 9. The construction of Fig. 3, therefore, provides a convenient way for determining the Bragg angle of any reflection

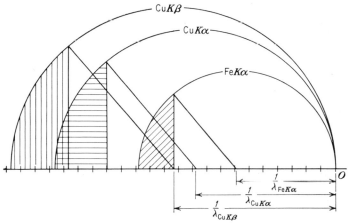

FIG. 4

for x-radiation of a given wavelength. After marking the various values of p on a horizontal line, a series of circles of radius $1/\lambda$ are drawn to the same scale using this line as a diameter, and having the point $p = 0$ on their circumferences. This construction is illustrated in Fig. 4. The shaded sections represent the fields of reflections that lie in the back-reflection region ($\theta = 65$ to $90°$) for CuKβ, CuKα, and FeKα radiations. A different construction like Fig. 4 is needed for each crystal being investigated, since the length of $p_{\min}$ changes with the value of a^{*2} according to (13).

This construction also can be used for crystals belonging to the tetragonal and hexagonal systems, particularly when it is desired to select reflections whose d values depend on the values of a^{*2} or c^{*2} exclusively.

In these cases two constructions are needed, one for

$$p_a = \frac{\lambda}{2}(Na^{*2}), \tag{18}$$

where N has the values of N_T or N_H, and the other for

$$p_c = \frac{\lambda}{2}(Nc^{*2}), \tag{19}$$

where N takes on the values of l^2. The two constructions can be combined in a single drawing by using different colors to mark off p_a and p_c, respectively.

It is important to remember that certain reflections may be missing if the crystal has a nonprimitive lattice. Thus, if the crystal is best described by a body-centered cubic lattice, only reflections having $h + k + l = 2n$ can arise and be recorded. Consequently, in a given case, only one or two actual reflections may fall in the back-reflection region. This number can be doubled by using unfiltered radiation so that the reflections due to the $K\beta$ component are recorded along with those due to $K\alpha$.

Carapella's method for determining appropriate wavelengths. A somewhat more elaborate construction devised by Carapella[1] has the advantage that a single chart can be used for different crystals belonging to the same crystal system. Combining the Bragg equation (2) with the interplanar-spacing equation for an isometric crystal,

$$d_{hkl} = \frac{a}{\sqrt{h^2 + k^2 + l^2}}, \tag{20}$$

one obtains

$$\frac{2\sin\theta}{h^2 + k^2 + l^2} = \frac{\lambda}{a}. \tag{21}$$

A chart may be prepared by first plotting a series of curves representing the left side of (21) as a function of θ (one curve for each possible value of hkl), followed by plotting on the same graph curves representing the right side of (21) as a function of a (similarly, a separate curve is drawn for each value of λ). The resulting charts are shown in Figs. 5 and 6, where Fig. 6B is essentially an enlargement and continuation of Fig. 6A for back-reflection lines of higher order.

The use of these charts is quite simple. A vertical line is drawn corresponding to the approximate value of a (dashed line in Fig. 5). At the intersection of this line with the curve representing the radiation deemed suitable, a horizontal line is drawn. The intersections of this horizontal line with the curves representing the planes (hkl) mark the glancing angle at which these planes reflect the chosen radiation. In the illustrative example, the cell edge of a crystal with a face-centered cubic lattice has the value 3.36 Å. If copper $K\alpha$ and $K\beta$ radiations are both used, five reflections can be recorded in the back-reflection region, as compared with only two reflections if $K\alpha$ radiation alone were used.

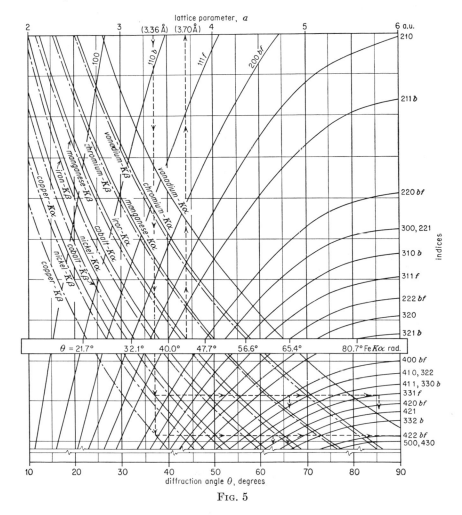

Fig. 5

If the crystals belong to the tetragonal or hexagonal systems, two lattice constants must be determined. In this case, a minimum of three back-reflection lines are necessary. For greater accuracy, even more lines are desirable, especially reflections from planes that permit the measurement of each lattice constant more or less independently. In selecting the x-radiation, therefore, it is necessary to note the indices as well as the number of available lines. A set of charts that allow one

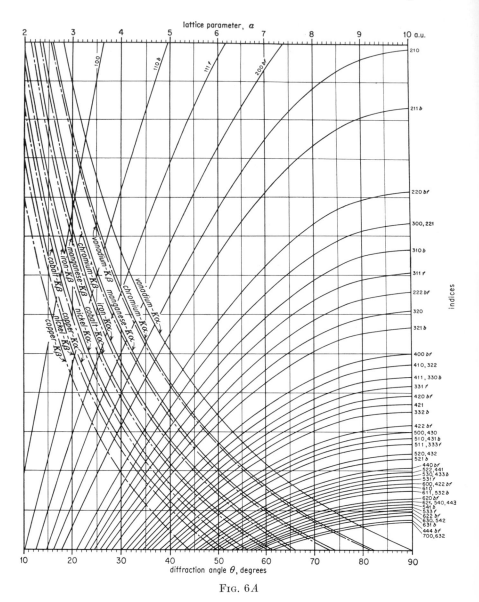

Fig. 6A

to do this graphically have been prepared by Carapella[2] (Figs. 7, 8, and 9).

Consider Fig. 7, which illustrates the use of a chart for the hexagonal lattice. The chart consists of two parts. In the lower half, curves representing (2) are first drawn, with d as a function of θ for radiations commonly used. Next, d/a is plotted against d, on a common ordinate, for

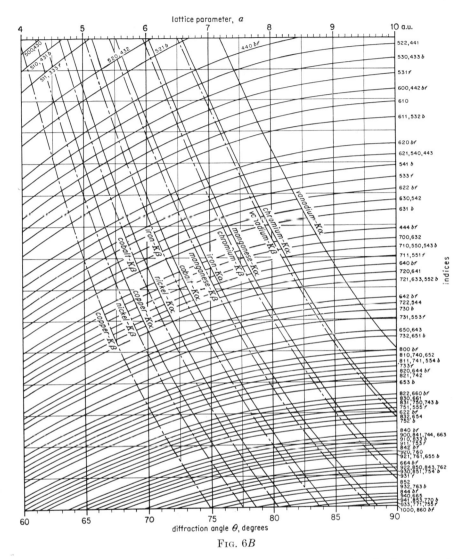

FIG. 6B

various values of a. The interplanar-spacing equation for the hexagonal system,

$$\frac{d_{hkl}}{a} = \frac{1}{\sqrt{\frac{4}{3}(h^2 + hk + k^2) + \frac{l^2}{(c/a)^2}}}, \tag{22}$$

is used to plot d/a as a function of c/a in the upper half of the chart, using the same abscissa for d/a as was used in the lower part. A separate curve must be drawn for each plane (hkl). Since rhombohedral crys-

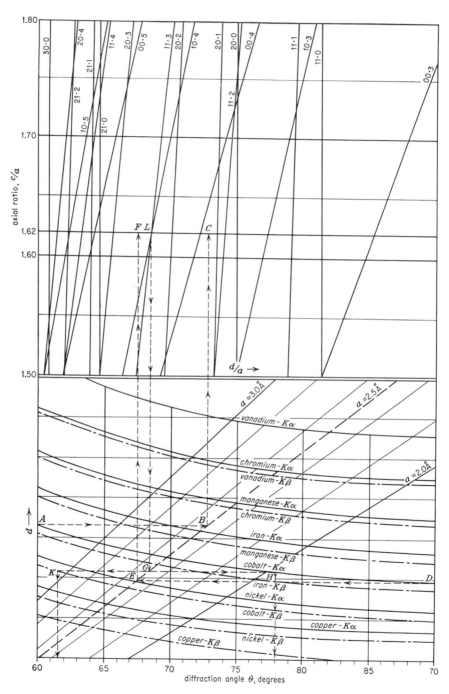

Fig. 7

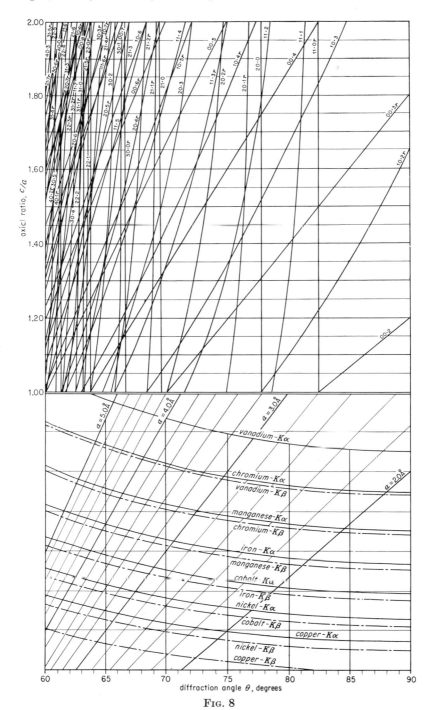

Fig. 8

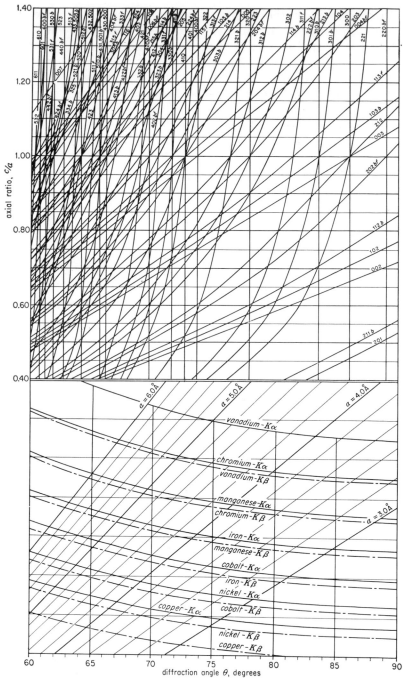

Fig. 9

tals can be indexed on the basis of a hexagonal lattice, the same chart can be used for rhombohedral crystals. The possible *hkl* values of a rhombohedral crystal indexed on a hexagonal lattice are marked by the letter *r* following the indices. Similarly, in Fig. 9, the indices belonging to a body-centered tetragonal lattice are followed by the letter *b*. (Occasionally, a larger, face-centered tetragonal unit cell is incorrectly chosen to describe the body-centered tetragonal lattice. The possible indices for such a choice are those followed by the letter *f*.)

The use of the chart is illustrated in Fig. 7 for a hexagonal crystal having $a = 2.50$ Å and $c/a = 1.62$. Suppose cobalt $K\alpha$ is selected. Two horizontal lines, A and D, are drawn from the extreme ends of the $CoK\alpha$ curve until they intersect the cell-edge curve at B and E, respectively. These points are then joined by vertical lines BC and EF to the horizontal line in the upper half of Fig. 7 representing $c/a = 1.62$. The curves representing the reflections lying in the back-reflection region for cobalt $K\alpha$ radiation intersect this line between the points F and C. By reversing the above procedures, it is possible to determine the value of the glancing angle θ for such a reflection. Suppose the θ value of the 11·3 and 10·4 reflections of this crystal is desired. A vertical line is dropped from L until it intersects the cell-edge curve at G. A horizontal line is next drawn until it intersects the $CoK\alpha$ curve at H. The θ value is now read directly from the abscissa; in this example $\theta = 78°$. Similarly, the glancing angle if $CoK\beta$ radiation were used is found to be 61.5°.

It is possible to construct other types of graphical aids. For example, Thomas[3] has devised a slide rule which allows one to determine rapidly the indices of back-reflection lines of cubic crystals. Alternatively, this slide rule can be used to determine the wavelength for which a reflection lies in the back-reflection region.

Practice of careful experimental technique

M. E. Straumanis[5] is the strongest exponent of attaining precision through careful experimental techniques. The specimen is centered very carefully with the aid of a microscope and a precision-made camera assembly. Errors due to the uncertainty of the film radius are eliminated by using his asymmetric film-mounting arrangement. Absorption of x-rays by the sample is decreased by diluting the sample, mounting it on the surface of a capillary tube, and holding the over-all diameter to less than 0.2 mm. This also has the advantage that the lines on the film are narrow. By using only back-reflection lines near $\theta = 90°$, other systematic errors are minimized. Finally, a specially constructed film-reading device is employed to assure the highest possible accuracy of reading.

An important consideration in the measurement of any physical quan-

Fig. 10*A*

Fig. 10*B*

tity is the maintenance of standard conditions of temperature, pressure, etc. For precision cell-edge determination, therefore, it is important to know the temperature at which the measurement is made. Since the duration of a single exposure may be several hours, it is necessary to take steps to maintain the temperature constant during this time. Changes in temperature not only change the value of the cell edge but may also introduce errors owing to relative movement of various parts of the camera. In order to eliminate these difficulties, Straumanis[7] has built an elaborate temperature-control system. The camera and its accessories are shown in Fig. 10*A*. The entire apparatus, consisting of the camera enclosed in its thermostat, and mounted on the x-ray unit, two separate baths for providing either heated or cooled liquids, and the necessary controls, is shown in Fig. 10*B*.

Minimizing errors by using an internal standard

A simple and rapid procedure for obtaining spacing values consists of mixing the material under investigation with another whose spacings are known to the desired accuracy. The resulting photograph contains the lines due to both substances on the same film. Since the accurate *d* values of one are known, those of the other can be obtained by simple interpolation provided certain precautions are observed. The first is that the particle sizes of both substances should be less than 1 micron. The mixture of the two should, of course, be quite intimate, and the two sets of reflections should not overlap. Finally, the temperature at which the spacings of the standard were determined must be taken into account provided that it differs from the temperature at which the photograph was made in the laboratory.

The *d* values thus determined can be used with the extrapolation methods described in a subsequent section. Alternatively, the values of $\Delta d/d \times 100$, where Δd is the difference between the known and the measured value of *d* of the standard, can be plotted against some extrapolation function of θ, say $\dfrac{1}{2}\left(\dfrac{\cos^2 \theta}{\sin \theta} + \dfrac{\cos^2 \theta}{\theta}\right)$ as discussed beyond. The percentage correction for any measured *d* value can then be read directly from this graph. The accuracy attainable by this procedure is directly comparable with the accuracy to which the spacings of the standard are known. Moreover, Bacon[9] has shown that this same procedure need not be limited to isometric crystals. Since the methods described in the later sections become more difficult to handle for nonisometric crystals, the internal-standard procedure is highly recommended for that purpose.

Matching absorption errors

In Chapter 5 it was shown that absorption of x-rays by the sample obeys the relation

$$I = I_0 e^{-\mu t}, \tag{23}$$

where μ is the linear absorption coefficient and t the thickness of the sample. Accordingly, if the radius of a powder specimen is kept constant, its absorption varies exponentially with changes in the linear absorption coefficient for different substances. On the other hand, if, by diluting a heavily absorbing substance, the effective absorption coefficient of the sample were made equal to that of a sample of another substance, then specimens prepared from both substances would have the same absorption. Consequently, both samples would introduce identical absorption errors in the measured glancing-angle values. Thus, if the product μr for different specimens is kept constant (the radius $r \sim t$), the absorption of different samples is the same. If identical procedures are followed in preparing the powder photographs of such samples, the spacing errors for all these samples should be the same. This effect can be utilized to develop a procedure for determining precise spacing values.

To make use of this scheme, the ratio

$$R = \frac{\sin \theta_{\text{theoretical}}}{\sin \theta_{\text{experimental}}} \tag{24}$$

is first determined for several substances whose unit-cell dimensions are accurately known. Next, the sample under investigation is fashioned into a specimen having the same μr value as one of the standards. Since a plot of R against $\frac{1}{2}\left(\dfrac{\cos^2 \theta}{\sin \theta} + \dfrac{\cos^2 \theta}{\theta}\right)$ for different values of θ is a straight line,† such a plot prepared for the standard can now be used to read off R values for corresponding reflections of the unknown. These R values, in turn, can be used to compute $\sin \theta_{\text{theor}}$ from which the accurate value of the unit-cell edges follows.

Scatturin, Tornati, and Zannetti[11] used samples of NaCl, KCl, $Pb(NO_3)_2$, TlCl, and TlO_3 as standards. These substances represent a range in linear absorption coefficients from 161 cm^{-1} to 2120 cm^{-1} for $CuK\alpha$ radiation. They diluted the unknown samples with starch to obtain the same value of the product μr as that of the nearest standard. Some of the diluents mentioned in Chapter 5 can also be used and may be preferable.

In order to determine the value of μ for such a mixture, the procedures discussed in Chapter 5 can be used. Alternatively, since the absorption

---

† The reasons for choosing the function $\frac{1}{2}\left(\dfrac{\cos^2 \theta}{\sin \theta} + \dfrac{\cos^2 \theta}{\theta}\right)$ will be explained in a later section of this chapter.

coefficient of the diluent is far less than that of the sample, an approximate relation

$$\mu_{\text{effective}} = \frac{\mu_{\text{sample}}}{v}, \tag{25}$$

where
$$v = \frac{\text{tot. vol. of specimen}}{\text{vol. of sample}},$$

can be used instead. It should be remembered that μ_{eff} is simply the desired value of μ, that is, the μ value of the nearest standard, so that relation (25) can be used to determine the fractional volume of the sample to be used in preparing the specimen.

Graphical extrapolation

Kettmann's extrapolation. It was pointed out in the last chapter that the systematic errors vanished at $\theta = 90°$. Although the d value of a reflection whose glancing angle is 90° cannot be measured, it is possible to extrapolate to this value from d values of reflections that can be

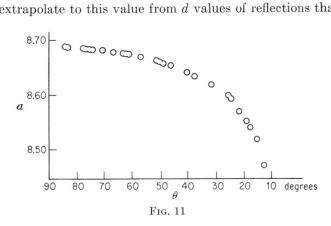

Fig. 11

recorded. Based on the assumption that the systematic errors decrease with increasing θ, a curve drawn through the plotted points, representing the experimentally determined values, should cross the axis corresponding to $\theta = 90°$ at the correct (error-free) value. The first to suggest such a plot was Kettmann.[12] The earliest plots were made against θ (also against $1 - \cot \theta$) directly. Such plots have the disadvantage of nonlinearity, making it difficult to extend the curve accurately to $\theta = 90°$ (Fig. 11).

Bradley and Jay's extrapolation. Bradley and Jay,[13] who were the first to study systematically the effects of the errors discussed in the last chapter, found that most errors depended on various functions of θ which can be combined into a single expression which is linear with $\cos^2 \theta$ for

$\theta > 60°$. The advantage of the use of $\cos^2 \theta$ is that it is easy to extrapolate along a straight line.   It should be noted that, since $\sin^2 \theta = 1 - \cos^2 \theta$, the errors also vary linearly with $\sin^2 \theta$.   If $\cos^2 \theta$ is used, the plotted d values are extrapolated to $\cos^2 \theta = 0$; if $\sin^2 \theta$ is used, the d values are

Table 1
The dependence of errors in d on functions of θ

Type of error	*Dependence of error in d*
Absorption error	$\dfrac{\cos^2 \theta}{\theta}$ or $\dfrac{1}{2}\left(\dfrac{\cos^2 \theta}{\sin \theta} + \dfrac{\cos^2 \theta}{\theta}\right)$
Specimen-eccentricity error	$\cos^2 \theta$
Radius error	$\left(\dfrac{\pi}{2} - \theta\right) \cot \theta$
Beam-divergence error (umbrella effect)	$\dfrac{\cos^2 \theta}{\sin^2 \theta} - 1$

extrapolated to $\sin^2 \theta = 1$.   Tables listing $\cos^2 \theta$ and $\sin^2 \theta$ values are given in Appendixes 4 and 5.

The dependence of the various errors in the measured d value on θ, described in the previous chapter, is summarized in Table 1.   If the

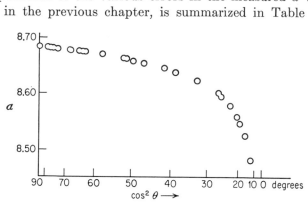

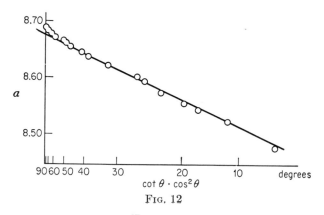

Fɪɢ. 12

radius error is eliminated by the asymmetric film-mounting arrangement and the beam-divergence error (umbrella effect) is ignored because it is much smaller than the other errors, then only two principal errors remain, namely, the absorption error and the specimen-eccentricity error. Both errors depend primarily on $\cos^2 \theta$ for large values of θ since $\sin \theta$ tends

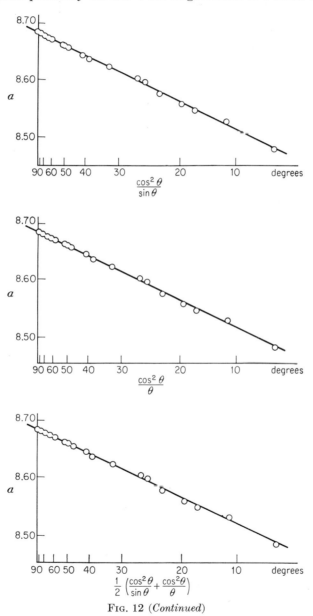

FIG. 12 (*Continued*)

to unity as θ approaches 90°. For large values of θ, therefore, the plot against cos² θ is justified.

Taylor, Sinclair, Nelson, and Riley's extrapolation. Recently, Taylor and Sinclair[15] and, independently, Nelson and Riley,[16] have shown that an even better extrapolation function to use is $\dfrac{1}{2}\left(\dfrac{\cos^2 \theta}{\sin \theta} + \dfrac{\cos^2 \theta}{\theta}\right)$. Numerical values for this function are listed in Appendix 3. By reference to Table 1, it can be seen that this represents a mean between the different functions which are proportional to the error caused by absorption and the error caused by beam divergence. Since modern powder cameras eliminate the radius error and minimize the eccentricity error, absorption and divergent radiation remain the chief source of error.

Nelson and Riley[16] used a very careful experimental technique in order to compare and evaluate the different extrapolation functions described

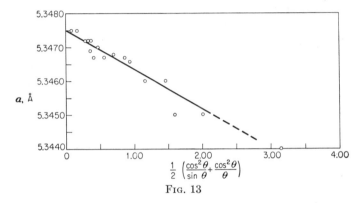

FIG. 13

above. They selected the gamma modification of Cu_9Al_4 for their specimen because this alloy is isometric and because it gives many strong reflections beginning at $\theta = 12.7°$ and ending at $\theta = 84.8°$. The specimen was placed in a camera carefully constructed to eliminate any eccentricity in the specimen. The camera was thermostatically maintained at a constant temperature during the exposure and a special film reader was employed to measure the glancing angles. The principal remaining source of error, therefore, was absorption of the x-ray beam by the specimen.

The data thus obtained were used to construct graphs showing the variation of the cell edge a as a function of the different extrapolation functions listed in Table 1 of Chapter 14. Some of Nelson and Riley's graphs are reproduced in Fig. 12. As can be seen in this figure, the plot against cos² θ is linear for θ values greater than 60°. The plots against either $\dfrac{\cos^2 \theta}{\sin \theta}$ or $\dfrac{\cos^2 \theta}{\theta}$ are linear for much smaller values of θ. It is for this reason that Nelson and Riley recommended the use of the arithmetic mean of these two functions whenever absorption is the chief source of error. They further confirmed this conclusion by varying the absorption in the sample by suitably diluting it and observing no significant changes in the linearity of the resulting graphs.

As an additional illustration of the use of this extrapolation procedure, Fig. 13 shows a graph for CaF_2. The data were obtained using a regular 114.6-mm-diameter powder camera under typical conditions, that is, no extraordinary attempts to achieve accuracy by mechanical means were made. Consequently, θ was measured to .01°, and the d values were determined to 1 part in 10,000. Despite the scatter among the experimental points, it is nevertheless possible to draw a long straight line because $\frac{1}{2}\left(\frac{\cos^2\theta}{\sin\theta} + \frac{\cos^2\theta}{\theta}\right)$ is linear over such a large range. This feature becomes even more important for nonisometric crystals where it is often difficult to find a close grouping of useful lines in the back-reflection region.

Least-squares method

Principles. It is well known that the repetition of a given set of measurements followed by an averaging of the measurements obtained tends to improve the accuracy of the determination. This procedure was first formally stated by Legendre in 1806, and later developed by Laplace and Gauss.† The latter formulated the problem in the following way: If a series of observations expressed by linear equations of the form

$$a_1 x + b_1 y + c_1 z + \cdots = n_1 + \epsilon_1,$$
$$a_2 x + b_2 y + c_2 z + \cdots = n_2 + \epsilon_2, \qquad (26)$$
$$a_3 x + b_3 y + c_3 z + \cdots = n_3 + \epsilon_3,$$

can be written to relate a set of observational parameters $x, y, z, \ldots$ with a set of observations $n_1, n_2, n_3, \ldots$ having associated with them the errors of measurement $\epsilon_1, \epsilon_2, \epsilon_3, \ldots$ then the most satisfactory set of observational parameters is the one that makes the sum of the squares of the errors a minimum. The name of this procedure is the *method of least squares*. The details of this method, as well as the theoretical justifications for this procedure, can be found in books on advanced calculus.

Cohen's method. In 1936, Cohen[20,21] adapted the method of least squares to the determination of the most satisfactory unit-cell dimensions from a single powder photograph. The chief virtue of this procedure is the elimination of subjective errors which arise when a curve or a straight

† Sir Edmund Whittaker and G. Robinson, *The calculus of observations* (D. Van Nostrand Company, Inc., London, 1949), 4th ed., Chapter IX.

line must be fitted to a set of points of a graph. In effect, the straight line is fitted by the method of least squares. On the other hand, therein also lies one of its chief limitations, namely, the removal of the observer's judgment in deciding which of the individual measurements can be deemed more reliable. Hess[23] has pointed out that this can be overcome by appropriately weighting the individual measurements, but not without an expenditure of additional computational labor. The ever-increasing availability of rapid automatic computers tends to minimize this drawback and to favor the use of this procedure in precision cell-edge determinations. Furthermore, the same procedure can be readily extended to deal with crystals that are not isometric, whereas graphical methods become less useful in such cases.

It has been shown in the preceding sections that the errors in measured d values caused by absorption and specimen eccentricity (the principal causes of systematic errors) can be set proportional to $\cos^2 \theta$, that is,

$$\frac{\Delta d}{d} \sim \cos^2 \theta. \tag{27}$$

Let the Bragg equation, written

$$\frac{\lambda}{2} = d \sin \theta, \tag{28}$$

be first squared:

$$\left(\frac{\lambda}{2}\right)^2 = d^2 \sin^2 \theta. \tag{29}$$

Taking logarithms of both sides gives

$$2 \log \frac{\lambda}{2} = 2 \log d + \log (\sin^2 \theta). \tag{30}$$

Differentiating (30) results in

$$2 \frac{\Delta \lambda}{\lambda} = 2 \frac{\Delta d}{d} + \frac{\Delta \sin^2 \theta}{\sin^2 \theta}. \tag{31}$$

Since it is assumed that the x-ray wavelength is known exactly, that is, that $\lambda = \text{const}$, and $\Delta \lambda = 0$, (31) can be written

$$- 2 \frac{\Delta d}{d} = \frac{\Delta \sin^2 \theta}{\sin^2 \theta}. \tag{32}$$

Equations (27) and (32) can be combined to give

$$\frac{\Delta \sin^2 \theta}{\sin^2 \theta} \sim \cos^2 \theta,$$

or $$\Delta \sin^2 \theta \sim \cos^2 \theta \sin^2 \theta. \tag{33}$$

Since $\sin \theta \cos \theta = \frac{1}{2} \sin 2\theta$, equation (33) can also be written

$$\Delta \sin^2 \theta = D \sin^2 2\theta, \tag{34}$$

where D is a proportionality constant having the same value for all angles θ of one film, but having different values for different films. It should be noted that, since $\Delta\lambda$ was assumed to be identically equal to zero, the values of θ in (34) must be referred to a single wavelength.

Application to isometric crystals. The relation determining the value of $\sin^2 \theta$ for isometric crystals can be written

$$\sin^2 \theta = A_0(h^2 + k^2 + l^2), \tag{35}$$

where $A_0 = \lambda^2/4a^2$. If the possibility of an error in the value of $\sin^2 \theta$ is postulated, (35) can be written†

$$(h^2 + k^2 + l^2)A_0 = \sin^2 \theta - \Delta \sin^2 \theta. \tag{36}$$

Substituting for $\Delta \sin^2 \theta$ from (34), and rearranging the terms, equation (36) becomes

$$(h^2 + k^2 + l^2)A_0 + D \sin^2 2\theta = \sin^2 \theta. \tag{37}$$

Let $\alpha = (h^2 + k^2 + l^2)$ and $\delta = 10 \sin^2 2\theta$ (the factor 10 is included to make the values of δ more nearly equal to the values of α) so that (37) can be written

$$\alpha A_0 + \delta D = \sin^2 \theta. \tag{38}$$

Equation (38) relates the known value of α to the measured value of θ provided only systematic errors occur. As was shown in the preceding sections, random observational errors also occur, with the result that the two sides of (38) are not identically equal; for a given reflection,

$$\alpha_1 A_0 + \delta_1 D - \sin^2 \theta_1 = \epsilon_1. \tag{39}$$

According to the method of least squares, the most accurate values of the observational parameters A_0 and D are those which make the sum of the squares of the random errors a minimum. The sum of such equations as (39) can be written

$$\sum_i{}' (\alpha_i A_0 + \delta_i D - \sin^2 \theta_i) = \sum_i \epsilon_i. \tag{40}$$

---

† Cohen's original notation is used in this derivation. This procedure could be equally well carried out using reciprocal-lattice coordinates. Specifically, multiplying both sides of (37) by $4/\lambda^2$ gives

$$(h^2 + k^2 + l^2)a^{*2} + \sin^2 2\theta D' = Q_{hkl},$$

where $D = (4/\lambda^2)D$.

Squaring each term on both sides, (40) becomes

$$\sum_i (\alpha_i A_0 + \delta_i D - \sin^2 \theta_i)^2 = \sum_i \epsilon_i^2. \tag{41}$$

In order that (41) be a minimum, the first derivatives of $\sum_i \epsilon_i^2$ with respect to A_0 and D must vanish. Therefore,

$$\frac{\partial}{\partial A_0} \sum_i \epsilon_i^2 = 0 = \sum_i \alpha_i(\alpha_i A_0 + \delta_i D - \sin^2 \theta_i), \tag{42}$$

and $$\frac{\partial}{\partial D} \sum_i \epsilon_i^2 = 0 = \sum_i \delta_i(\alpha_i A_0 + \delta_i D - \sin^2 \theta_i). \tag{43}$$

Multiplying through by α_i in (42) and by δ_i in (43) and rearranging the terms, the so-called *normal equations* are obtained:

$$\begin{aligned}
\left(\sum_i \alpha_i^2 \right) A_0 + \left(\sum_i \alpha_i \delta_i \right) D &= \sum_i \alpha_i \sin^2 \theta_i, \\
\left(\sum_i \delta_i \alpha_i \right) A_0 + \left(\sum_i \delta_i^2 \right) D &= \sum_i \delta_i \sin^2 \theta_i.
\end{aligned} \tag{44}$$

These two simultaneous equations can be solved in the usual way to give the value of A_0 and D. Actually, it is only necessary to determine the value of A_0, since $a = \lambda^2/4A_0$.

The application of Cohen's procedure consists of first forming the products α_i^2, $\alpha_i \delta_i$, δ_i^2, $\alpha_i \sin^2 \theta_i$, and $\delta_i \sin^2 \theta_i$, for each reflection i. These terms are then summed for all reflections in the back-reflection region, and the appropriate sums are substituted in (43). Finally, the values of A_0 and a are determined.

Extension to uniaxial crystals. This procedure is readily extended to nonisometric crystals. For uniaxial crystals (38) becomes

$$\alpha_i A_0 + \gamma_i C_0 + \delta_i D = \sin^2 \theta_i, \tag{45}$$

where the values of the symbols for the tetragonal and hexagonal systems are given in Table 2.

Table 2
Values of symbols in (45) and (46) for tetragonal and hexagonal systems

Tetragonal system	Hexagonal system
$A_0 = \lambda^2/4a^2$	$A_0 = \lambda^2/3a^2$
$C_0 = \lambda^2/4c^2$	$C_0 = \lambda^2/4c^2$
$\alpha_i = (h^2 + k^2)_i$	$\alpha_i = (h^2 + hk + k^2)_i$
$\gamma_i = (l^2)_i$	$\gamma_i = (l^2)_i$
$\delta_i = 10 \sin^2 2\theta_i$	$\delta_i = 10 \sin^2 2\theta_i$

The normal equations (obtained in a manner similar to that of the isometric case) are

$$\left(\sum_i \alpha_i^2 \right) A_0 + \left(\sum_i \alpha_i \gamma_i \right) C_0 + \left(\sum_i \alpha_i \delta_i \right) D = \sum_i \alpha_i \sin^2 \theta_i$$

$$\left(\sum_i \gamma_i \alpha_i \right) A_0 + \left(\sum_i \gamma_i^2 \right) C_0 + \left(\sum_i \gamma_i \delta_i \right) D = \sum_i \gamma_i \sin^2 \theta_i \qquad (46)$$

$$\left(\sum_i \delta_i \alpha_i \right) A_0 + \left(\sum_i \delta_i \gamma_i \right) C_0 + \left(\sum_i \delta_i^2 \right) D = \sum_i \delta_i \sin^2 \theta_i.$$

As before, the products are first formed for each reflection i, and then summed and substituted in the normal equations, which, in turn, are solved for A_0 and C_0 by using determinants.

Biaxial crystals can be treated similarly. The number of parameters to be determined and the number of normal equations to be formed, however, considerably increase the computational labor.

Practical considerations

In the least-squares method the value of δ need be computed to only two significant figures, whereas the value of $\sin^2 \theta$ must be known to as many figures as are required in the final value of the lattice constant. The number of significant figures needed in computations can be decreased by adopting an approximate value of A_0 (and C_0, etc.) and finding the corrections to these values from the differences between $\sin^2 \theta_{(exp)}$ and $\sin^2 \theta$ computed from the adopted A_0. Letting $v = \sin^2 \theta_{(adopted)} - \sin^2 \theta_{(exp)}$ the terms in the normal equations are $\sum_i \alpha_i v_i$ instead of $\sum_i \alpha_i \sin^2 \theta_i$, and $\sum_i \delta_i v_i$ instead of $\sum_i \delta_i \sin^2 \theta_i$, and ΔA_0 instead of A_0, for isometric crystals. Similar substitutions can be made in the other systems also. When these difference equations are used, ΔA_0 usually has only three significant figures and v only two.

The graphical methods give the correct value of a cell edge because this value is determined from an imaginary reflection at $\theta = 90°$, the angle at which all systematic errors vanish. Similarly, the least-squares method fits an imaginary straight line to the observed d values and extrapolates it analytically to $\theta = 90°$. If a real choice between these two procedures exists, therefore, it must be resolved on the basis of whether the *random* errors are large or whether they are negligible. In the graphical method, they manifest themselves in the scatter of observational points about the straight line that must be drawn to determine the extrapolated value. When such scatter is small, the line can be drawn easily and the graphical method is very accurate. On the other hand, when the scatter is great, the least-squares method should give better results since it is expressly designed to eliminate random errors. In most instances, however, intelligent handling of either procedure results in equally high accuracy.

Literature

Determination of appropriate wavelengths

[1] Louis A. Carapella, *A graphical method for selecting suitable targets for precision determination of cubic lattice constants and for solving cubic powder patterns*, J. Appl. Phys. **11** (1940) 510–514.

[2] Louis A. Carapella, *A graphical method for selecting suitable radiations for the precision determination of noncubic lattice constants and for indexing back-reflection lines in powder x-ray photographs*, J. Appl. Phys. **11** (1940) 800–805.

[3] D. E. Thomas, *A slide rule for x-ray diffraction by cubic crystals*, J. Sci. Instr. **18** (1941) 205.

[4] M. E. Straumanis, *Graphical indexing of powder patterns of cubic substances and the choice of radiation for precision measurements of lattice parameters*, Am. Mineralogist **37** (1952) 48–52.

Practice of careful technique

[5] M. Straumanis and A. Ieviņš, *Die Präzisionsbestimmung von Gitterkonstanten nach der asymmetrischen Methode* (Julius Springer, Berlin, 1940). (Reprinted by Edwards Bros., Inc., Ann Arbor, Mich., 1948.)

[6] D. E. Thomas, *Precision measurement of crystal-lattice parameters*, J. Sci. Instr. **25** (1948) 440–444.

[7] M. E. Straumanis, *Lattice parameters, expansion coefficients, and atomic and molecular weights*, Anal. Chem. **25** (1953) 700–704.

Use of internal standard

[8] W. P. Davey, *A study of crystal structure and its applications* (McGraw-Hill Book Company, Inc., New York, 1934) 113–114, 158–163.

[9] G. E. Bacon, *The determination of the unit-cell dimensions of non-cubic substances*, Acta Cryst. **1** (1948) 337.

[10] K. W. Andrews, *The determination of interplanar spacings and cell dimensions from powder photographs using an internal standard*, Acta Cryst. **4** (1951) 562–563.

[11] Vladimiro Scatturin, Maria Tornati, and Roberto Zannetti, *A system for the elimination of absorption errors in x-ray investigations with Debye-Scherrer cylindrical cameras*, Ricerca Sci. **25** (1955) 1447–1460.

Graphical extrapolation

[12] Gustav Kettmann, *Beiträge zur Auswertung von Debye-Scherrer Aufnahmen*, Z. Physik **53** (1929) 198–209.

[13] A. J. Bradley and A. H. Jay, *A method for deducing accurate values of the lattice spacing from x-ray powder photographs taken by the Debye-Scherrer method*, Proc. Phys. Soc. (London) **44** (1932) 563–579.

[14] A. J. Bradley and A. Taylor, *The crystal structures of Ni_2Al_3 and $NiAl_3$*, London, Edinburgh and Dublin Phil. Mag. and J. Sci. **23** (1937) 1049–1067.

[15] A. Taylor and H. Sinclair, *On the determination of lattice parameters by the Debye-Scherrer method*, Proc. Phys. Soc. (London) **57** (1945) 126–135.

[16] J. B. Nelson and D. P. Riley, *An experimental investigation of extrapolation methods in the derivation of accurate unit-cell dimensions of crystals*, Proc. Phys. Soc. (London) **57** (1945) 160–177.

[17] J. B. Nelson and D. P. Riley, *The thermal expansion of graphite from 15°C to 800°C: Part I. Experimental*, Proc. Phys. Soc. (London) **57** (1945) 477–486.

[18] A. Taylor and R. W. Floyd, *Precision measurements of lattice parameters of non-cubic crystals*, Acta Cryst. **3** (1950) 285–289.

[19] A. F. Ieviņš and Ya K. Ozol, *A precision determination of unit cell parameters of triclinic crystals*, Doklady Akad. Nauk. S.S.S.R. **98** (1954) 589–592.

Least-squares methods

[20] M. U. Cohen, *Precision lattice constants from x-ray powder photographs*, Rev. Sci. Instr. **6** (1935) 68–74.

[21] M. U. Cohen, *Errata; Precision lattice constants from x-ray powder photographs*, Rev. Sci. Instr. **7** (1936) 155.

[22] Eric R. Jette and Frank Foote, *Precision determination of lattice constants*, J. Chem. Phys. **3** (1935) 605–616.

[23] James B. Hess, *A modification of the Cohen procedure for computing precision lattice constants from powder data*, Acta Cryst. **4** (1951) 209–215.

16

Appearance of powder photographs

Ideally, a powder photograph that has been correctly prepared and processed should consist of sharply outlined arcs superimposed on a uniform transparent background. In practice, this state of perfection is rarely, if ever, attained for many different reasons. These reasons can be classified into three groups:

1. Physical effects inherent in the experimental arrangement
2. Experimental effects subject to the investigator's control
3. Characteristics of the sample whose recognition may significantly affect the interpretation of the results

It is the purpose of this chapter to describe briefly these three possible causes of irregularities appearing in the final photograph, their origin, interpretation, and means of elimination (whenever possible). By thus acquainting the reader with effects frequently observed in actual practice, it is hoped to stimulate further investigation of their origins and meaning. A special collection of photographs reproduced in this chapter illustrates some of the more frequently observed effects.

Physical effects

One of the most glaring departures from the ideal powder photographs described in the first sentence of this chapter is the nonuniformity of the background. It may appear at first that this is undesirable and that means should be sought for eliminating the background. It will become clear, however, that this is not entirely possible and that it is necessary to accept this condition. Moreover, it turns out that certain kinds of irregularities in the background can be used to divulge information about the nature of the sample under investigation. What first appeared as a nuisance can thus be used as a means for studying the properties of matter.

Some of the physical causes contributing to the background intensity are inherent in the powder method and, hence, are present in all powder photographs. It is important to know what they are so that they can be recognized and differentiated from other effects that may be due to the characteristics of the sample. To this end, the most important physical causes are listed and briefly discussed below. A more complete exposition of these phenomena can be found in textbooks on the physics of x-rays.†

Air scattering. The direct beam is scattered by the gas molecules comprising the gas through which it passes. Since the density of the gas is low, the total x-radiation scattered is relatively slight. The scattering takes place most intensely in a forward direction, causing a darkening in the film around the direct-beam exit hole. It can be minimized by surrounding the beam with a tunnel near the exit port as described in Chapter 4. With this precaution, air scattering is rarely a serious problem, but if it is still troublesome, it can be removed entirely by evacuating the camera.

Fluorescent scattering. It has been pointed out in Chapter 5 that, whenever the sample contains atoms whose atomic number is slightly less than that of the atoms comprising the x-ray target, the sample absorbs and badly scatters the incident beam. This is explained by the fact that the energy of the incident x-radiation is sufficiently greater than the energy required to knock out an electron from the K shell of such atoms in the sample. The ejection of a K electron permits another electron in the same atom to fall into the K shell. When this occurs, energy in the form of x-rays is emitted by the atom. This phenomenon is called *fluorescence,* and the emitted radiation is called *fluorescent* or *characteristic radiation* since its wavelength is characteristic of the atom emitting it. Fluorescent radiation is independent of angle and tends to increase the background intensity uniformly. A simple, but not always very effective, means for decreasing this effect is to place an absorbing foil of metal between the sample and the film. This foil may be the metal that is being used to filter the incident beam.

The best procedure for eliminating fluorescence of the sample is to choose another x ray target. In this connection it should be noted that, if higher-energy x-rays are used, their energy is sufficient to knock out K electrons also, but the probability that such a collision will occur is inversely proportional to the third power of the wavelength used. If

† A concise treatment of the physics of x-rays is given in F. K. Richtmyer, E. H. Kennard, and T. Lauritsen, *Introduction to modern physics,* 5th ed. (McGraw-Hill Book Company, Inc., New York, 1947). A comprehensive treatment of this subject is given in Arthur H. Compton and Samuel K. Allison, *X-Rays in theory and experiment* (D. Van Nostrand and Company, Inc., New York, 1935).

lower-energy x-radiation is used, the problem is entirely eliminated.   For example, if Cu$K\alpha$ radiation is used in making a powder photograph of a material containing iron, the iron atoms fluoresce strongly, producing a heavy background in the photograph.   This is because Cu has an atomic number greater than Fe by 3.   The Fe fluorescence can be avoided by using radiation from a target which has the same atomic number (that is, Fe) or one unit higher (that is, Co†).   Alternatively, a target of a metal of a much higher atomic number can be used, such as Mo.

General radiation scattering.   Despite the use of filters to enhance the relative intensity of the *characteristic K radiation* from an x-ray tube, a certain amount of noncharacteristic radiation is contained in the direct beam also.   This so-called *general radiation* represents a continuous spectrum of energies.   X-rays having a continuous range of wavelengths are present in this spectrum and are reflected by a crystal plane through a continuous range of angles.   The most intense portion of the general radiation which passes through a filter has wavelengths considerably shorter than the characteristic wavelength.   Corresponding to each powder line, therefore, there is a band which is recorded at smaller angles. For several reasons the intensities of these bands are enhanced in the small-angle region.

Such build-up of background intensity in the front-reflection region has a very real importance in the powder method.   The major constituents of the film emulsion, silver and bromine, have the property of strongly absorbing x-rays having wavelengths up to 0.485 Å for silver, and up to 0.918 Å for bromine.   The result is that the film is greatly darkened by x-rays having these wavelengths, and becomes suddenly lighter for longer wavelengths.   These so-called *absorption edges* are particularly apparent when the direct beam contains a large portion of general radiation (see Fig. 1A).

Occasionally the sample contains a species of atoms that has an absorption edge close to that of either silver or bromine.   This produces a very interesting, and sometimes confusing effect.   If a sample contains cadmium, for instance, the cadmium atoms, having an absorption edge at 0.464 Å, absorb virtually all radiation having wavelengths shorter than this.   The silver in the film, on the other hand, strongly absorbs the transmitted x-radiation up to the wavelength of its absorption edge. The result is that relatively intense lines appear for apparent Bragg reflections of x-rays having an average wavelength of $\frac{1}{2}(0.464 + 0.485)$ Å. This effect is illustrated in Fig. 1B.

Incoherent scattering.   An x-ray beam passing through a substance transfers energy to the electrons in that substance by accelerating these

† A Co target is preferable to an Fe target because it can be operated at a higher current.   It also has a somewhat shorter wavelength, which is usually desirable.

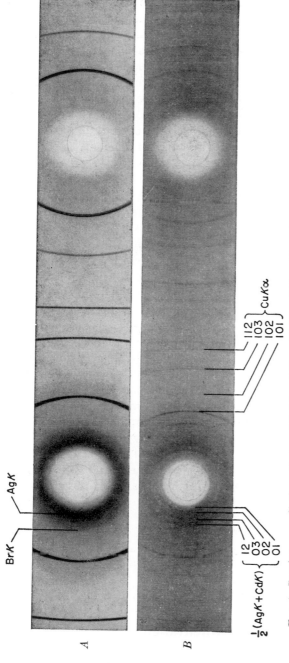

FIG. 1. Continuous radiation effects. (Unfiltered Cu radiation, 40 kv, 18 ma.)

A. Diamond The *K* absorption edges of Ag and Br for the general radiation scattered by the (111) planes of diamond are indicated.

B. Cadmium. The lines appearing at very small angles appear to correspond to Bragg reflections but are actually absorption effects. The normal lines due to Cu*Kα* appear at larger angles, as indicated.

electrons.   When an electron absorbs all the energy of the incident x-ray quantum, it subsequently loses this newly acquired energy by radiating x-rays of like energy in all directions.   This process appears, therefore, as if the electrons were "scattering" the incident quanta and is called *coherent scattering*.   On the other hand, if an electron absorbs only part of the incident quantum's energy, the electron, in turn, radiates x-radiations which differ in energy from that of the incident beam.   This process is called *incoherent scattering* or *Compton modified scattering*, after its discoverer.   As described in Chapter 3, the observed diffraction maxima are formed by the constructive interference of coherently scattered x-rays, whereas incoherent scattering strikes the film at random positions, thereby fogging the film.

Temperature diffuse scattering.   Diffraction theory assumes that the electrons are restricted to small volumes near an atom's nucleus. The background between the diffraction maxima of a perfect crystalline substance, therefore, should contain no contributions from coherently scattered x-rays.   Debye recognized at an early date that actual crystals are not perfect, in that the atoms undergo thermal vibrations at room temperatures.   The thermal motion consists of random displacements of the atoms from their ideal positions.   This has the effect of diffusing the diffraction maxima and increasing the background on which these maxima are superimposed.   This effect is not unlike other effects occasioned by departures from crystalline perfection, and can be distinguished from these other effects by observing the diffuse maxima as a function of temperature.   Temperature diffuse scattering is a useful tool for studying the elastic properties of crystals.

The Lorentz and polarization effects.   Another pair of factors affecting the relative intensities of coherently scattered radiation on different parts of the film are the geometrical Lorentz factor L and the Thomson polarization factor p.   The combined Lorentz-polarization factor Lp is a function of the glancing angle θ and for the powder method has the form

$$Lp = \frac{1 + \cos^2 2\theta}{\sin^2 \theta \cos \theta}. \tag{1}$$

This factor reaches its maximum value as θ approaches $0°$ and $90°$, and has a minimum value at $\theta = 49°$.   It has the effect, therefore, of increasing the intensities at very small and very large angles.

Experimental effects

The effects classified in this section are chiefly occasioned by faulty experimental technique.   Consequently the discussion given here can be used as a practical guide in daily laboratory procedure. Probably the most important subject, described at the close of this section, is the procedure to be followed in properly aligning a camera. In this connection, the reader should review the subject matter of Chapter 4. Although the alignment procedure for powder cameras only is described, the same principles apply to the alignment of any diffraction arrangement.

Line doubling caused by resolution of the $K\alpha$ doublet. As discussed in Chapter 5 the characteristic $K\alpha$ radiation actually consists of two parts, α_1 and α_2. These two radiations usually have wavelengths very nearly alike, and so are diffracted to virtually the same place on the film. Occasionally, the crystallites comprising the powder under investigation are very nearly perfect and resolve these two wavelengths from each other. This resolving power increases with glancing angle, and the resolved doublet can usually be observed in the back-reflection region.

Line doubling caused by absorption. If a very thick, or otherwise heavily absorbing, sample specimen is used, the diffracted x-ray beam is more highly absorbed by the thick central portion of the sample. In extreme cases this results in an absence of the central part of a diffraction line, the line then appearing as a doublet. This effect is more likely to occur when the more readily absorbed long-wavelength radiations are used.

It is important that this effect be recognized when it occurs. Since a doublet could indicate that two of the three unit-cell dimensions are very nearly of the same length (for example, a slight tetragonal distortion of a cubic cell), it is possible to entertain a false interpretation if the line is regarded as a true doublet. Fortunately it is easy to recognize this effect when it occurs. As discussed in Chapter 4, absorption primarily affects those reflections produced by rays which must pass entirely through the sample, namely, reflections having small glancing angles. Consequently, in such a case the forward-reflection lines are split but the back-reflection lines are not. On the other hand, doublets occurring in the back-reflection region may be due to a different cause since back-reflection lines are very sensitive to small changes in the lattice spacings. If line splitting is observed at small glancing angles only, it can be eliminated most readily by using more care in preparing the sample (see Chapter 5).

Diffraction effects due to impurities in the target. An x-ray tube may contain small amounts of impurities in addition to the desired metal in the tube target. This condition was common in the past and is still encountered occasionally. As a consequence of the contamination, the direct beam contains an appreciable amount of characteristic radiations having a wavelength different from the desired one. This produces Bragg reflections at unexpected angles. Manufacturers of x-ray tubes now usually supply a spectral analysis of the target with each tube sold.

These should be scrutinized, therefore, before a new tube is placed in operation.

After a tube has been in use for a long time, or if a tube partially loses its vacuum, the target may become contaminated by tungsten evaporated from the filament. Although normal operating potentials of less than 50 kv are incapable of exciting the K spectrum of tungsten, it is possible to excite the L spectrum. The most intense radiation of the tungsten L spectrum is the α_1 line having a wavelength of 1.476 Å. If an extra line is observed, the substitution of this wavelength value in the Bragg equation will disclose whether the additional reflection is caused by tungsten contamination of the target. If the extra reflection cannot be explained by this procedure, one should test for target contamination as a possible cause by repeating the photograph using another tube, preferably having a different target material. Unless the same impurity is present in both tubes, the line should either disappear or occur at a different glancing angle. If the latter occurs, it is safe to assume that the extra reflection is inherently due to the sample.

Maladjustment of the camera. Spurious diffraction effects are caused when the direct beam improperly strikes one of the metal parts of the camera. This situation can arise when the different parts of a camera are not in correct adjustment. If curious diffraction effects are observed, therefore, one should check the components of the arrangement for possible maladjustment.

Maladjustment can occur from injury to the parts of a powder camera during the handling of the camera in the darkroom. One of the most common causes of trouble arises from bending the collimator or direct-beam trap, or from crimping the ends of their cones. As a consequence of such damage, the direct beam is likely to graze these parts and produce a diffraction pattern of the metal, usually brass, from which these parts are made. Since the position of such an obstruction differs from the sample position, the diffraction lines are displaced from where they would occur were they part of the sample's pattern. Their recognition is aided by their fuzziness, by their occurrence in the front-reflection region, by their different curvatures, and by their tendency to appear on one side of the central hole only.

Another source of trouble may lie in the contamination of the sample itself. If, because of poor adjustment of the sample, a part of the sample's support is struck by the direct beam, its diffraction pattern is obtained in addition to the desired one. Use of glass capillaries or foreign materials to fashion the sample mount can also contribute undesirable reflections. Means of avoiding this particular difficulty have already been discussed in Chapter 5.

It is obvious from the above discussion that much is lost by using an improperly adjusted instrument. The camera itself should need no internal adjustment if it has been properly constructed by the manufacturer, unless it is accidentally damaged. Correct construction of the camera requires that the collimator and direct-beam trap form a straight line bisecting the camera, and containing the center of the specimen rotation axis. This line must be parallel to the base of the camera that rests on the track. If it is suspected that this is not so, it is advisable to have the beam system realigned either by a skilled machinist or by the camera's manufacturer. It is not advisable to make such internal adjustments in the laboratory. A simple test of the alignment of the beam system can be made in the laboratory, however, after the camera track has been correctly aligned. This is done by sliding the camera along the track and observing whether the direct beam continues to strike the fluorescent screen in the direct-beam trap at the same place. If this spot moves as the camera is displaced along the track, it indicates that the collimator-sample-beam-trap axis is not parallel to the surface of the track. Since the x-ray beam passes through an open space when the camera is withdrawn from the x-ray tube during this test, care should be taken to safeguard the operator from exposure.

Misalignment of the track. The track must be aligned so that the pinhole system of the camera points at the focal spot of the x-ray tube target. Every time a new tube is inserted, therefore, the track alignment should be checked and, if necessary, adjusted. If the following simple procedure is observed, an arrangement yielding optimum results is obtained.

1. The angle of inclination of the track with respect to the horizontal should be fixed. For maximum intensity and optimum resolution this angle should be 7°. (Somewhat sharper lines are obtained if a smaller angle, say 3°, is used. This increased resolution, however, is paid for in a decrease in intensity of the direct beam.)

2. Place a properly adjusted camera on the track and raise and lower the track with the x-ray beam turned on, until the brightest spot is observed in the center of the fluorescent screen located in the direct-beam trap.

3. Remove the camera, insert a highly absorbing sample in the camera, and center it accurately. It is helpful for such purposes to use a brass pin that has been turned down to a diameter approximately equal to half the width of the incident beam. Replace the camera on the track and repeat the alignment procedure of step 2 until the pin bisects the direct-beam spot seen on the fluorescent screen in the direct-beam trap.

Following such an alignment of the track, the camera's adjustment can be checked as described above. If the camera is damaged it can be readjusted properly only by a machinist equipped with the proper gauges, etc. It usually happens that two cameras are not exactly alike in that the distance from the base to the center of the specimen axis differs. For this reason, a track that is properly aligned for one camera may be slightly out of alignment for another. It is good laboratory procedure, therefore, to use a particular camera always with the same track.

Characteristics of samples

The ideal powder photograph described in the first sentence of this chapter can be partially attained by the elimination of the experimental effects described in the preceding section through careful technique and by minimizing the physical effects through use of suitable, truly mono-chromatized x-radiation. Two more conditions, however, must be sat-isfied: The powder must contain particles of appropriate size, and the irradiated materials must be truly crystalline. Departures from these two conditions form the subject of the remainder of this chapter. Some of the different effects constituting such departures are illustrated by actual powder photographs: Figs. 2, 3, and 4.

Particle-size effects. It has been indicated in the introductory chapters that, in order for a crystal to diffract at all, the reflecting planes must meet the incident x-ray beam at one of a set of specified angles. This is necessary in order that the x-rays reflected from different points on these planes reach the film in phase, that is, that their path lengths differ by integral multiples of one wavelength. When the diffracting crystal is large, containing thousands of parallel planes, this condition must be satisfied very precisely, and when it is, the diffraction maxima are sharp. As the crystallites become smaller, however, the smaller num-ber of cooperating planes relaxes this condition somewhat. Finally, when the crystallites are so small that they contain only a few planes, in-phase diffraction by these planes is no longer capable of producing sharp dif-fraction maxima. As the crystallite size decreases, the normally sharp diffraction maxima first become broader at their base, then broaden uni-formly throughout (concomitantly decreasing in height in order that the area under the peak remains constant) until, finally, they become so broad that they are no longer clearly visible.

Scherrer was the first to study *particle-size broadening* (as this is called) in a quantitative way. He derived the following expression in 1918:

$$B = \frac{K\lambda}{L \cos \theta}, \tag{2}$$

where B is the broadening of the line expressed in units of 2θ; K is a con-stant approximately equal to 1; L is the average length of the crystallite; and λ and θ have their usual meaning. The actual value of K depends on the experimental arrangement used and on the definition of the *average length L*.   The *breadth* of the line *B* must be differentiated from the natural line breadth due to the experimental arrangement used, such as slit sizes, sample thickness, beam divergence, x-ray line width, etc.   The latter factor requires that radiation of only one wavelength, say $K\alpha_1$, be used.   These factors are further discussed below.

It is convenient to discuss the effects of particle-size broadening in four categories, depending on the average size of the crystallites producing the effect, as follows:

1. Very small crystallites, average diameter less than 10^{-6} cm
2. Small crystallites, average diameter in the range 2×10^{-5} to 10^{-6} cm
3. Crystallites of proper size, average diameter in the range 10^{-3} to 5×10^{-5} cm
4. Crystallites which are too large, average diameter greater than 10^{-3} cm

The ranges of these classifications are approximate, and depend on various properties of the crystals.

Very small crystallites.   When the crystallites are much smaller than 100 Å in their average length, they are too small to yield normal diffraction maxima.   In fact, the diffraction maxima corresponding to lines are so broad that it is not possible to distinguish them from the general background.   Consequently it is not possible to measure line breadth, and the Scherrer equation (2) does not apply.   The main feature of the powder diagram from such very small crystallites consists of a maximum occurring near the undeviated beam direction, accompanied by one or more weaker satellite bands.   (This can be regarded as broadening of the reflection 000.)   This region of the film is called the *small-angle* region, and special experimental arrangements employing long specimen-to-film distances for increased resolution are used to study it.   Small-angle studies are particularly effective in studying particle sizes of colloids, organic substances, or substances suspended in liquid matrices.   The actual range of sizes that can be studied by small-angle cameras includes the very small particles and extends partially into the next category of the classification employed here.

Small crystallites.   The particle size of crystallites in this category is large enough to produce resolved diffraction maxima.   Figure 2*A* is a powder photograph illustrating the effects typical of this size range.

It is in this range that particle size can be studied by means of the Scherrer equation (2).   The upper limit of particle sizes that can be studied is approximately $L = 2 \times 10^{-5}$ cm.   Crystallites much larger than this produce diffraction broadening effects that cannot be clearly resolved from line broadening due to the experimental arrangement used. This so-called *instrumental broadening* must be taken into account if (2) is used to determine the actual particle size.

Warren has shown that this can be done by mixing the unknown with a powder whose particle size is known to be large enough to cause no particle-size broadening. In this *method of mixtures*, the width of the peak of the known substance B_{inst} is first determined. Either the width of the peak at half maximum intensity, or the integral peak width can be used. The broadening due to the particle-size effect B_{part} can then be determined from the measured peak width B_{meas} by the relation

$$B_{meas} = \sqrt{B_{inst}^2 + B_{part}^2}. \tag{3}$$

The value of K in (2) depends on the definition of what is meant by the average length L, and whether the half-maximum or integral breadth of the peak is measured. It also depends, therefore, on the crystallite shape, which is generally not known precisely, and on the particular hkl values of the reflections being studied. Consequently it is not possible to determine the absolute size of the crystallites with very great precision, except in very rare instances when the particle shape is known or can be readily determined. The fact that one usually does not have particles of uniform size, but rather some type of probability distribution, means that the accuracy of particle-size determinations is usually no better than 20 to 40 per cent. In this connection it should be remembered that microscopic examination, whether optical or electron microscopes are used, can do much to help the accuracy of such investigations. The advantages of the x-ray method lie chiefly in its ability to examine large quantities of sample more rapidly.

Crystallites of proper size. The crystallites in this category come closest to giving the ideal powder photograph already mentioned (Fig. 2*B*). The range of sizes given above, 10^{-3} to 2×10^{-5} cm, represents the range normally encountered in the laboratory. Departures from sharp diffraction lines in this case, therefore, must be due to other causes. These are discussed in later sections in this chapter.

Crystallites which are too large. A crystallite whose average length is 10^{-2} cm is normally considered to be a single crystal as distinct from a powder particle, and is best handled by single-crystal methods described elsewhere.† The diffraction spectrum of a single crystal consists of spots (corresponding to the individual reciprocal-lattice points described in Chapter 9) rather than arcs or circles. Consequently, a powder photograph of a sample containing crystallites larger than 10^{-3} cm contains

† For the handling of single crystals see M. J. Buerger, *X-ray crystallography* (John Wiley & Sons, Inc., New York, 1942).

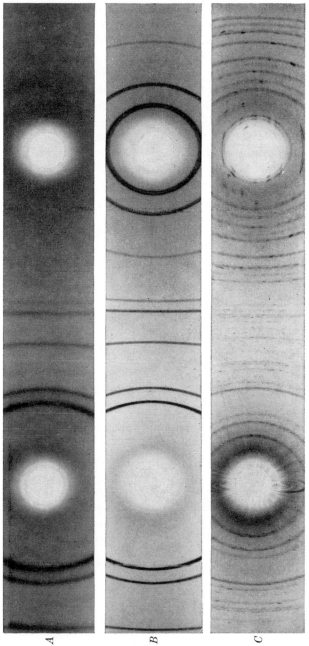

Fig. 2. Particle size effects. (CuKα, 40 kv, 18 ma.) (About 10^{-5} cm.)
A. Nickel catalyst. Note line broadening due to small particle size.
B. Nickel powder of proper size for powder photography. Note $\alpha_1\alpha_2$ resolution in back-reflection region.
C. Titanium dioxide (rutile) powder insufficiently ground. Note spottiness of lines.

spots superimposed on the arcs. In fact, if no particles smaller than 10^{-3} cm are present, the arcs are not continuous but rather consist of closely spaced discrete spots. X-ray diffraction diagrams showing this effect are said to have *spotty lines* (see Fig. 2C).

Since the size of a spot on the film is related to the size of the crystal producing it, it is possible to get some idea of crystallite sizes from direct size measurements on the film. A possible procedure for doing this consists of counting the total number of spots on the complete ring. A flat-cassette method must be used, usually in the transmission or front-reflection arrangement, in order that the complete ring may be recorded. Provided that the sample is stationary during the exposure, the total number of spots counted is related to the total number of diffracting crystals in the irradiated portion of the sample. Accordingly, it is possible, in principle, to determine the average crystallite size. The accuracy of such schemes is very limited for various reasons. When dealing with crystallites in this size range it is far easier to examine the film qualitatively and deduce their sizes from the following observations:

10^{-3} to 4×10^{-3} cm: clearly discernible diffraction lines consisting of very many spots closely spaced

4×10^{-3} to 8×10^{-3} cm: lines still visible, but a scattering of spots deviating from the true line position

8×10^{-3} to 2×10^{-2} cm: lines no longer visible, many spots randomly scattered

2×10^{-2} cm and larger: few scattered spots

Preferred orientation. It has been assumed that the particles of a powder are oriented completely at random. Ordinarily the preparation of the powder sample produces substantially this ideal condition. There are, however, kinds of samples in which the condition is not realized. One is a sample composed of grains of special shape. Another is a sample in which plastic deformation has taken place.

Shape orientation. If the powder consists of needle-shaped particles, these tend to become aligned parallel to the specimen axis in the preparation of the sample. The crystal planes containing the needle axis thus have a greater chance of reflecting to the equatorial line of the powder photograph. This causes an intensification of the arcs of these lines near the equator, and a weakening of them elsewhere. This is illustrated in Fig. 3A.

In a similar manner, plate-shaped particles tend to become aligned so that the plane of the plate is parallel to the axis of the specimen. The crystal plane parallel to the plate, therefore, has the greatest chance of reflecting to the equatorial line of the photograph. This causes the orders of reflection from this plane to be intensified near the equator and weakened away from it. This is illustrated in Fig. 3B.

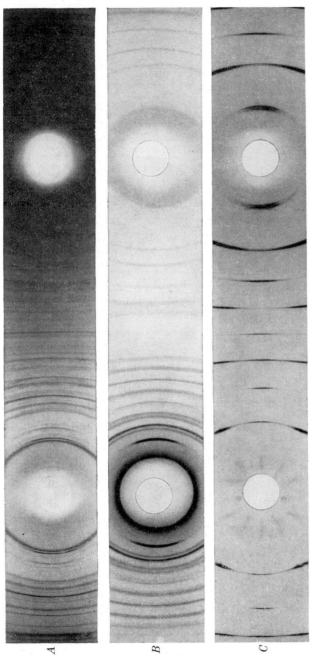

Fig. 3. Preferred orientation effects in powder photographs.   (CuKα, 40 kv, 18 m.a.)

A. Shape orientation.   Needle-like crystals of PbO.
B. Shape orientation.   Plate-shaped crystals of talc.
C. Deformation orientation.   Drawn tungsten wire.

Planes normal to the specimen axis never have a chance to reflect.   In the case of needle-shaped particles this requires that reflections from planes normal to the needle axis are missing.   In the case of plate-shaped particles, all planes normal to the plate have nearly equal opportunities to reflect and hence give rise to powder lines which have normal appearance.   In case the plate is also lath-shaped, the sample gives a diffraction diagram approaching that of a single crystal rotated about the lath axis.   The diagram thus tends to approximate a rotating-crystal diagram.

Deformation orientation.   Many crystals are plastic.   This property arises in certain crystals because of the possibility of shear displacement (called *slip* by metallurgists and *translation gliding* by mineralogists) along rational planes of the crystal.   If an aggregate of crystals is plastically deformed, this process leads to changes of the orientations of the crystals of the aggregate.   When metals are "worked," as, for instance, in rolling a sheet or drawing a wire, a crystallographic direction tends to orient itself parallel to the drawing direction in the case of a wire, or parallel to the plane of rolling in the case of a sheet.   A preferential alignment occurs during each operation of shaping the metal.   Since in almost all instances such preferred orientation of the grains tends to change the physical properties of the metal, the detection of preferred orientation is an oft-recurring problem in metallurgy (see Fig. 3C).

Many substances, metals being notable examples, occur as polycrystalline aggregates.   Even though they are not "powders" in the conventional sense of the word, these materials can be studied by the powder method without requiring that they be first ground up.   When preferred orientation occurs simultaneously for more than one direction, as in rolled sheets, for instance, it is possible to determine the different directions by special studies.   For example, the metal sheet is mounted in transmission normal to the x-ray beam.   Successive powder photographs are recorded on a flat cassette.   Between each photograph, the metal sheet is rotated by a few degrees about a direction parallel to the plane of the cassette.   The intensity distributions on each photograph are studied from the point of view of both location and absolute value.   This makes it possible to determine the relative number of grains having each orientation, and therefore to obtain an idea of the nature of the preferred orientations.   The interpretation of such photographs is greatly aided by a special construction called a *pole figure diagram.*   To construct such a figure, radii of a sphere are drawn along directions of preferred orientation.   The intersections of these radii with the sphere are called *poles.*   The poles are ordinarily projected onto a plane by means of the stereographic projection.   The resulting pole figure is a representation of the three-dimensional orientations of the grains in a sheet.

In a similar way, it is possible to determine the grain orientations in different metal parts such as wires, ribbons, tapes, and springs. Mineralogists use similar procedures for determining the orientations of specific minerals, say mica, in different parts of a rock specimen. A thin section, bearing a known orientation to the rock, is first cut out and then examined in the manner described above.

Crystal texture. In one sense, all matter can be considered as having varying degrees of crystallinity. If a crystal is defined as a state of aggregation of atoms or molecules in a three-dimensional periodic array, then a glass can be thought of as a completely disorganized crystal. Thus, in solids, the range of aggregations extends from ideal crystals such as Iceland spar (a particularly perfect variety of calcite) down to glasses. Even crystals exhibit a range in perfection of the state of aggregation of their constituent atoms. Thus, it is possible to speak of *crystal texture* in describing departures from ideal crystals. In terms of the classical definition of a crystal, this means that departures from true periodicity can occur. In a broader sense, a crystalline substance may have only one-dimensional or two-dimensional periodicity. Some of these departures are discussed below and illustrated in Fig. 4.

Strains in crystals. There are two kinds of strains that can be studied by the powder method. The first kind is produced when a substance is deformed without exceeding its elastic limit. This type of strain is called a *macro strain*. The nature of the deformation usually consists of stretching (or compressing) the sample along one direction. On an atomic scale this is accompanied by dilation (or contraction) of some crystallographic dimension of the unit cell, say the c axis. Simultaneously, one or both of the two directions normal to this axis, say a and b, are contracted (or dilated). The effect observed on a powder photograph in such a case is a shift in the diffraction-line positions in accordance with the changes in the respective d values. Thus, in the above illustration, elongation along c requires that the glancing angle decrease for $00l$ reflections and increase for $hk0$ reflections. Since individual crystallites do not usually deform uniformly, the lines are slightly broadened.

Because these variations in the glancing angle are expected to be small, back-reflection lines must be measured to detect such variations. Commercial equipment is available with which the x-ray tube can be brought directly to the specimen which might be, for instance, a large piece of machinery, or a truss in a bridge. Such an arrangement has the advantage that the stress in a very small area can be measured, thus permitting the determination of stress gradients along the entire member. It is not necessary to compare such measurements with cell-dimension determinations in the unstressed state. This means that not only can localized stresses be measured, but also that the measurements can be made without relieving the stresses. The limited penetrability of the x-ray beam, however, restricts such measurements to the surface of the test specimen.

Another kind of strain, the so-called *micro strain*, is produced when the elastic limit is exceeded. This occurs, for instance, in the cold working of metals. On an atomic scale this type of strain is pictured as a collection of local imperfections randomly distributed throughout the sample. Such random imperfections have the effect of broadening the diffraction lines and they usually also affect the line profiles.

Another feature of powder photographs of strained materials is the presence of radial streaks emanating from the direct-beam exit position on the film. This is called *asterism* and is due to components of non-characteristic radiation present in the incident beam which are reflected by the strained portions of the crystals. Asterism can also be due to other causes and cannot be used, therefore, as anything more than an indication of the possible presence of strains.

It will be recalled from Chapter 5 that the reader was cautioned that, in the preparation of specimens from metal filings, they must be annealed after completing the filing process. Annealing relieves the strains by imparting sufficient thermal motion to the atoms so that the filings can return to the unstressed state.

Stacking faults. In recent years, as the nature of crystals as well as their structures have become better understood, increasing attention has been devoted to the study of imperfections in crystals. A special kind of imperfection occurs for certain crystals whose structures consist of densely populated atomic planes stacked in a prescribed manner along a direction normal to these planes. A simple illustration of this type of structure is cobalt, which consists of (0001) sheets of cobalt atoms in a close-packed hexagonal array. There are two ways of stacking such sheets: hexagonal close packing (requiring two sheets to complete the translation period) and cubic close packing (requiring three sheets to complete the period). Under certain conditions it is possible to obtain a mixture of both kinds of stacking sequences in the same crystal. The discontinuities arising in these stacking sequences are called *stacking faults*.

In addition to cobalt, such stacking faults are found to occur in zinc sulfide, cadmium sulfide, silicon carbide, and other structures where both hexagonal close-packed and cubic close-packed structures are possible. The effect of stacking faults on the powder photograph is to broaden certain diffraction lines in a characteristic fashion. The small-angle side of the line is sharp, whereas for larger angles the intensity of the line declines very gradually. Not all reflections show these so-called "tails." For instance, reflections from planes normal to the stacking direction are unaffected and remain sharp since the spacing between the planes is not changed by the stacking sequence.

Random layer structures. Many crystals having layer structures, such as certain clays, micas, and graphites, are composed of atomic layers that are perfectly periodic only in the two dimensions of the layer. The stacking of these layers in the third dimension is not periodic. Although the spacing between layers may remain constant, there is no fixed relation between corresponding points of successive layers. The difference between these structures and those described in the previous section is the complete randomness of their stacking sequence.

The effect on the powder photographs of such substances is similar to, and yet different from that due to simple stacking faults. Assuming c to be the stacking direction, usually no general hkl reflections are observed at all, the $hk0$ reflections show intensity decline, and the $00l$ reflections may be sharp or broad depending on whether the spacings between successive layers are uniform or not. The $hk0$ reflections may decline uniformly or may show undulations.

Disordered crystals. In a normal crystal composed of more than one kind of atom, the different atoms are arranged so that each kind occupies a prescribed position in the structure. Under certain conditions it is possible that the atoms are not in their correct positions; for example, in a crystal of composition AB, some or all sites of atom A may be occupied by atoms B, whose sites, in turn, may be partially or wholly occupied by atom A, or by another kind of atom C. Such a crystal is said to be *disordered* to distinguish it from its *ordered* counterpart. Disorder is common in crystals composed of like atoms (for example, alloys) although not necessarily restricted to such compounds. Thus, many minerals, notably sulfides, exhibit this type of disorder as regards the positions of the metal atoms.

A disordered crystal is periodic only in a statistical sense, that is, a given atomic site is occupied by some kind of atom. In such cases, some period is usually a submultiple of the corresponding period in the ordered structure. Quite frequently the disordered crystal has a higher symmetry, again because atom A cannot be uniquely distinguished from atom B as belonging to any given site. Thus the disordered crystal, by virtue of its higher symmetry, or its smaller cell size, or both, yields a powder photograph having a small number of well-separated lines.

Disordered crystals can usually be ordered by annealing. When this is accompanied by an increase in one or more of the three periods, new lines appear between those formerly present on the photograph. These new reflections are called *superlattice* reflections. As the crystal goes from the disordered to the ordered state, these lines increase in intensity, until order is complete. Meanwhile, the relative intensities of the original set of lines, the so-called *fundamental* reflections, do not change.

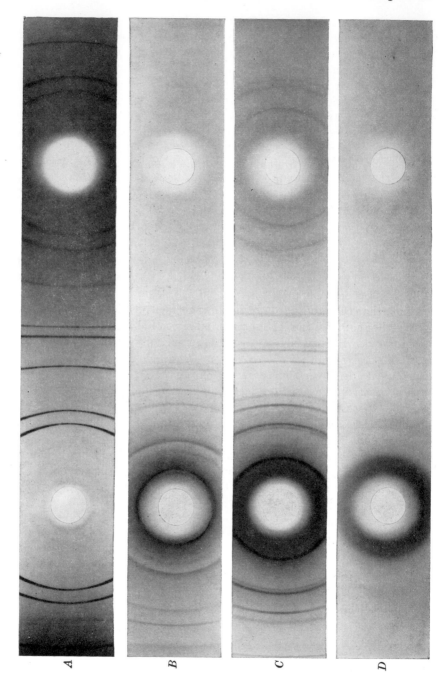

Fig. 4. Crystal texture effects in powder photographs. (Cu$K\alpha$, 40 kv, 18 ma.)

A. Cold-worked brass pin. Note slight asterism at small angles and line broadening at large angles caused by micro strains. Also note weak superstructure lines.

B. Montmorillonite having randomness in stacking of layers. Note that the intensity distribution in certain reflections is uneven (intense on the small angle side of the line, gradually decreasing with increasing angle) while other reflections appear normal. Also note virtual absence of back-reflection lines, in this case due to disorder in the structure.

C. Graphite. Note broadening of some reflection lines and absence of $\alpha_1\alpha_2$ resolution in the back-reflection region, characteristic of imperfect crystals. Also note intense halo at small angles due to presence of amorphous carbon.

D. Lamp black. Note that only broad halos at small angles appear, since this is a noncrystalline form of carbon.

Liquids. The atomic arrangement in liquids is random, not unlike that of a gas. The greater density of the liquid as compared with that of a gas results in a certain ordering among the atoms due to packing considerations. It is usual to describe a liquid as consisting of atoms whose immediate environs do not differ greatly from those of other like atoms. As a consequence of more uniform packing, the coherent scattering of x-rays by a liquid more nearly approximates that of a crystal than that of a gas. Several distinct maxima at different scattering angles can usually be detected. These peaks can be interpreted by discovering a radial-distribution curve which has several peaks at radial values corresponding to interatomic distances of nearest neighbors, next-nearest neighbors, second-next-nearest neighbors, etc. Since liquids differ from amorphous solids or glasses essentially by being more fluid, the x-ray diffraction effects observed for these substances are similar.

Certain substances that appear to be amorphous may not be. For example, most waxes are crystalline despite their softness and pliability. It is usually possible to determine the state of aggregation of a substance only by the powder method.

Literature

The references listed below are primarily intended to serve as a guide to further reading on the applications of the powder method.

[1] J. T. Randall, *The diffraction of x-rays and electrons by amorphous solids, liquids and gases* (John Wiley & Sons, Inc., New York, 1931).

[2] A. Taylor, *X-ray metallography* (Chapman & Hall, Ltd., London, 1945).

[3] H. R. Isenberger, *Bibliography on x-ray stress analysis* (St. John X-ray Laboratory, Califon, N.J., 1949).

[4] A. J. C. Wilson, *X-ray optics* (Methuen & Co., Ltd., London, 1949).

[5] Charles S. Barrett, *Structure of metals*, 2d ed. (McGraw-Hill Book Company, Inc., New York, 1952).

[6] André Guinier, *X-ray crystallographic technology*, English translation by T. L. Tippell, edited by Kathleen Lonsdale (Hilger and Watts Ltd., London, 1952).

[7] Harold P. Klug and Leroy E. Alexander, *X-ray diffraction procedures for polycrystalline and amorphous materials* (John Wiley & Sons, Inc., New York, 1954).

[8] H. S. Peiser, H. P. Rooksby, and A. J. C. Wilson, *X-ray diffraction by polycrystalline materials* (Institute of Physics, London, 1955).

[9] André Guinier and Gérard Fournet, *Small-angle scattering of x-rays*, English translation by Christopher B. Walker with bibliography by Kenneth L. Yudovitch (John Wiley & Sons, Inc., New York, 1955).

Appendixes

Appendix 1
Quadratic forms

Quadratic forms for the isometric, tetragonal, and hexagonal systems

N	hkl	hk0	
	Isometric $(N_C = h^2 + k^2 + l^2)$	Tetragonal $(N_T = h^2 + k^2)$	Hexagonal $(N_H = h^2 + hk + k^2)$
1	1,0,0	1,0	1,0
2	1,1,0	1,1	
3	1,1,1		1,1
4	2,0,0	2,0	2,0
5	2,1,0	2,1	
6	2,1,1		
7			2,1
8	2,2,0	2,2	
9	3,0,0; 2,2,1	3,0	3,0
10	3,1,0	3,1	
11	3,1,1		
12	2,2,2		2,2
13	3,2,0	3,2	3,1
14	3,2,1		
15			
16	4,0,0	4,0	4,0
17	4,1,0; 3,2,2	4,1	
18	4,1,1; 3,3,0	3,3	
19	3,3,1		3,2
20	4,2,0	4,2	
21	4,2,1		4,1
22	3,3,2		
23			
24	4,2,2		
25	5,0,0; 4,3,0	5,0	5,0
26	5,1,0; 4,3,1	5,1	
27	5,1,1; 3,3,3		3,3
28			4,2
29	5,2,0; 4,3,2	5,2	
30	5,2,1		
31			5,1
32	4,4,0	4,4	
33	5,2,2; 4,4,1		
34	5,3,0; 4,3,3	5,3	
35	5,3,1		
36	6,0,0; 4,4,2	6,0	6,0
37	6,1,0	6,1	4,3
38	6,1,1; 5,3,2		
39			5,2

N	hkl Isometric $(N_C = h^2 + k^2 + l^2)$	$hk0$ Tetragonal $(N_T = h^2 + k^2)$	$hk0$ Hexagonal $(N_H = h^2 + hk + k^2)$
40	6,2,0	6,2	
41	6,2,1; 5,4,0; 4,4,3	5,4	
42	5,4,1		
43	5,3,3		6,1
44	6,2,2		
45	6,3,0; 5,4,2	6,3	
46	6,3,1		
47			
48	4,4,4		4,4
49	7,0,0; 6,3,2	7,0	7,0; 5,3
50	7,1,0; 5,5,0; 5,4,3	7,1; 5,5	
51	7,1,1; 5,5,1		
52	6,4,0	6,4	6,2
53	7,2,0; 6,4,1	7,2	
54	7,2,1; 6,3,3; 5,5,2		
55			
56	6,4,2		
57	7,2,2; 5,4,4		7,1
58	7,3,0	7,3	
59	7,3,1; 5,5,3		
60			
61	6,5,0; 6,4,3	6,5	5,4
62	7,3,2; 6,5,1		
63			6,3
64	8,0,0	8,0	8,0
65	8,1,0; 7,4,0; 6,5,2	8,1; 7,4	
66	8,1,1; 7,4,1; 5,5,4		
67	7,3,3		7,2
68	8,2,0; 6,4,4	8,2	
69	8,2,1; 7,4,2		
70	6,5,3		
71			
72	8,2,2; 6,6,0	6,6	
73	8,3,0; 6,6,1	8,3	8,1
74	8,3,1; 7,5,0; 7,4,3	7,5	
75	7,5,1; 5,5,5		5,5
76	6,6,2		6,4
77	8,3,2; 6,5,4		
78	7,5,2		
79			7,3

N	hkl		$hk0$	
	Isometric $(N_C = h^2 + k^2 + l^2)$	Tetragonal $(N_T = h^2 + k^2)$	Hexagonal $(N_H = h^2 + hk + k^2)$
80	8,4,0	8,4	
81	9,0,0; 8,4,1; 7,4,4; 6,6,3	9,0	9,0
82	9,1,0; 8,3,3	9,1	
83	9,1,1; 7,5,3		
84	8,4,2		8,2
85	9,2,0; 7,6,0	9,2; 7,6	
86	9,2,1; 7,6,1; 6,5,5		
87			
88	6,6,4		
89	9,2,2; 8,5,0; 8,4,3; 7,6,2	8,5	
90	9,3,0; 8,5,1; 7,5,4	9,3	
91	9,3,1		6,5; 9,1
92			
93	8,5,2		7,4
94	9,3,2; 7,6,3		
95			
96	8,4,4		
97	9,4,0; 6,6,5	9,4	8,3
98	9,4,1; 8,5,3; 7,7,0	7,7	
99	9,3,3; 7,7,1; 7,5,5		
100	10,0,0; 8,6,0	10,0; 8,6	10,0
101	10,1,0; 9,4,2; 8,6,1; 7,6,4	10,1	
102	10,1,1; 7,7,2		
103			9,2
104	10,2,0; 8,6,2	10,2	
105	10,2,1; 8,5,4		
106	9,5,0; 9,4,3	9,5	
107	9,5,1; 7,7,3		
108	10,2,2; 6,6,6		6,6
109	10,3,0; 8,6,3	10,3	7,5
110	10,3,1; 9,5,2; 7,6,5		
111			10,1
112			8,4
113	10,3,2; 9,4,4; 8,7,0	8,7	
114	8,7,1; 8,5,5; 7,7,4		
115	9,5,3		
116	10,4,0	10,4	
117	10,4,1; 9,6,0; 8,7,2	9,6	9,3
118	10,3,3; 9,6,1		
119			

N	hkl		hk0	
	Isometric $(N_C = h^2 + k^2 + l^2)$	Tetragonal $(N_T = h^2 + k^2)$	Hexagonal $(N_H = h^2 + hk + k^2)$	
120	10,4,2			
121	11,0,0; 9,6,2; 7,6,6	11,0	11,0	
122	11,1,0; 9,5,4; 8,7,3	11,1	10,2	
123	11,1,1; 7,7,5			
124				
125	11,2,0; 10,5,0; 10,4,3; 8,6,5	11,2; 10,5		
126	11,2,1; 10,5,1; 9,6,3			
127			7,6	
128	8,8,0	8,8		
129	11,2,2; 10,5,2; 8,8,1; 8,7,4		8,5	
130	11,3,0; 9,7,0	11,3; 9,7		
131	11,3,1; 9,7,1; 9,5,5			
132	10,4,4; 8,8,2			
133	9,6,4		11,1; 9,4	
134	11,3,2; 10,5,3; 9,7,2; 7,7,6			
135				
136	10,6,0; 8,6,6	10,6		
137	11,4,0; 10,6,1; 8,8,3	11,4		
138	11,4,1; 8,7,5			
139	11,3,3; 9,7,3		10,3	
140	10,6,2			
141	11,4,2; 10,5,4			
142	9,6,5			
143				
144	12,0,0; 8,8,4	12,0	12,0	
145	12,1,0; 10,6,3; 9,8,0	12,1; 9,8		
146	12,1,1; 11,5,0; 11,4,3; 9,8,1; 9,7,4	11,5		
147	11,5,1; 7,7,7		11,2; 7,7	
148	12,2,0	12,2	8,6	
149	12,2,1; 10,7,0; 9,8,2; 8,7,6	10,7		
150	11,5,2; 10,7,1; 10,5,5			
151			9,5	
152	12,2,2; 10,6,4			
153	12,3,0; 11,4,4; 10,7,2; 9,6,6; 8,8,5	12,3		
154	12,3,1; 9,8,3			
155	11,5,3; 9,7,5			
156			10,4	
157	12,3,2; 11,6,0	11,6	12,1	
158	11,6,1; 10,7,3			
159				

N	hkl	hk0	
	Isometric $(N_C = h^2 + k^2 + l^2)$	Tetragonal $(N_T = h^2 + k^2)$	Hexagonal $(N_H = h^2 + hk + k^2)$
160	12,4,0	12,4	
161	12,4,1; 11,6,2; 10,6,5; 9,8,4		
162	12,3,3; 11,5,4; 9,9,0; 8,7,7	9,9	
163	9,9,1		11,3
164	12,4,2; 10,8,0; 8,8,6	10,8	
165	10,8,1; 10,7,4		
166	11,6,3; 9,9,2; 9,7,6		
167			
168	10,8,2		
169	13,0,0; 12,5,0; 12,4,3	13,0; 12,5	13,0; 8,7
170	13,1,0; 12,5,1; 11,7,0; 9,8,5	13,1; 11,7	
171	13,1,1; 11,7,1; 11,5,5; 9,9,3		9,6
172	10,6,6		12,2
173	13,2,0; 12,5,2; 11,6,4; 10,8,3	13,2	
174	13,2,1; 11,7,2; 10,7,5		
175			10,5
176	12,4,4		
177	13,2,2; 8,8,7		
178	13,3,0; 12,5,3; 9,9,4	13,3	
179	13,3,1; 11,7,3; 9,7,7		
180	12,6,0; 10,8,4	12,6	
181	12,6,1; 10,9,0; 9,8,6	10,9	11,4
182	13,3,2; 11,6,5; 10,9,1		
183			13,1
184	12,6,2		
185	13,4,0; 12,5,4; 11,8,0; 10,9,2; 10,7,6	13,4; 11,8	
186	13,4,1; 11,8,1; 11,7,4		
187	13,3,3; 9,9,5		
188			
189	13,4,2; 12,6,3; 11,8,2; 10,8,5		12,3
190	10,9,3		
191			
192	8,8,8		8,8
193	12,7,0; 11,6,6	12,7	9,7
194	13,5,0; 13,4,3; 12,7,1; 12,5,5; 11,8,3; 9,8,7	13,5	
195	13,5,1; 11,7,5		
196	14,0,0; 12,6,4	14,0	14,0; 10,6
197	14,1,0; 12,7,2; 10,9,4	14,1	
198	14,1,1; 13,5,2; 10,7,7; 9,9,6		
199			13,2
200	14,2,0; 10,10,0; 10,8,6	14,2; 10,10	

Appendix 2
Conversion of d to Q

This appendix contains a table for converting d to $1/d^2 = Q$ for all values of d from 1.000 to 10.000. The values of Q are given to six significant numbers. This permits the determination of Q for d values greater than 10.000 and less than 1.000 to within at least four significant figures, an accuracy sufficient for all possible uses discussed in this book.

In order to use this table for d values lying outside the range of 1.000 to 10.000, shift the decimal point under d by the necessary number of places and, in the corresponding value under $1/d^2$, by twice that number of places, making the shifts in opposite directions. As an example of the use of this table, three representative d values and their respective $1/d^2$ values are tabulated below (to nearest .0001)

d	$1/d^2$
0.6642	2.2667
6.642	0.0227
66.42	0.0002

274

d	1/d²	d	1/d²	d	1/d²	d	1/d²
1.000	1.0000	1.050	.9070	1.100	.8264	1.150	.7561
1	.9980	1	.9053	1	.8249	1	.7548
2	.9960	2	.9036	2	.8235	2	.7535
3	.9940	3	.9019	3	.8220	3	.7522
4	.9920	4	.9002	4	.8205	4	.7509
5	.9901	5	.8985	5	.8190	5	.7596
6	.9881	6	.8968	6	.8175	6	.7483
7	.9861	7	.8951	7	.8160	7	.7470
8	.9842	8	.8934	8	.8145	8	.7457
9	.9822	9	.8917	9	.8131	9	.7444
1.010	.9803	1.060	.8900	1.110	.8116	1.160	.7432
1	.9784	1	.8883	1	.8102	1	.7419
2	.9762	2	.8866	2	.8087	2	.7406
3	.9745	3	.8850	3	.8073	3	.7393
4	.9726	4	.8833	4	.8058	4	.7381
5	.9707	5	.8817	5	.8044	5	.7368
6	.9688	6	.8800	6	.8029	6	.7355
7	.9668	7	.8784	7	.8015	7	.7343
8	.9650	8	.8767	8	.8000	8	.7330
9	.9631	9	.8751	9	.7986	9	.7318
1.020	.9612	1.070	.8734	1.120	.7972	1.170	.7305
1	.9593	1	.8718	1	.7958	1	.7293
2	.9574	2	.8702	2	.7944	2	.7280
3	.9555	3	.8686	3	.7930	3	.7268
4	.9537	4	.8669	4	.7915	4	.7255
5	.9518	5	.8653	5	.7901	5	.7243
6	.9499	6	.8637	6	.7887	6	.7231
7	.9481	7	.8621	7	.7873	7	.7219
8	.9463	8	.8605	8	.7859	8	.7206
9	.9444	9	.8589	9	.7845	9	.7194
1.030	.9426	1.080	.8573	1.130	.7831	1.180	.7182
1	.9408	1	.8557	1	.7818	1	.7169
2	.9390	2	.8542	2	.7804	2	.7158
3	.9371	3	.8526	3	.7790	3	.7145
4	.9353	4	.8510	4	.7776	4	.7133
5	.9335	5	.8495	5	.7763	5	.7121
6	.9317	6	.8479	6	.7749	6	.7109
7	.9299	7	.8463	7	.7735	7	.7097
8	.9281	8	.8448	8	.7722	8	.7085
9	.9263	9	.8432	9	.7708	9	.7074
1.040	.9246	1.090	.8417	1.140	.7695	1.190	.7062
1	.9228	1	.8401	1	.7681	1	.7050
2	.9210	2	.8386	2	.7668	2	.7038
3	.9192	3	.8370	3	.7654	3	.7026
4	.9175	4	.8356	4	.7641	4	.7014
5	.9157	5	.8340	5	.7628	5	.7003
6	.9140	6	.8325	6	.7614	6	.6991
7	.9122	7	.8310	7	.7601	7	.6979
8	.9105	8	.8295	8	.7588	8	.6968
9	.9088	9	.8280	9	.7575	9	.6956

d	$1/d^2$	d	$1/d^2$	d	$1/d^2$	d	$1/d^2$
1.200	.6944	1.250	.6400	1.300	.5917	1.350	.5487
1	.6933	1	.6390	1	.5908	1	.5479
2	.6921	2	.6380	2	.5899	2	.5471
3	.6910	3	.6369	3	.5890	3	.5463
4	.6898	4	.6359	4	.5881	4	.5455
5	.6887	5	.6349	5	.5872	5	.5447
6	.6876	6	.6339	6	.5863	6	.5439
7	.6864	7	.6329	7	.5854	7	.5431
8	.6853	8	.6319	8	.5845	8	.5423
9	.6841	9	.6309	9	.5836	9	.5415
1.210	.6830	1.260	.6299	1.310	.5827	1.360	.5407
1	.6819	1	.6289	1	.5818	1	.5399
2	.6808	2	.6279	2	.5809	2	.5391
3	.6796	3	.6269	3	.5800	3	.5383
4	.6785	4	.6259	4	.5792	4	.5375
5	.6774	5	.6249	5	.5783	5	.5367
6	.6763	6	.6239	6	.5774	6	.5359
7	.6752	7	.6229	7	.5765	7	.5351
8	.6741	8	.6220	8	.5757	8	.5344
9	.6730	9	.6210	9	.5748	9	.5336
1.220	.6719	1.270	.6200	1.320	.5739	1.370	.5328
1	.6708	1	.6190	1	.5731	1	.5320
2	.6697	2	.6181	2	.5722	2	.5312
3	.6686	3	.6171	3	.5713	3	.5305
4	.6675	4	.6161	4	.5705	4	.5297
5	.6664	5	.6151	5	.5696	5	.5289
6	.6653	6	.6142	6	.5687	6	.5282
7	.6642	7	.6132	7	.5679	7	.5274
8	.6631	8	.6123	8	.5670	8	.5266
9	.6621	9	.6113	9	.5662	9	.5259
1.230	.6610	1.280	.6104	1.330	.5653	1.380	.5251
1	.6599	1	.6094	1	.5645	1	.5243
2	.6588	2	.6084	2	.5636	2	.5236
3	.6578	3	.6075	3	.5628	3	.5228
4	.6567	4	.6066	4	.5619	4	.5221
5	.6556	5	.6056	5	.5611	5	.5213
6	.6546	6	.6047	6	.5603	6	.5206
7	.6535	7	.6037	7	.5594	7	.5198
8	.6525	8	.6028	8	.5586	8	.5191
9	.6514	9	.6019	9	.5577	9	.5183
1.240	.6504	1.290	.6009	1.340	.5569	1.390	.5176
1	.6493	1	.6000	1	.5561	1	.5168
2	.6483	2	.5991	2	.5553	2	.5161
3	.6472	3	.5981	3	.5544	3	.5153
4	.6462	4	.5972	4	.5536	4	.5146
5	.6452	5	.5963	5	.5528	5	.5139
6	.6441	6	.5954	6	.5520	6	.5131
7	.6431	7	.5945	7	.5511	7	.5124
8	.6421	8	.5935	8	.5503	8	.5117
9	.6410	9	.5926	9	.5495	9	.5109

d	$1/d^2$	d	$1/d^2$	d	$1/d^2$	d	$1/d^2$
1.400	.5102	1.450	.4756	1.500	.4444	1.550	.4162
1	.5095	1	.4750	1	.4439	1	.4157
2	.5087	2	.4743	2	.4433	2	.4152
3	.5080	3	.4737	3	.4427	3	.4146
4	.5073	4	.4730	4	.4421	4	.4141
5	.5066	5	.4724	5	.4415	5	.4136
6	.5095	6	.4717	6	.4409	6	.4130
7	.5051	7	.4711	7	.4403	7	.4125
8	.5044	8	.4704	8	.4397	8	.4120
9	.5037	9	.4698	9	.4392	9	.4114
1.410	.5030	1.460	.4691	1.510	.4386	1.560	.4109
1	.5023	1	.4685	1	.4380	1	.4104
2	.5016	2	.4678	2	.4374	2	.4099
3	.5009	3	.4672	3	.4368	3	.4093
4	.5002	4	.4666	4	.4363	4	.4088
5	.4994	5	.4659	5	.4357	5	.4083
6	.4987	6	.4653	6	.4351	6	.4078
7	.4980	7	.4647	7	.4345	7	.4073
8	.4973	8	.4640	8	.4340	8	.4067
9	.4966	9	.4634	9	.4334	9	.4062
1.420	.4959	1.470	.4628	1.520	.4328	1.570	.4057
1	.4952	1	.4621	1	.4323	1	.4052
2	.4945	2	.4615	2	.4317	2	.4047
3	.4938	3	.4609	3	.4311	3	.4041
4	.4932	4	.4603	4	.4306	4	.4036
5	.4925	5	.4596	5	.4300	5	.4031
6	.4918	6	.4590	6	.4294	6	.4026
7	.4911	7	.4584	7	.4289	7	.4021
8	.4904	8	.4578	8	.4283	8	.4016
9	.4897	9	.4572	9	.4277	9	.4011
1.430	.4890	1.480	.4565	1.530	.4272	1.580	.4006
1	.4883	1	.4559	1	.4266	1	.4001
2	.4877	2	.4553	2	.4261	2	.3996
3	.4870	3	.4547	3	.4255	3	.3991
4	.4863	4	.4541	4	.4250	4	.3986
5	.4856	5	.4535	5	.4244	5	.3981
6	.4849	6	.4529	6	.4239	6	.3976
7	.4843	7	.4522	7	.4233	7	.3971
8	.4836	8	.4516	8	.4228	8	.3966
9	.4829	9	.4510	9	.4222	9	.3961
1.440	.4823	1.490	.4504	1.540	.4217	1.590	.3956
1	.4816	1	.4498	1	.4211	1	.3951
2	.4809	2	.4492	2	.4206	2	.3946
3	.4802	3	.4486	3	.4200	3	.3941
4	.4796	4	.4480	4	.4195	4	.3936
5	.4789	5	.4474	5	.4189	5	.3931
6	.4783	6	.4468	6	.4184	6	.3926
7	.4776	7	.4462	7	.4178	7	.3921
8	.4769	8	.4456	8	.4173	8	.3916
9	.4763	9	.4450	9	.4168	9	.3911

d	$1/d^2$	d	$1/d^2$	d	$1/d^2$	d	$1/d^2$
1.600	.3906	1.650	.3673	1.700	.3460	1.750	.3265
1	.3901	1	.3669	1	.3456	1	.3262
2	.3897	2	.3664	2	.3452	2	.3258
3	.3892	3	.3660	3	.3448	3	.3254
4	.3887	4	.3655	4	.3444	4	.3250
5	.3882	5	.3651	5	.3440	5	.3247
6	.3877	6	.3647	6	.3436	6	.3243
7	.3872	7	.3642	7	.3432	7	.3239
8	.3867	8	.3638	8	.3428	8	.3236
9	.3863	9	.3633	9	.3424	9	.3232
1.610	.3858	1.660	.3629	1.710	.3420	1.760	.3228
1	.3853	1	.3625	1	.3416	1	.3225
2	.3848	2	.3620	2	.3412	2	.3221
3	.3844	3	.3616	3	.3408	3	.3217
4	.3839	4	.3612	4	.3404	4	.3214
5	.3834	5	.3607	5	.3400	5	.3210
6	.3829	6	.3603	6	.3396	6	.3206
7	.3825	7	.3599	7	.3392	7	.3203
8	.3820	8	.3594	8	.3388	8	.3199
9	.3815	9	.3590	9	.3384	9	.3196
1.620	.3810	1.670	.3586	1.720	.3380	1.770	.3192
1	.3806	1	.3581	1	.3376	1	.3188
2	.3801	2	.3577	2	.3372	2	.3185
3	.3796	3	.3573	3	.3368	3	.3181
4	.3792	4	.3569	4	.3365	4	.3178
5	.3787	5	.3564	5	.3361	5	.3174
6	.3782	6	.3560	6	.3357	6	.3170
7	.3778	7	.3556	7	.3353	7	.3167
8	.3773	8	.3552	8	.3349	8	.3163
9	.3768	9	.3547	9	.3345	9	.3160
1.630	.3764	1.680	.3543	1.730	.3341	1.780	.3156
1	.3759	1	.3539	1	.3337	1	.3153
2	.3755	2	.3535	2	.3334	2	.3149
3	.3750	3	.3530	3	.3330	3	.3146
4	.3745	4	.3526	4	.3326	4	.3142
5	.3741	5	.3522	5	.3322	5	.3139
6	.3736	6	.3518	6	.3318	6	.3135
7	.3732	7	.3514	7	.3314	7	.3131
8	.3727	8	.3510	8	.3311	8	.3128
9	.3723	9	.3505	9	.3307	9	.3124
1.640	.3718	1.690	.3501	1.740	.3303	1.790	.3121
1	.3713	1	.3497	1	.3299	1	.3118
2	.3709	2	.3493	2	.3295	2	.3114
3	.3704	3	.3489	3	.3292	3	.3111
4	.3700	4	.3485	4	.3288	4	.3107
5	.3695	5	.3481	5	.3284	5	.3104
6	.3691	6	.3477	6	.3280	6	.3100
7	.3686	7	.3472	7	.3277	7	.3097
8	.3682	8	.3468	8	.3273	8	.3093
9	.3678	9	.3464	9	.3269	9	.3090

d	$1/d^2$	d	$1/d^2$	d	$1/d^2$	d	$1/d^2$
1.800	.3086	1.850	.2922	1.900	.2770	1.950	.2630
1	.3083	1	.2919	1	.2767	1	.2627
2	.3080	2	.2916	2	.2764	2	.2624
3	.3076	3	.2912	3	.2761	3	.2622
4	.3073	4	.2909	4	.2758	4	.2619
5	.3069	5	.2906	5	.2756	5	.2616
6	.3066	6	.2903	6	.2753	6	.2614
7	.3063	7	.2900	7	.2750	7	.2611
8	.3059	8	.2897	8	.2747	8	.2608
9	.3056	9	.2894	9	.2744	9	.2606
1.810	.3052	1.860	.2891	1.910	.2741	1.960	.2603
1	.3049	1	.2887	1	.2738	1	.2600
2	.3046	2	.2884	2	.2735	2	.2598
3	.3042	3	.2881	3	.2733	3	.2595
4	.3039	4	.2878	4	.2730	4	.2592
5	.3036	5	.2875	5	.2727	5	.2590
6	.3032	6	.2872	6	.2724	6	.2587
7	.3029	7	.2869	7	.2721	7	.2585
8	.3026	8	.2866	8	.2718	8	.2582
9	.3022	9	.2863	9	.2716	9	.2579
1.820	.3019	1.870	.2860	1.920	.2713	1.970	.2577
1	.3016	1	.2857	1	.2710	1	.2574
2	.3012	2	.2854	2	.2707	2	.2571
3	.3009	3	.2851	3	.2704	3	.2569
4	.3006	4	.2847	4	.2701	4	.2566
5	.3002	5	.2844	5	.2699	5	.2564
6	.2999	6	.2841	6	.2696	6	.2561
7	.2996	7	.2838	7	.2693	7	.2559
8	.2993	8	.2835	8	.2690	8	.2556
9	.2989	9	.2832	9	.2687	9	.2553
1.830	.2986	1.880	.2829	1.930	.2685	1.980	.2551
1	.2983	1	.2826	1	.2682	1	.2548
2	.2980	2	.2823	2	.2679	2	.2546
3	.2976	3	.2820	3	.2676	3	.2543
4	.2973	4	.2817	4	.2674	4	.2540
5	.2970	5	.2814	5	.2671	5	.2538
6	.2967	6	.2811	6	.2668	6	.2535
7	.2963	7	.2808	7	.2665	7	.2533
8	.2960	8	.2805	8	.2663	8	.2530
0	2957	9	.2802	9	.2660	9	.2528
1.840	.2954	1.890	.2799	1.940	.2657	1.990	.2525
1	.2950	1	.2797	1	.2654	1	.2523
2	.2947	2	.2794	2	.2652	2	.2520
3	.2944	3	.2791	3	.2649	3	.2518
4	.2941	4	.2788	4	.2646	4	.2515
5	.2938	5	.2785	5	.2643	5	.2513
6	.2935	6	.2782	6	.2641	6	.2510
7	.2931	7	.2779	7	.2638	7	.2508
8	.2928	8	.2776	8	.2635	8	.2505
9	.2925	9	.2773	9	.2633	9	.2503

d	$1/d^2$	d	$1/d^2$	d	$1/d^2$	d	$1/d^2$
2.000	.2500	2.050	.2380	2.100	.2268	2.150	.2163
1	.2498	1	.2377	1	.2265	1	.2161
2	.2495	2	.2375	2	.2263	2	.2159
3	.2493	3	.2373	3	.2261	3	.2157
4	.2490	4	.2370	4	.2259	4	.2155
5	.2488	5	.2368	5	.2257	5	.2153
6	.2485	6	.2366	6	.2255	6	.2151
7	.2483	7	.2363	7	.2253	7	.2149
8	.2480	8	.2361	8	.2250	8	.2147
9	.2478	9	.2359	9	.2248	9	.2145
2.010	.2475	2.060	.2356	2.110	.2246	2.160	.2143
1	.2473	1	.2354	1	.2244	1	.2141
2	.2470	2	.2352	2	.2242	2	.2139
3	.2468	3	.2350	3	.2240	3	.2137
4	.2465	4	.2347	4	.2238	4	.2135
5	.2463	5	.2345	5	.2236	5	.2133
6	.2460	6	.2343	6	.2233	6	.2131
7	.2458	7	.2341	7	.2231	7	.2130
8	.2456	8	.2338	8	.2229	8	.2128
9	.2453	9	.2336	9	.2227	9	.2126
2.020	.2451	2.070	.2334	2.120	.2225	2.170	.2124
1	.2448	1	.2332	1	.2223	1	.2122
2	.2446	2	.2329	2	.2221	2	.2120
3	.2443	3	.2327	3	.2219	3	.2118
4	.2441	4	.2325	4	.2217	4	.2116
5	.2439	5	.2323	5	.2215	5	.2114
6	.2436	6	.2320	6	.2212	6	.2112
7	.2434	7	.2318	7	.2210	7	.2110
8	.2431	8	.2316	8	.2208	8	.2108
9	.2429	9	.2314	9	.2206	9	.2106
2.030	.2427	2.080	.2311	2.130	.2204	2.180	.2104
1	.2424	1	.2309	1	.2202	1	.2102
2	.2422	2	.2307	2	.2200	2	.2100
3	.2419	3	.2305	3	.2198	3	.2098
4	.2417	4	.2303	4	.2196	4	.2096
5	.2415	5	.2300	5	.2194	5	.2095
6	.2412	6	.2298	6	.2192	6	.2093
7	.2410	7	.2296	7	.2190	7	.2091
8	.2408	8	.2294	8	.2188	8	.2089
9	.2405	9	.2292	9	.2186	9	.2087
2.040	.2403	2.090	.2289	2.140	.2184	2.190	.2085
1	.2401	1	.2287	1	.2182	1	.2083
2	.2398	2	.2285	2	.2180	2	.2081
3	.2396	3	.2283	3	.2177	3	.2079
4	.2394	4	.2281	4	.2175	4	.2077
5	.2391	5	.2278	5	.2173	5	.2076
6	.2389	6	.2276	6	.2171	6	.2074
7	.2387	7	.2274	7	.2169	7	.2072
8	.2384	8	.2272	8	.2167	8	.2070
9	.2382	9	.2270	9	.2165	9	.2068

d	$1/d^2$	d	$1/d^2$	d	$1/d^2$	d	$1/d^2$
2.200	.2066	2.250	.1975	2.300	.1890	2.350	.1811
1	.2064	1	.1974	1	.1889	1	.1809
2	.2062	2	.1972	2	.1887	2	.1808
3	.2060	3	.1970	3	.1885	3	.1806
4	.2059	4	.1968	4	.1884	4	.1805
5	.2057	5	.1967	5	.1882	5	.1803
6	.2055	6	.1965	6	.1881	6	.1802
7	.2053	7	.1963	7	.1879	7	.1800
8	.2051	8	.1961	8	.1877	8	.1799
9	.2049	9	.1960	9	.1876	9	.1797
2.210	.2047	2.260	.1958	2.310	.1874	2.360	.1795
1	.2046	1	.1956	1	.1872	1	.1794
2	.2044	2	.1954	2	.1871	2	.1792
3	.2042	3	.1953	3	.1869	3	.1791
4	.2040	4	.1951	4	.1868	4	.1789
5	.2038	5	.1949	5	.1866	5	.1788
6	.2036	6	.1948	6	.1864	6	.1786
7	.2035	7	.1946	7	.1863	7	.1785
8	.2033	8	.1944	8	.1861	8	.1783
9	.2031	9	.1942	9	.1860	9	.1782
2.220	.2029	2.270	.1941	2.320	.1858	2.370	.1780
1	.2027	1	.1939	1	.1856	1	.1779
2	.2025	2	.1937	2	.1855	2	.1777
3	.2024	3	.1936	3	.1853	3	.1776
4	.2022	4	.1934	4	.1852	4	.1774
5	.2020	5	.1932	5	.1850	5	.1773
6	.2018	6	.1930	6	.1848	6	.1771
7	.2016	7	.1929	7	.1847	7	.1770
8	.2015	8	.1927	8	.1845	8	.1768
9	.2013	9	.1925	9	.1844	9	.1767
2.230	.2011	2.280	.1924	2.330	.1842	2.380	.1765
1	.2009	1	.1922	1	.1840	1	.1764
2	.2007	2	.1920	2	.1839	2	.1762
3	.2005	3	.1919	3	.1837	3	.1761
4	.2004	4	.1917	4	.1836	4	.1759
5	.2002	5	.1915	5	.1834	5	.1758
6	.2000	6	.1914	6	.1833	6	.1757
7	.1998	7	.1912	7	.1831	7	.1755
8	.1997	8	.1910	8	.1829	8	.1754
9	.1995	9	.1909	9	.1828	9	.1752
2.240	.1993	2.290	.1907	2.340	.1826	2.390	.1751
1	.1991	1	.1905	1	.1825	1	.1749
2	.1989	2	.1904	2	.1823	2	.1748
3	.1988	3	.1902	3	.1822	3	.1746
4	.1986	4	.1900	4	.1820	4	.1745
5	.1984	5	.1899	5	.1819	5	.1743
6	.1982	6	.1897	6	.1817	6	.1742
7	.1981	7	.1895	7	.1815	7	.1740
8	.1979	8	.1894	8	.1814	8	.1739
9	.1977	9	.1892	9	.1812	9	.1738

d	$1/d^2$	d	$1/d^2$	d	$1/d^2$	d	$1/d^2$
2.400	.1736	2.450	.1666	2.500	.1600	2.550	.1538
1	.1735	1	.1665	1	.1599	1	.1537
2	.1733	2	.1663	2	.1597	2	.1535
3	.1732	3	.1662	3	.1596	3	.1534
4	.1730	4	.1661	4	.1595	4	.1533
5	.1729	5	.1659	5	.1594	5	.1532
6	.1727	6	.1658	6	.1592	6	.1531
7	.1726	7	.1656	7	.1591	7	.1529
8	.1725	8	.1655	8	.1590	8	.1528
9	.1723	9	.1654	9	.1589	9	.1527
2.410	.1722	2.460	.1652	2.510	.1587	2.560	.1526
1	.1720	1	.1651	1	.1586	1	.1525
2	.1719	2	.1650	2	.1585	2	.1523
3	.1717	3	.1648	3	.1583	3	.1522
4	.1716	4	.1647	4	.1582	4	.1521
5	.1715	5	.1646	5	.1581	5	.1520
6	.1713	6	.1644	6	.1580	6	.1519
7	.1712	7	.1643	7	.1578	7	.1518
8	.1710	8	.1642	8	.1577	8	.1516
9	.1709	9	.1640	9	.1576	9	.1515
2.420	.1708	2.470	.1639	2.520	.1575	2.570	.1514
1	.1706	1	.1638	1	.1573	1	.1513
2	.1705	2	.1636	2	.1572	2	.1512
3	.1703	3	.1635	3	.1571	3	.1510
4	.1702	4	.1634	4	.1570	4	.1509
5	.1700	5	.1632	5	.1568	5	.1508
6	.1699	6	.1631	6	.1567	6	.1507
7	.1698	7	.1630	7	.1566	7	.1506
8	.1696	8	.1629	8	.1565	8	.1505
9	.1695	9	.1627	9	.1564	9	.1503
2.430	.1694	2.480	.1626	2.530	.1562	2.580	.1502
1	.1692	1	.1625	1	.1561	1	.1501
2	.1691	2	.1623	2	.1560	2	.1500
3	.1689	3	.1622	3	.1559	3	.1499
4	.1688	4	.1621	4	.1557	4	.1498
5	.1687	5	.1619	5	.1556	5	.1497
6	.1685	6	.1618	6	.1555	6	.1495
7	.1684	7	.1617	7	.1554	7	.1494
8	.1682	8	.1615	8	.1552	8	.1493
9	.1681	9	.1614	9	.1551	9	.1492
2.440	.1680	2.490	.1613	2.540	.1550	2.590	.1491
1	.1678	1	.1612	1	.1549	1	.1490
2	.1677	2	.1610	2	.1548	2	.1488
3	.1676	3	.1609	3	.1546	3	.1487
4	.1674	4	.1608	4	.1545	4	.1486
5	.1673	5	.1606	5	.1544	5	.1485
6	.1671	6	.1605	6	.1543	6	.1484
7	.1670	7	.1604	7	.1541	7	.1483
8	.1669	8	.1603	8	.1540	8	.1482
9	.1667	9	.1601	9	.1539	9	.1480

d	$1/d^2$	d	$1/d^2$	d	$1/d^2$	d	$1/d^2$
2.600	.1479	2.650	.1424	2.700	.1372	2.750	.1322
1	.1478	1	.1423	1	.1371	1	.1321
2	.1477	2	.1422	2	.1370	2	.1320
3	.1476	3	.1421	3	.1369	3	.1319
4	.1475	4	.1420	4	.1368	4	.1318
5	.1474	5	.1419	5	.1367	5	.1318
6	.1472	6	.1418	6	.1366	6	.1317
7	.1471	7	.1417	7	.1365	7	.1316
8	.1470	8	.1415	8	.1364	8	.1315
9	.1469	9	.1414	9	.1363	9	.1314
2.610	.1468	2.660	.1413	2.710	.1362	2.760	.1313
1	.1467	1	.1412	1	.1361	1	.1312
2	.1466	2	.1411	2	.1360	2	.1311
3	.1465	3	.1410	3	.1359	3	.1310
4	.1463	4	.1409	4	.1358	4	.1309
5	.1462	5	.1408	5	.1357	5	.1308
6	.1461	6	.1407	6	.1356	6	.1307
7	.1460	7	.1406	7	.1355	7	.1306
8	.1459	8	.1405	8	.1354	8	.1305
9	.1458	9	.1404	9	.1353	9	.1304
2.620	.1457	2.670	.1403	2.720	.1352	2.770	.1303
1	.1456	1	.1402	1	.1351	1	.1302
2	.1455	2	.1401	2	.1350	2	.1301
3	.1453	3	.1400	3	.1349	3	.1300
4	.1452	4	.1399	4	.1348	4	.1300
5	.1451	5	.1398	5	.1347	5	.1299
6	.1450	6	.1396	6	.1346	6	.1298
7	.1449	7	.1395	7	.1345	7	.1297
8	.1448	8	.1394	8	.1344	8	.1296
9	.1447	9	.1393	9	.1343	9	.1295
2.630	.1446	2.680	.1392	2.730	.1342	2.780	.1294
1	.1445	1	.1391	1	.1341	1	.1293
2	.1444	2	.1390	2	.1340	2	.1292
3	.1442	3	.1389	3	.1339	3	.1291
4	.1441	4	.1388	4	.1338	4	.1290
5	.1440	5	.1387	5	.1337	5	.1289
6	.1439	6	.1386	6	.1336	6	.1288
7	.1438	7	.1385	7	.1335	7	.1287
8	.1437	8	.1384	8	.1334	8	.1287
9	.1436	9	.1383	9	.1333	9	.1286
2.640	.1435	2.690	.1382	2.740	.1332	2.790	.1285
1	.1434	1	.1381	1	.1331	1	.1284
2	.1433	2	.1380	2	.1330	2	.1283
3	.1432	3	.1379	3	.1329	3	.1282
4	.1430	4	.1378	4	.1328	4	.1281
5	.1429	5	.1377	5	.1327	5	.1280
6	.1428	6	.1376	6	.1326	6	.1279
7	.1427	7	.1375	7	.1325	7	.1278
8	.1426	8	.1374	8	.1324	8	.1277
9	.1425	9	.1373	9	.1323	9	.1276

d	$1/d^2$	d	$1/d^2$	d	$1/d^2$	d	$1/d^2$
2.800	.1276	2.850	.1231	2.900	.1189	2.950	.1149
1	.1275	1	.1230	1	.1188	1	.1148
2	.1274	2	.1229	2	.1187	2	.1148
3	.1273	3	.1229	3	.1187	3	.1147
4	.1272	4	.1228	4	.1186	4	.1146
5	.1271	5	.1227	5	.1185	5	.1145
6	.1270	6	.1226	6	.1184	6	.1144
7	.1269	7	.1225	7	.1183	7	.1144
8	.1268	8	.1224	8	.1183	8	.1143
9	.1267	9	.1223	9	.1182	9	.1142
2.810	.1266	2.860	.1223	2.910	.1181	2.960	.1141
1	.1266	1	.1222	1	.1180	1	.1141
2	.1265	2	.1221	2	.1179	2	.1140
3	.1264	3	.1220	3	.1178	3	.1139
4	.1263	4	.1219	4	.1178	4	.1138
5	.1262	5	.1218	5	.1177	5	.1137
6	.1261	6	.1217	6	.1176	6	.1137
7	.1260	7	.1217	7	.1175	7	.1136
8	.1259	8	.1216	8	.1174	8	.1135
9	.1258	9	.1215	9	.1174	9	.1134
2.820	.1257	2.870	.1214	2.920	.1173	2.970	.1134
1	.1257	1	.1213	1	.1172	1	.1133
2	.1256	2	.1212	2	.1171	2	.1132
3	.1255	3	.1212	3	.1170	3	.1131
4	.1254	4	.1211	4	.1170	4	.1131
5	.1253	5	.1210	5	.1169	5	.1130
6	.1252	6	.1209	6	.1168	6	.1129
7	.1251	7	.1208	7	.1167	7	.1128
8	.1250	8	.1207	8	.1166	8	.1128
9	.1249	9	.1206	9	.1166	9	.1127
2.830	.1249	2.880	.1206	2.930	.1165	2.980	.1126
1	.1248	1	.1205	1	.1164	1	.1125
2	.1247	2	.1204	2	.1163	2	.1125
3	.1246	3	.1203	3	.1162	3	.1124
4	.1245	4	.1202	4	.1162	4	.1123
5	.1244	5	.1201	5	.1161	5	.1122
6	.1243	6	.1201	6	.1160	6	.1122
7	.1242	7	.1200	7	.1159	7	.1121
8	.1242	8	.1199	8	.1159	8	.1120
9	.1241	9	.1198	9	.1158	9	.1119
2.840	.1240	2.890	.1197	2.940	.1157	2.990	.1119
1	.1239	1	.1196	1	.1156	1	.1118
2	.1238	2	.1196	2	.1155	2	.1117
3	.1237	3	.1195	3	.1155	3	.1116
4	.1236	4	.1194	4	.1154	4	.1116
5	.1235	5	.1193	5	.1153	5	.1115
6	.1235	6	.1192	6	.1152	6	.1114
7	.1234	7	.1192	7	.1151	7	.1113
8	.1233	8	.1191	8	.1151	8	.1113
9	.1232	9	.1190	9	.1150	9	.1112

d	$1/d^2$	d	$1/d^2$	d	$1/d^2$	d	$1/d^2$
3.000	.1111	3.050	.1075	3.100	.1041	3.150	.1008
1	.1110	1	.1074	1	.1040	1	.1007
2	.1110	2	.1074	2	.1039	2	.1007
3	.1109	3	.1073	3	.1039	3	.1006
4	.1108	4	.1072	4	.1038	4	.1005
5	.1107	5	.1071	5	.1037	5	.1005
6	.1107	6	.1071	6	.1037	6	.1004
7	.1106	7	.1070	7	.1036	7	.1003
8	.1105	8	.1069	8	.1035	8	.1003
9	.1104	9	.1069	9	.1035	9	.1002
3.010	.1104	3.060	.1068	3.110	.1034	3.160	.1001
1	.1103	1	.1067	1	.1033	1	.1001
2	.1102	2	.1067	2	.1033	2	.1000
3	.1102	3	.1066	3	.1032	3	.1000
4	.1101	4	.1065	4	.1031	4	.0999
5	.1100	5	.1064	5	.1031	5	.0998
6	.1099	6	.1064	6	.1030	6	.0998
7	.1099	7	.1063	7	.1029	7	.0997
8	.1098	8	.1062	8	.1029	8	.0996
9	.1097	9	.1062	9	.1028	9	.0996
3.020	.1096	3.070	.1061	3.120	.1027	3.170	.0995
1	.1096	1	.1060	1	.1027	1	.0995
2	.1095	2	.1060	2	.1026	2	.0994
3	.1094	3	.1059	3	.1025	3	.0993
4	.1094	4	.1058	4	.1025	4	.0993
5	.1093	5	.1058	5	.1024	5	.0992
6	.1092	6	.1057	6	.1023	6	.0991
7	.1091	7	.1056	7	.1023	7	.0991
8	.1091	8	.1056	8	.1022	8	.0990
9	.1090	9	.1055	9	.1021	9	.0990
3.030	.1089	3.080	1054	3.130	.1021	3.180	.0989
1	.1088	1	.1053	1	.1020	1	.0988
2	.1088	2	.1053	2	.1019	2	.0988
3	.1087	3	.1052	3	.1019	3	.0987
4	.1086	4	.1051	4	.1018	4	.0986
5	.1086	5	.1051	5	.1017	5	.0986
6	.1085	6	.1050	6	.1017	6	.0985
7	.1084	7	.1049	7	.1016	7	.0985
8	.1083	8	.1049	8	.1016	8	.0984
9	.1083	9	1048	9	.1015	9	.0983
3.040	.1082	3.090	.1047	3.140	.1014	3.190	.0983
1	.1081	1	.1047	1	.1014	1	.0982
2	.1081	2	.1046	2	.1013	2	.0981
3	.1080	3	.1045	3	.1012	3	.0981
4	.1079	4	.1045	4	.1012	4	.0980
5	.1079	5	.1044	5	.1011	5	.0980
6	.1078	6	.1043	6	.1010	6	.0979
7	.1077	7	.1043	7	.1010	7	.0978
8	.1076	8	.1042	8	.1009	8	.0978
9	.1076	9	.1041	9	.1008	9	.0977

d	$1/d^2$	d	$1/d^2$	d	$1/d^2$	d	$1/d^2$
3.200	.0977	3.250	.0947	3.300	.0918	3.350	.0891
1	.0976	1	.0946	1	.0918	1	.0891
2	.0975	2	.0946	2	.0917	2	.0890
3	.0975	3	.0945	3	.0917	3	.0889
4	.0974	4	.0944	4	.0916	4	.0889
5	.0974	5	.0944	5	.0915	5	.0888
6	.0973	6	.0943	6	.0915	6	.0888
7	.0972	7	.0943	7	.0914	7	.0887
8	.0972	8	.0942	8	.0914	8	.0887
9	.0971	9	.0942	9	.0913	9	.0886
3.210	.0970	3.260	.0941	3.310	.0913	3.360	.0886
1	.0970	1	.0940	1	.0912	1	.0885
2	.0969	2	.0940	2	.0912	2	.0885
3	.0969	3	.0939	3	.0911	3	.0884
4	.0968	4	.0939	4	.0911	4	.0884
5	.0967	5	.0938	5	.0910	5	.0883
6	.0967	6	.0937	6	.0909	6	.0883
7	.0966	7	.0937	7	.0909	7	.0882
8	.0966	8	.0936	8	.0908	8	.0882
9	.0965	9	.0936	9	.0908	9	.0881
3.220	.0964	3.270	.0935	3.320	.0907	3.370	.0881
1	.0964	1	.0935	1	.0907	1	.0880
2	.0963	2	.0934	2	.0906	2	.0879
3	.0963	3	.0933	3	.0906	3	.0879
4	.0962	4	.0933	4	.0905	4	.0878
5	.0961	5	.0932	5	.0905	5	.0878
6	.0961	6	.0932	6	.0904	6	.0877
7	.0960	7	.0931	7	.0903	7	.0877
8	.0960	8	.0931	8	.0903	8	.0876
9	.0959	9	.0930	9	.0902	9	.0876
3.230	.0959	3.280	.0930	3.330	.0902	3.380	.0875
1	.0958	1	.0929	1	.0901	1	.0875
2	.0957	2	.0928	2	.0901	2	.0874
3	.0957	3	.0928	3	.0900	3	.0874
4	.0956	4	.0927	4	.0900	4	.0873
5	.0956	5	.0927	5	.0899	5	.0873
6	.0955	6	.0926	6	.0899	6	.0872
7	.0954	7	.0926	7	.0898	7	.0872
8	.0954	8	.0925	8	.0897	8	.0871
9	.0953	9	.0924	9	.0897	9	.0871
3.240	.0953	3.290	.0924	3.340	.0896	3.390	.0870
1	.0952	1	.0923	1	.0896	1	.0870
2	.0951	2	.0923	2	.0895	2	.0869
3	.0951	3	.0922	3	.0895	3	.0869
4	.0950	4	.0922	4	.0894	4	.0868
5	.0950	5	.0921	5	.0894	5	.0868
6	.0949	6	.0921	6	.0893	6	.0867
7	.0948	7	.0920	7	.0893	7	.0867
8	.0948	8	.0919	8	.0892	8	.0866
9	.0947	9	.0919	9	.0892	9	.0866

d	$1/d^2$	d	$1/d^2$	d	$1/d^2$	d	$1/d^2$
3.400	.0865	3.450	.0840	3.500	.0816	3.550	.0793
1	.0865	1	.0840	1	.0816	1	.0793
2	.0864	2	.0839	2	.0815	2	.0793
3	.0864	3	.0839	3	.0815	3	.0792
4	.0863	4	.0838	4	.0814	4	.0792
5	.0863	5	.0838	5	.0814	5	.0791
6	.0862	6	.0837	6	.0814	6	.0791
7	.0862	7	.0837	7	.0813	7	.0790
8	.0861	8	.0836	8	.0813	8	.0790
9	.0860	9	.0836	9	.0812	9	.0789
3.410	.0860	3.460	.0835	3.510	.0812	3.560	.0789
1	.0859	1	.0835	1	.0811	1	.0789
2	.0859	2	.0834	2	.0811	2	.0788
3	.0858	3	.0834	3	.0810	3	.0788
4	.0858	4	.0833	4	.0810	4	.0787
5	.0857	5	.0833	5	.0809	5	.0787
6	.0857	6	.0832	6	.0809	6	.0786
7	.0856	7	.0832	7	.0808	7	.0786
8	.0856	8	.0831	8	.0808	8	.0786
9	.0855	9	.0831	9	.0808	9	.0785
3.420	.0855	3.470	.0831	3.520	.0807	3.570	.0785
1	.0854	1	.0830	1	.0807	1	.0784
2	.0854	2	.0830	2	.0806	2	.0784
3	.0853	3	.0829	3	.0806	3	.0783
4	.0853	4	.0829	4	.0805	4	.0783
5	.0852	5	.0828	5	.0805	5	.0782
6	.0852	6	.0828	6	.0804	6	.0782
7	.0851	7	.0827	7	.0804	7	.0782
8	.0851	8	.0827	8	.0803	8	.0781
9	.0850	9	.0826	9	.0803	9	.0781
3.430	.0850	3.480	.0826	3.530	.0803	3.580	.0780
1	.0849	1	.0825	1	.0802	1	.0780
2	.0849	2	.0825	2	.0802	2	.0779
3	.0849	3	.0824	3	.0801	3	.0779
4	.0848	4	.0824	4	.0801	4	.0779
5	.0848	5	.0823	5	.0800	5	.0778
6	.0847	6	.0823	6	.0800	6	.0778
7	.0847	7	.0822	7	.0799	7	.0777
8	.0846	8	.0822	8	.0799	8	.0777
9	.0846	9	.0821	0	.0798	9	.0776
3.440	.0845	3.490	.0821	3.540	.0798	3.590	.0776
1	.0845	1	.0821	1	.0798	1	.0775
2	.0844	2	.0820	2	.0797	2	.0775
3	.0844	3	.0820	3	.0797	3	.0775
4	.0843	4	.0819	4	.0796	4	.0774
5	.0843	5	.0819	5	.0796	5	.0774
6	.0842	6	.0818	6	.0795	6	.0773
7	.0842	7	.0818	7	.0795	7	.0773
8	.0841	8	.0817	8	.0794	8	.0772
9	.0841	9	.0817	9	.0794	9	.0772

d	$1/d^2$	d	$1/d^2$	d	$1/d^2$	d	$1/d^2$
3.600	.0772	3.650	.0751	3.700	.0730	3.750	.0711
1	.0771	1	.0750	1	.0730	1	.0711
2	.0771	2	.0750	2	.0730	2	.0710
3	.0770	3	.0749	3	.0729	3	.0710
4	.0770	4	.0749	4	.0729	4	.0710
5	.0769	5	.0749	5	.0728	5	.0709
6	.0769	6	.0748	6	.0728	6	.0709
7	.0769	7	.0748	7	.0728	7	.0708
8	.0768	8	.0747	8	.0727	8	.0708
9	.0768	9	.0747	9	.0727	9	.0708
3.610	.0767	3.660	.0747	3.710	.0727	3.760	.0707
1	.0767	1	.0746	1	.0726	1	.0707
2	.0766	2	.0746	2	.0726	2	.0707
3	.0766	3	.0745	3	.0725	3	.0706
4	.0766	4	.0745	4	.0725	4	.0706
5	.0765	5	.0744	5	.0725	5	.0705
6	.0765	6	.0744	6	.0724	6	.0705
7	.0764	7	.0744	7	.0724	7	.0705
8	.0764	8	.0743	8	.0723	8	.0704
9	.0764	9	.0743	9	.0723	9	.0704
3.620	.0763	3.670	.0742	3.720	.0723	3.770	.0704
1	.0763	1	.0742	1	.0722	1	.0703
2	.0762	2	.0742	2	.0722	2	.0703
3	.0762	3	.0741	3	.0721	3	.0702
4	.0761	4	.0741	4	.0721	4	.0702
5	.0761	5	.0740	5	.0721	5	.0702
6	.0761	6	.0740	6	.0720	6	.0701
7	.0760	7	.0740	7	.0720	7	.0701
8	.0760	8	.0739	8	.0720	8	.0701
9	.0759	9	.0739	9	.0719	9	.0700
3.630	.0759	3.680	.0738	3.730	.0719	3.780	.0700
1	.0758	1	.0738	1	.0718	1	.0699
2	.0758	2	.0738	2	.0718	2	.0699
3	.0758	3	.0737	3	.0718	3	.0699
4	.0757	4	.0737	4	.0717	4	.0698
5	.0757	5	.0736	5	.0717	5	.0698
6	.0756	6	.0736	6	.0716	6	.0698
7	.0756	7	.0736	7	.0716	7	.0697
8	.0756	8	.0735	8	.0716	8	.0697
9	.0755	9	.0735	9	.0715	9	.0697
3.640	.0755	3.690	.0734	3.740	.0715	3.790	.0696
1	.0754	1	.0734	1	.0715	1	.0696
2	.0754	2	.0734	2	.0714	2	.0695
3	.0753	3	.0733	3	.0714	3	.0695
4	.0753	4	.0733	4	.0713	4	.0695
5	.0753	5	.0732	5	.0713	5	.0694
6	.0752	6	.0732	6	.0713	6	.0694
7	.0752	7	.0732	7	.0712	7	.0694
8	.0751	8	.0731	8	.0712	8	.0693
9	.0751	9	.0731	9	.0711	9	.0693

d	$1/d^2$	d	$1/d^2$	d	$1/d^2$	d	$1/d^2$
3.800	.0693	3.850	.0675	3.900	.0657	3.950	.0641
1	.0692	1	.0674	1	.0657	1	.0641
2	.0692	2	.0674	2	.0657	2	.0640
3	.0691	3	.0674	3	.0656	3	.0640
4	.0691	4	.0673	4	.0656	4	.0640
5	.0691	5	.0673	5	.0656	5	.0639
6	.0690	6	.0673	6	.0655	6	.0639
7	.0690	7	.0672	7	.0655	7	.0639
8	.0690	8	.0672	8	.0655	8	.0638
9	.0689	9	.0672	9	.0654	9	.0638
3.810	.0689	3.860	.0671	3.910	.0654	3.960	.0638
1	.0689	1	.0671	1	.0654	1	.0637
2	.0688	2	.0670	2	.0653	2	.0637
3	.0688	3	.0670	3	.0053	3	.0637
4	.0687	4	.0670	4	.0653	4	.0636
5	.0687	5	.0669	5	.0652	5	.0636
6	.0687	6	.0669	6	.0652	6	.0636
7	.0686	7	.0669	7	.0652	7	.0635
8	.0686	8	.0668	8	.0651	8	.0635
9	.0686	9	.0668	9	.0651	9	.0635
3.820	.0685	3.870	.0668	3.920	.0651	3.970	.0634
1	.0685	1	.0667	1	.0650	1	.0634
2	.0685	2	.0667	2	.0650	2	.0634
3	.0684	3	.0667	3	.0650	3	.0634
4	.0684	4	.0666	4	.0649	4	.0633
5	.0683	5	.0666	5	.0649	5	.0633
6	.0683	6	.0666	6	.0649	6	.0633
7	.0683	7	.0665	7	.0648	7	.0632
8	.0682	8	.0665	8	.0648	8	.0632
9	.0682	9	.0665	9	.0648	9	.0632
3.830	.0682	3.880	.0664	3.930	.0647	3.980	.0631
1	.0681	1	.0664	1	.0647	1	.0631
2	.0681	2	.0664	2	.0647	2	.0631
3	.0681	3	.0663	3	.0646	3	.0630
4	.0680	4	.0663	4	.0646	4	.0630
5	.0680	5	.0663	5	.0646	5	.0630
6	.0680	6	.0662	6	.0645	6	.0629
7	.0679	7	.0662	7	.0645	7	.0629
8	.0679	8	.0662	8	.0645	8	.0629
9	.0679	9	.0661	9	.0645	9	.0628
3.840	.0678	3.890	.0661	3.940	.0644	3.990	.0628
1	.0678	1	.0661	1	.0644	1	.0628
2	.0677	2	.0660	2	.0644	2	.0628
3	.0677	3	.0660	3	.0643	3	.0627
4	.0677	4	.0659	4	.0643	4	.0627
5	.0676	5	.0659	5	.0643	5	.0627
6	.0676	6	.0659	6	.0642	6	.0626
7	.0676	7	.0658	7	.0642	7	.0626
8	.0675	8	.0658	8	.0642	8	.0626
9	.0675	9	.0658	9	.0641	9	.0625

d	$1/d^2$	d	$1/d^2$	d	$1/d^2$	d	$1/d^2$
4.000	.062500	4.050	.060966	4.100	.059488	4.150	.058064
1	.062469	1	.060936	1	.059459	1	.058036
2	.062438	2	.060906	2	.059430	2	.058008
3	.062406	3	.060876	3	.059401	3	.057980
4	.062375	4	.060846	4	.059372	4	.057952
5	.062344	5	.060816	5	.059344	5	.057924
6	.062313	6	.060786	6	.059315	6	.057896
7	.062282	7	.060756	7	.059286	7	.057868
8	.062251	8	.060726	8	.059257	8	.057840
9	.062220	9	.060696	9	.059228	9	.057813
4.010	.062189	4.060	.060666	4.110	.059199	4.160	.057785
1	.062158	1	.060636	1	.059170	1	.057757
2	.062127	2	.060607	2	.059142	2	.057729
3	.062096	3	.060577	3	.059113	3	.057702
4	.062065	4	.060547	4	.059084	4	.057674
5	.062034	5	.060517	5	.059056	5	.057646
6	.062003	6	.060487	6	.059027	6	.057618
7	.061972	7	.060458	7	.058998	7	.057591
8	.061941	8	.060428	8	.058969	8	.057563
9	.061910	9	.060398	9	.058941	9	.057536
4.020	.061880	4.070	.060369	4.120	.058912	4.170	.057508
1	.061849	1	.060339	1	.058884	1	.057480
2	.061818	2	.060309	2	.058855	2	.057453
3	.061787	3	.060280	3	.058827	3	.057425
4	.061757	4	.060250	4	.058798	4	.057398
5	.061726	5	.060221	5	.058770	5	.057370
6	.061695	6	.060191	6	.058741	6	.057343
7	.061665	7	.060161	7	.058713	7	.057315
8	.061634	8	.060132	8	.058684	8	.057288
9	.061604	9	.060103	9	.058656	9	.057261
4.030	.061573	4.080	.060073	4.130	.058627	4.180	.057233
1	.061542	1	.060044	1	.058599	1	.057206
2	.061512	2	.060014	2	.058571	2	.057178
3	.061481	3	.059985	3	.058542	3	.057151
4	.061451	4	.059955	4	.058514	4	.057124
5	.061420	5	.059926	5	.058486	5	.057096
6	.061390	6	.059897	6	.058457	6	.057069
7	.061360	7	.059867	7	.058429	7	.057042
8	.061329	8	.059838	8	.058401	8	.057015
9	.061299	9	.059809	9	.058373	9	.056987
4.040	.061269	4.090	.059780	4.140	.058344	4.190	.056960
1	.061238	1	.059750	1	.058316	1	.056933
2	.061208	2	.059721	2	.058288	2	.056906
3	.061178	3	.059692	3	.058260	3	.056879
4	.061147	4	.059663	4	.058232	4	.056852
5	.061117	5	.059634	5	.058204	5	.056825
6	.061087	6	.059605	6	.058176	6	.056797
7	.061057	7	.059576	7	.058148	7	.056770
8	.061027	8	.059546	8	.058120	8	.056743
9	.060996	9	.059517	9	.058092	9	.056716

d	$1/d^2$	d	$1/d^2$	d	$1/d^2$	d	$1/d^2$
4.200	.056689	4.250	.055363	4.300	.054083	4.350	.052847
1	.056662	1	.055337	1	.054058	1	.052823
2	.056635	2	.055311	2	.054033	2	.052799
3	.056608	3	.055285	3	.054008	3	.052774
4	.056582	4	.055259	4	.053983	4	.052750
5	.056555	5	.055233	5	.053958	5	.052726
6	.056528	6	.055207	6	.053933	6	.052702
7	.056501	7	.055181	7	.053908	7	.052677
8	.056474	8	.055155	8	.053883	8	.052653
9	.056447	9	.055130	9	.053858	9	.052629
4.210	.056420	4.260	.055104	4.310	.053833	4.360	.052605
1	.056394	1	.055078	1	.053808	1	.052581
2	.056367	2	.055052	2	.053783	2	.052557
3	.056340	3	.055026	3	.053758	3	.052533
4	.056313	4	.055000	4	.053733	4	.052509
5	.056287	5	.054975	5	.053708	5	.052485
6	.056260	6	.054949	6	.053683	6	.052461
7	.056233	7	.054923	7	.053658	7	.052436
8	.056207	8	.054897	8	.053633	8	.052412
9	.056180	9	.054872	9	.053608	9	.052388
4.220	.056153	4.270	.054846	4.320	.053584	4.370	.052365
1	.056127	1	.054820	1	.053559	1	.052341
2	.056100	2	.054795	2	.053534	2	.052317
3	.056074	3	.054769	3	.053509	3	.052293
4	.056047	4	.054743	4	.053485	4	.052269
5	.056020	5	.054718	5	.053460	5	.052245
6	.055994	6	.054692	6	.053435	6	.052221
7	.055967	7	.054667	7	.053410	7	.052197
8	.055941	8	.054641	8	.053386	8	.052173
9	.055915	9	.054615	9	.053361	9	.052150
4.230	.055888	4.280	.054590	4.330	.053336	4.380	.052126
1	.055862	1	.054564	1	.053312	1	.052102
2	.055835	2	.054539	2	.053287	2	.052078
3	.055809	3	.054513	3	.053263	3	.052054
4	.055783	4	.054488	4	.053238	4	.052031
5	.055756	5	.054463	5	.053214	5	.052007
6	.055730	6	.054437	6	.053189	6	.051983
7	.055704	7	.054412	7	.053164	7	.051959
8	.055677	8	.054386	8	.053140	8	.051936
9	.055651	0	.054361	9	.053115	9	.051912
4.240	.055625	4.290	.054336	4.340	.053091	4.390	.051888
1	.055599	1	.054310	1	.053067	1	.051865
2	.055572	2	.054285	2	.053042	2	.051841
3	.055546	3	.054260	3	.053018	3	.051818
4	.055520	4	.054235	4	.052993	4	.051794
5	.055494	5	.054209	5	.052969	5	.051770
6	.055468	6	.054184	6	.052944	6	.051747
7	.055442	7	.054159	7	.052920	7	.051723
8	.055415	8	.054134	8	.052896	8	.051700
9	.055389	9	.054108	9	.052871	9	.051676

d	$1/d^2$	d	$1/d^2$	d	$1/d^2$	d	$1/d^2$
4.400	.051653	4.450	.050499	4.500	.049383	4.550	.048303
1	.051629	1	.050476	1	.049361	1	.048282
2	.051606	2	.050453	2	.049339	2	.048261
3	.051583	3	.050431	3	.049317	3	.048240
4	.051559	4	.050408	4	.049295	4	.048219
5	.051536	5	.050385	5	.049273	5	.048197
6	.051512	6	.050363	6	.049251	6	.048176
7	.051489	7	.050340	7	.049229	7	.048155
8	.051466	8	.050318	8	.049208	8	.048134
9	.051442	9	.050295	9	.049186	9	.048113
4.410	.051419	4.460	.050272	4.510	.049164	4.560	.048092
1	.051396	1	.050250	1	.049142	1	.048071
2	.051372	2	.050227	2	.049120	2	.048050
3	.051349	3	.050205	3	.049099	3	.048029
4	.051326	4	.050182	4	.049077	4	.048007
5	.051303	5	.050160	5	.049055	5	.047986
6	.051279	6	.050137	6	.049033	6	.047965
7	.051256	7	.050115	7	.049012	7	.047944
8	.051233	8	.050093	8	.048990	8	.047923
9	.051210	9	.050070	9	.048968	9	.047902
4.420	.051187	4.470	.050048	4.520	.048947	4.570	.047881
1	.051163	1	.050025	1	.048925	1	.047861
2	.051140	2	.050003	2	.048903	2	.047840
3	.051117	3	.049981	3	.048882	3	.047819
4	.051094	4	.049958	4	.048860	4	.047798
5	.051071	5	.049936	5	.048839	5	.047777
6	.051048	6	.049914	6	.048817	6	.047756
7	.051025	7	.049891	7	.048795	7	.047735
8	.051002	8	.049869	8	.048774	8	.047714
9	.050979	9	.049847	9	.048752	9	.047693
4.430	.050956	4.480	.049825	4.530	.048731	4.580	.047673
1	.050933	1	.049802	1	.048709	1	.047652
2	.050910	2	.049780	2	.048688	2	.047631
3	.050887	3	.049758	3	.048666	3	.047610
4	.050864	4	.049736	4	.048645	4	.047589
5	.050841	5	.049714	5	.048623	5	.047569
6	.050818	6	.049691	6	.048602	6	.047548
7	.050795	7	.049669	7	.048581	7	.047527
8	.050772	8	.049647	8	.048559	8	.047507
9	.050749	9	.049625	9	.048538	9	.047486
4.440	.050726	4.490	.049603	4.540	.048516	4.590	.047465
1	.050704	1	.049581	1	.048495	1	.047444
2	.050681	2	.049559	2	.048474	2	.047424
3	.050658	3	.049537	3	.048452	3	.047403
4	.050635	4	.049515	4	.048431	4	.047383
5	.050612	5	.049493	5	.048410	5	.047362
6	.050590	6	.049471	6	.048388	6	.047341
7	.050567	7	.049449	7	.048367	7	.047321
8	.050544	8	.049427	8	.048346	8	.047300
9	.050521	9	.049405	9	.048325	9	.047280

d	$1/d^2$	d	$1/d^2$	d	$1/d^2$	d	$1/d^2$
4.600	.047259	4.650	.046248	4.700	.045269	4.750	.044321
1	.047238	1	.046228	1	.045250	1	.044303
2	.047218	2	.046208	2	.045231	2	.044284
3	.047197	3	.046189	3	.045212	3	.044265
4	.047177	4	.046169	4	.045192	4	.044247
5	.047156	5	.046149	5	.045173	5	.044228
6	.047136	6	.046129	6	.045154	6	.044210
7	.047115	7	.046109	7	.045135	7	.044191
8	.047095	8	.046089	8	.045116	8	.044172
9	.047075	9	.046070	9	.045096	9	.044154
4.610	.047054	4.660	.046050	4.710	.045077	4.760	.044135
1	.047034	1	.046030	1	.045058	1	.044117
2	.047013	2	.046010	2	.045039	2	.044098
3	.046993	3	.045991	3	.045020	3	.044080
4	.046973	4	.045971	4	.045001	4	.044061
5	.046952	5	.045951	5	.044982	5	.044043
6	.046932	6	.045931	6	.044963	6	.044024
7	.046912	7	.045912	7	.044944	7	.044006
8	.046891	8	.045892	8	.044925	8	.043987
9	.046871	9	.045872	9	.044906	9	.043969
4.620	.046851	4.670	.045853	4.720	.044887	4.770	.043950
1	.046830	1	.045833	1	.044868	1	.043932
2	.046810	2	.045814	2	.044849	2	.043914
3	.046790	3	.045794	3	.044830	3	.043895
4	.046770	4	.045774	4	.044811	4	.043877
5	.046749	5	.045755	5	.044792	5	.043858
6	.046729	6	.045735	6	.044773	6	.043840
7	.046709	7	.045716	7	.044754	7	.043822
8	.046669	8	.045696	8	.044735	8	.043803
9	.046669	9	.045677	9	.044716	9	.043785
4.630	.046649	4.680	.045657	4.730	.044697	4.780	.043767
1	.046628	1	.045638	1	.044678	1	.043748
2	.046608	2	.045618	2	.044659	2	.043730
3	.046588	3	.045599	3	.044640	3	.043712
4	.046568	4	.045579	4	.044621	4	.043694
5	.046548	5	.045560	5	.044603	5	.043675
6	.046528	6	.045540	6	.044584	6	.043657
7	.046508	7	.045521	7	.044565	7	.043639
8	.046488	8	.045501	8	.044546	8	.043621
9	.046468	9	.045482	9	.044527	9	.043602
4.640	.046448	4.690	.045463	4.740	.044509	4.790	.043584
1	.046428	1	.045443	1	.044490	1	.043566
2	.046408	2	.045424	2	.044471	2	.043548
3	.046388	3	.045405	3	.044452	3	.043530
4	.046368	4	.045385	4	.044434	4	.043511
5	.046348	5	.045366	5	.044415	5	.043493
6	.046328	6	.045347	6	.044396	6	.043475
7	.046308	7	.045327	7	.044377	7	.043457
8	.046288	8	.045308	8	.044359	8	.043439
9	.046268	9	.045289	9	.044340	9	.043421

d	$1/d^2$	d	$1/d^2$	d	$1/d^2$	d	$1/d^2$
4.800	.043403	4.850	.042512	4.900	.041649	4.950	.040812
1	.043385	1	.042495	1	.041632	1	.040796
2	.043367	2	.042477	2	.041615	2	.040779
3	.043349	3	.042460	3	.041598	3	.040763
4	.043331	4	.042442	4	.041581	4	.040746
5	.043313	5	.042425	5	.041564	5	.040730
6	.043294	6	.042408	6	.041548	6	.040713
7	.043276	7	.042390	7	.041531	7	.040697
8	.043258	8	.042373	8	.041514	8	.040681
9	.043240	9	.042355	9	.041497	9	.040664
4.810	.043223	4.860	.042338	4.910	.041480	4.960	.040648
1	.043205	1	.042320	1	.041463	1	.040631
2	.043187	2	.042303	2	.041446	2	.040615
3	.043169	3	.042286	3	.041429	3	.040599
4	.043151	4	.042268	4	.041412	4	.040582
5	.043133	5	.042251	5	.041395	5	.040566
6	.043115	6	.042233	6	.041379	6	.040550
7	.043097	7	.042216	7	.041362	7	.040533
8	.043079	8	.042199	8	.041345	8	.040517
9	.043061	9	.042181	9	.041328	9	.040501
4.820	.043043	4.870	.042164	4.920	.041311	4.970	.040484
1	.043025	1	.042147	1	.041295	1	.040468
2	.043008	2	.042129	2	.041278	2	.040452
3	.042990	3	.042112	3	.041261	3	.040436
4	.042972	4	.042095	4	.041244	4	.040419
5	.042954	5	.042078	5	.041228	5	.040403
6	.042936	6	.042060	6	.041211	6	.040387
7	.042919	7	.042043	7	.041194	7	.040371
8	.042901	8	.042026	8	.041177	8	.040354
9	.042883	9	.042009	9	.041161	9	.040338
4.830	.042865	4.880	.041991	4.930	.041144	4.980	.040322
1	.042848	1	.041974	1	.041127	1	.040306
2	.042830	2	.041957	2	.041111	2	.040290
3	.042812	3	.041940	3	.041094	3	.040273
4	.042794	4	.041923	4	.041077	4	.040257
5	.042777	5	.041905	5	.041061	5	.040241
6	.042759	6	.041888	6	.041044	6	.040225
7	.042741	7	.041871	7	.041027	7	.040209
8	.042724	8	.041854	8	.041011	8	.040193
9	.042706	9	.041837	9	.040994	9	.040177
4.840	.042688	4.890	.041820	4.940	.040978	4.990	.040160
1	.042671	1	.041803	1	.040961	1	.040144
2	.042653	2	.041786	2	.040944	2	.040128
3	.042635	3	.041769	3	.040928	3	.040112
4	.042618	4	.041752	4	.040911	4	.040096
5	.042600	5	.041734	5	.040895	5	.040080
6	.042583	6	.041717	6	.040878	6	.040064
7	.042565	7	.041700	7	.040862	7	.040048
8	.042548	8	.041683	8	.040845	8	.040032
9	.042530	9	.041666	9	.040829	9	.040016

d	$1/d^2$	d	$1/d^2$	d	$1/d^2$	d	$1/d^2$
5.000	.040000	5.050	.039212	5.100	.038447	5.150	.037704
1	.039984	1	.039196	1	.038432	1	.037689
2	.039968	2	.039181	2	.038417	2	.037675
3	.039952	3	.039165	3	.038402	3	.037660
4	.039936	4	.039150	4	.038387	4	.037645
5	.039920	5	.039134	5	.038371	5	.037631
6	.039904	6	.039119	6	.038356	6	.037616
7	.039888	7	.039103	7	.038341	7	.037602
8	.039872	8	.039088	8	.038326	8	.037587
9	.039856	9	.039072	9	.038311	9	.037572
5.010	.039840	5.060	.039057	5.110	.038296	5.160	.037558
1	.039825	1	.039042	1	.038281	1	.037543
2	.039809	2	.039026	2	.038266	2	.037529
3	.039793	3	.039011	3	.038252	3	.037514
4	.039777	4	.038995	4	.038237	4	.037500
5	.039761	5	.038980	5	.038222	5	.037485
6	.039745	6	.038965	6	.038207	6	.037471
7	.039729	7	.038949	7	.038192	7	.037456
8	.039714	8	.038934	8	.038177	8	.037442
9	.039698	9	.038918	9	.038162	9	.037427
5.020	.039682	5.070	.038903	5.120	.038147	5.170	.037413
1	.039666	1	.038888	1	.038132	1	.037398
2	.039650	2	.038872	2	.038117	2	.037384
3	.039635	3	.038857	3	.038102	3	.037369
4	.039619	4	.038842	4	.038087	4	.037355
5	.039603	5	.038826	5	.038073	5	.037340
6	.039587	6	.038811	6	.038058	6	.037326
7	.039571	7	.038796	7	.038043	7	.037312
8	.039556	8	.038781	8	.038028	8	.037297
9	.039540	9	.038765	9	.038013	9	.037283
5.030	.039524	5.080	.038750	5.130	.037998	5.180	.037268
1	.039509	1	.038735	1	.037984	1	.037254
2	.039493	2	.038720	2	.037969	2	.037240
3	.039477	3	.038704	3	.037954	3	.037225
4	.039462	4	.038689	4	.037939	4	.037211
5	.039446	5	.038674	5	.037924	5	.037197
6	.039430	6	.038659	6	.037910	6	.037182
7	.039415	7	.038644	7	.037895	7	.037168
8	.039399	8	.038628	8	.037880	8	.037154
9	.039383	9	.038613	9	.037865	9	.037139
5.040	.039368	5.090	.038598	5.140	.037851	5.190	.037125
1	.039352	1	.038583	1	.037836	1	.037111
2	.039336	2	.038568	2	.037821	2	.037096
3	.039321	3	.038553	3	.037807	3	.037082
4	.039305	4	.038537	4	.037792	4	.037068
5	.039290	5	.038522	5	.037777	5	.037053
6	.039274	6	.038507	6	.037762	6	.037039
7	.039258	7	.038492	7	.037748	7	.037025
8	.039243	8	.038477	8	.037733	8	.037011
9	.039227	9	.038462	9	.037718	9	.036996

d	$1/d^2$	d	$1/d^2$	d	$1/d^2$	d	$1/d^2$
5.200	.036982	5.250	.036281	5.300	.035600	5.350	.034938
1	.036968	1	.036267	1	.035586	1	.034924
2	.036954	2	.036254	2	.035573	2	.034911
3	.036940	3	.036240	3	.035560	3	.034898
4	.036925	4	.036226	4	.035546	4	.034885
5	.036911	5	.036212	5	.035533	5	.034872
6	.036897	6	.036198	6	.035519	6	.034859
7	.036883	7	.036185	7	.035506	7	.034846
8	.036869	8	.036171	8	.035493	8	.034833
9	.036855	9	.036157	9	.035479	9	.034820
5.210	.036840	5.260	.036143	5.310	.035466	5.360	.034807
1	.036826	1	.036130	1	.035453	1	.034794
2	.036812	2	.036116	2	.035439	2	.034781
3	.036798	3	.036102	3	.035426	3	.034768
4	.036784	4	.036088	4	.035413	4	.034755
5	.036770	5	.036075	5	.035399	5	.034742
6	.036756	6	.036061	6	.035386	6	.034730
7	.036742	7	.036047	7	.035373	7	.034717
8	.036728	8	.036034	8	.035359	8	.034704
9	.036713	9	.036020	9	.035346	9	.034691
5.220	.036699	5.270	.036006	5.320	.035333	5.370	.034678
1	.036685	1	.035993	1	.035319	1	.034665
2	.036671	2	.035979	2	.035306	2	.034652
3	.036657	3	.035965	3	.035293	3	.034639
4	.036643	4	.035952	4	.035280	4	.034626
5	.036629	5	.035938	5	.035266	5	.034613
6	.036615	6	.035924	6	.035253	6	.034600
7	.036601	7	.035911	7	.035240	7	.034588
8	.036587	8	.035897	8	.035227	8	.034575
9	.036573	9	.035884	9	.035213	9	.034562
5.230	.036559	5.280	.035870	5.330	.035200	5.380	.034549
1	.036545	1	.035856	1	.035187	1	.034536
2	.036531	2	.035843	2	.035174	2	.034523
3	.036517	3	.035829	3	.035161	3	.034511
4	.036503	4	.035816	4	.035147	4	.034498
5	.036489	5	.035802	5	.035134	5	.034485
6	.036475	6	.035789	6	.035121	6	.034472
7	.036462	7	.035775	7	.035108	7	.034459
8	.036448	8	.035762	8	.035095	8	.034446
9	.036434	9	.035748	9	.035082	9	.034434
5.240	.036420	5.290	.035735	5.340	.035069	5.390	.034421
1	.036406	1	.035721	1	.035055	1	.034408
2	.036392	2	.035708	2	.035042	2	.034395
3	.036378	3	.035694	3	.035029	3	.034383
4	.036364	4	.035681	4	.035016	4	.034370
5	.036350	5	.035667	5	.035003	5	.034357
6	.036337	6	.035654	6	.034990	6	.034344
7	.036323	7	.035640	7	.034977	7	.034332
8	.036309	8	.035627	8	.034964	8	.034319
9	.036295	9	.035613	9	.034951	9	.034306

d	$1/d^2$	d	$1/d^2$	d	$1/d^2$	d	$1/d^2$
5.400	.034294	5.450	.033667	5.500	.033058	5.550	.032465
1	.034281	1	.033655	1	.033046	1	.032453
2	.034268	2	.033643	2	.033034	2	.032442
3	.034255	3	.033630	3	.033022	3	.032430
4	.034243	4	.033618	4	.033010	4	.032418
5	.034230	5	.033606	5	.032998	5	.032406
6	.034217	6	.033593	6	.032986	6	.032395
7	.034205	7	.033581	7	.032974	7	.032383
8	.034192	8	.033569	8	.032962	8	.032372
9	.034180	9	.033556	9	.032950	9	.032360
5.410	.034167	5.460	.033544	5.510	.032938	5.560	.032348
1	.034154	1	.033532	1	.032926	1	.032337
2	.034142	2	.033519	2	.032914	2	.032325
3	.034129	3	.033507	3	.032902	3	.032313
4	.034116	4	.033495	4	.032890	4	.032302
5	.034104	5	.033483	5	.032878	5	.032290
6	.034091	6	.033470	6	.032866	6	.032279
7	.034079	7	.033458	7	.032854	7	.032267
8	.034066	8	.033446	8	.032843	8	.032255
9	.034054	9	.033434	9	.032831	9	.032244
5.420	.034041	5.470	.033421	5.520	.032819	5.570	.032232
1	.034028	1	.033409	1	.032807	1	.032221
2	.034016	2	.033397	2	.032795	2	.032209
3	.034003	3	.033385	3	.032783	3	.032197
4	.033991	4	.033373	4	.032771	4	.032186
5	.033978	5	.033360	5	.032759	5	.032174
6	.033966	6	.033348	6	.032748	6	.032163
7	.033953	7	.033336	7	.032736	7	.032151
8	.033941	8	.033324	8	.032724	8	.032140
9	.033928	9	.033312	9	.032712	9	.032128
5.430	.033916	5.480	.033300	5.530	.032700	5.580	.032117
1	.033903	1	.033287	1	.032688	1	.032105
2	.033891	2	.033275	2	.032677	2	.032094
3	.033878	3	.033263	3	.032665	3	.032082
4	.033866	4	.033251	4	.032653	4	.032071
5	.033853	5	.033239	5	.032641	5	.032059
6	.033841	6	.033227	6	.032629	6	.032048
7	.033828	7	.033215	7	.032618	7	.032036
8	.033816	8	.033203	8	.032606	8	.032025
9	.033804	9	.033190	9	.032594	9	.032013
5.440	.033791	5.490	.033178	5.540	.032582	5.590	.032002
1	.033779	1	.033166	1	.032570	1	.031991
2	.033766	2	.033154	2	.032559	2	.031979
3	.033754	3	.033142	3	.032547	3	.031968
4	.033741	4	.033130	4	.032535	4	.031956
5	.033729	5	.033118	5	.032523	5	.031945
6	.033717	6	.033106	6	.032512	6	.031933
7	.033704	7	.033094	7	.032500	7	.031922
8	.033692	8	.033082	8	.032488	8	.031911
9	.033680	9	.033070	9	.032477	9	.031899

d	$1/d^2$	d	$1/d^2$	d	$1/d^2$	d	$1/d^2$
5.600	.031888	5.650	.031326	5.700	.030779	5.750	.030246
1	.031876	1	.031315	1	.030768	1	.030235
2	.031865	2	.031304	2	.030757	2	.030225
3	.031854	3	.031293	3	.030746	3	.030214
4	.031842	4	.031282	4	.030736	4	.030204
5	.031831	5	.031270	5	.030725	5	.030193
6	.031820	6	.031259	6	.030714	6	.030183
7	.031808	7	.031248	7	.030703	7	.030172
8	.031797	8	.031237	8	.030692	8	.030162
9	.031786	9	.031226	9	.030682	9	.030151
5.610	.031774	5.660	.031215	5.710	.030671	5.760	.030141
1	.031763	1	.031204	1	.030660	1	.030130
2	.031752	2	.031193	2	.030650	2	.030120
3	.031740	3	.031182	3	.030639	3	.030109
4	.031729	4	.031171	4	.030628	4	.030099
5	.031718	5	.031160	5	.030617	5	.030089
6	.031706	6	.031149	6	.030607	6	.030078
7	.031695	7	.031138	7	.030596	7	.030068
8	.031684	8	.031127	8	.030585	8	.030057
9	.031672	9	.031116	9	.030575	9	.030047
5.620	.031661	5.670	.031105	5.720	.030564	5.770	.030036
1	.031650	1	.031094	1	.030553	1	.030026
2	.031639	2	.031083	2	.030542	2	.030016
3	.031627	3	.031072	3	.030532	3	.030005
4	.031616	4	.031061	4	.030521	4	.029995
5	.031605	5	.031050	5	.030510	5	.029984
6	.031594	6	.031040	6	.030500	6	.029974
7	.031582	7	.031029	7	.030489	7	.029964
8	.031571	8	.031018	8	.030479	8	.029953
9	.031560	9	.031007	9	.030468	9	.029943
5.630	.031549	5.680	.030996	5.730	.030457	5.780	.029933
1	.031538	1	.030985	1	.030447	1	.029922
2	.031526	2	.030974	2	.030436	2	.029912
3	.031515	3	.030963	3	.030425	3	.029902
4	.031504	4	.030952	4	.030415	4	.029891
5	.031493	5	.030941	5	.030404	5	.029881
6	.031482	6	.030930	6	.030394	6	.029871
7	.031471	7	.030920	7	.030383	7	.029860
8	.031459	8	.030909	8	.030372	8	.029850
9	.031448	9	.030898	9	.030362	9	.029840
5.640	.031437	5.690	.030887	5.740	.030351	5.790	.029829
1	.031426	1	.030876	1	.030341	1	.029819
2	.031415	2	.030865	2	.030330	2	.029809
3	.031404	3	.030854	3	.030320	3	.029798
4	.031393	4	.030844	4	.030309	4	.029788
5	.031381	5	.030833	5	.030298	5	.029778
6	.031370	6	.030822	6	.030288	6	.029768
7	.031359	7	.030811	7	.030277	7	.029757
8	.031348	8	.030800	8	.030267	8	.029747
9	.031337	9	.030790	9	.030256	9	.029737

d	$1/d^2$	d	$1/d^2$	d	$1/d^2$	d	$1/d^2$
5.800	.029727	5.850	.029221	5.900	.028727	5.950	.028247
1	.029716	1	.029211	1	.028718	1	.028237
2	.029706	2	.029201	2	.028708	2	.028228
3	.029696	3	.029191	3	.028698	3	.028218
4	.029686	4	.029181	4	.028688	4	.028209
5	.029675	5	.029171	5	.028679	5	.028199
6	.029665	6	.029161	6	.028669	6	.028190
7	.029655	7	.029151	7	.028659	7	.028180
8	.029645	8	.029141	8	.028650	8	.028171
9	.029634	9	.029131	9	.028640	9	.028161
5.810	.029624	5.860	.029121	5.910	.028630	5.960	.028152
1	.029614	1	.029111	1	.028621	1	.028142
2	.029604	2	.029101	2	.028611	2	.028133
3	.029594	3	.029091	3	.028601	3	.028124
4	.029584	4	.029081	4	.028592	4	.028114
5	.029573	5	.029071	5	.028582	5	.028105
6	.029563	6	.029061	6	.028572	6	.028095
7	.029553	7	.029051	7	.028563	7	.028086
8	.029543	8	.029042	8	.028553	8	.028076
9	.029533	9	.029032	9	.028543	9	.028067
5.820	.029523	5.870	.029022	5.920	.028534	5.970	.028058
1	.029512	1	.029012	1	.028524	1	.028048
2	.029502	2	.029002	2	.028514	2	.028039
3	.029492	3	.028992	3	.028505	3	.028029
4	.029482	4	.028982	4	.028495	4	.028020
5	.029472	5	.028972	5	.028485	5	.028011
6	.029462	6	.028963	6	.028476	6	.028001
7	.029452	7	.028953	7	.028466	7	.027992
8	.029442	8	.028943	8	.028457	8	.027983
9	.029431	9	.028933	9	.028447	9	.027973
5.830	.029421	5.880	.028923	5.930	.028437	5.980	.027964
1	.029411	1	.028913	1	.028428	1	.027955
2	.029401	2	.028903	2	.028418	2	.027945
3	.029391	3	.028894	3	.028409	3	.027936
4	.029381	4	.028884	4	.028399	4	.027927
5	.029371	5	.028874	5	.028390	5	.027917
6	.029361	6	.028864	6	.028380	6	.027908
7	.029351	7	.028854	7	.028370	7	.027899
8	.029341	8	.028845	8	.028361	8	.027889
9	.029331	9	.028835	9	.028351	9	.027880
5.840	.029321	5.890	.028825	5.940	.028342	5.990	.027871
1	.029311	1	.028815	1	.028332	1	.027861
2	.029301	2	.028805	2	.028323	2	.027852
3	.029291	3	.028796	3	.028313	3	.027843
4	.029281	4	.028786	4	.028304	4	.027833
5	.029271	5	.028776	5	.028294	5	.027824
6	.029261	6	.028766	6	.028285	6	.027815
7	.029251	7	.028757	7	.028275	7	.027806
8	.029241	8	.028747	8	.028266	8	.027796
9	.029231	9	.028737	9	.028256	9	.027787

d	$1/d^2$	d	$1/d^2$	d	$1/d^2$	d	$1/d^2$
6.000	.027778	6.050	.027321	6.100	.026874	6.150	.026439
1	.027769	1	.027312	1	.026866	1	.026431
2	.027759	2	.027302	2	.026857	2	.026422
3	.027750	3	.027293	3	.026848	3	.026414
4	.027741	4	.027284	4	.026839	4	.026405
5	.027732	5	.027275	5	.026830	5	.026396
6	.027722	6	.027266	6	.026822	6	.026388
7	.027713	7	.027257	7	.026813	7	.026379
8	.027704	8	.027248	8	.026804	8	.026371
9	.027695	9	.027239	9	.026795	9	.026362
6.010	.027685	6.060	.027230	6.110	.026787	6.160	.026354
1	.027676	1	.027221	1	.026778	1	.026345
2	.027667	2	.027212	2	.026769	2	.026336
3	.027659	3	.027204	3	.026760	3	.026328
4	.027649	4	.027195	4	.026752	4	.026319
5	.027639	5	.027186	5	.026743	5	.026311
6	.027630	6	.027177	6	.026734	6	.026302
7	.027621	7	.027168	7	.026725	7	.026294
8	.027612	8	.027159	8	.026717	8	.026285
9	.027603	9	.027150	9	.026717	9	.026277
6.020	.027594	6.070	.027141	6.120	.026699	6.170	.026268
1	.027584	1	.027132	1	.026690	1	.026260
2	.027575	2	.027123	2	.026682	2	.026251
3	.027566	3	.027114	3	.026673	3	.026243
4	.027557	4	.027105	4	.026664	4	.026234
5	.027548	5	.027096	5	.026656	5	.026226
6	.027539	6	.027087	6	.026647	6	.026217
7	.027529	7	.027078	7	.026638	7	.026209
8	.027520	8	.027069	8	.026629	8	.026200
9	.027511	9	.027060	9	.026621	9	.026192
6.030	.027502	6.080	.027052	6.130	.026612	6.180	.026183
1	.027493	1	.027043	1	.026603	1	.026175
2	.027484	2	.027034	2	.026595	2	.026166
3	.027475	3	.027025	3	.026586	3	.026158
4	.027466	4	.027016	4	.026577	4	.026149
5	.027457	5	.027007	5	.026569	5	.026141
6	.027447	6	.026998	6	.026560	6	.026132
7	.027438	7	.026989	7	.026551	7	.026125
8	.027429	8	.026981	8	.026543	8	.026116
9	.027420	9	.026972	9	.026534	9	.026107
6.040	.027411	6.090	.026963	6.140	.026525	6.190	.026099
1	.027402	1	.026954	1	.026517	1	.026090
2	.027393	2	.026945	2	.026508	2	.026082
3	.027384	3	.026936	3	.026500	3	.026073
4	.027375	4	.026927	4	.026491	4	.026065
5	.027366	5	.026919	5	.026482	5	.026057
6	.027357	6	.026910	6	.026474	6	.026048
7	.027348	7	.026901	7	.026465	7	.026040
8	.027339	8	.026982	8	.026456	8	.026031
9	.027330	9	.026883	9	.026448	9	.026023

d	$1/d^2$	d	$1/d^2$	d	$1/d^2$	d	$1/d^2$
6.200	.026015	6.250	.025600	6.300	.025195	6.350	.024800
1	.026006	1	.025592	1	.025187	1	.024792
2	.025998	2	.025584	2	.025179	2	.024784
3	.025989	3	.025575	3	.025171	3	.024777
4	.025981	4	.025567	4	.025163	4	.024769
5	.025973	5	.025559	5	.025155	5	.024761
6	.025964	6	.025551	6	.025147	6	.024753
7	.025956	7	.025543	7	.025139	7	.024745
8	.025948	8	.025535	8	.025131	8	.024738
9	.025939	9	.025527	9	.025123	9	.024730
6.210	.025931	6.260	.025518	6.310	.025115	6.360	.024722
1	.025923	1	.025510	1	.025108	1	.024714
2	.025914	2	.025502	2	.025100	2	.024707
3	.025906	3	.025494	3	.025092	3	.024699
4	.025897	4	.025486	4	.025084	4	.024691
5	.025889	5	.025478	5	.025076	5	.024683
6	.025881	6	.025469	6	.025068	6	.024676
7	.025872	7	.025461	7	.025060	7	.024668
8	.025864	8	.025453	8	.025052	8	.024660
9	.025856	9	.025445	9	.025044	9	.024652
6.220	.025848	6.270	.025437	6.320	.025036	6.370	.024645
1	.025839	1	.025429	1	.025028	1	.024637
2	.025831	2	.025421	2	.025020	2	.024629
3	.025823	3	.025413	3	.025012	3	.024621
4	.025814	4	.025405	4	.025004	4	.024614
5	.025806	5	.025396	5	.024996	5	.024606
6	.025798	6	.025388	6	.024989	6	.024598
7	.025789	7	.025380	7	.024981	7	.024590
8	.025781	8	.025372	8	.024973	8	.024583
9	.025773	9	.025364	9	.024965	9	.024575
6.230	.025765	6.280	.025356	6.330	.024957	6.380	.024567
1	.025756	1	.025348	1	.024949	1	.024560
2	.025748	2	.025340	2	.024941	2	.024552
3	.025740	3	.025332	3	.024933	3	.024544
4	.025732	4	.025324	4	.024926	4	.024537
5	.025723	5	.025316	5	.024918	5	.024529
6	.025715	6	.025308	6	.024910	6	.024521
7	.025707	7	.025300	7	.024902	7	.024514
8	.025699	8	.025292	8	.024894	8	.024506
9	.025690	9	.025283	9	.024886	9	.024498
6.240	.025682	6.290	.025275	6.340	.024878	6.390	.024491
1	.025674	1	.025267	1	.024870	1	.024483
2	.025666	2	.025259	2	.024863	2	.024475
3	.025657	3	.025251	3	.024855	3	.024468
4	.025649	4	.025243	4	.024847	4	.024460
5	.025641	5	.025235	5	.024839	5	.024452
6	.025633	6	.025227	6	.024831	6	.024445
7	.025625	7	.025219	7	.024823	7	.024437
8	.025616	8	.025211	8	.024816	8	.024429
9	.025608	9	.025203	9	.024808	9	.024422

d	$1/d^2$	d	$1/d^2$	d	$1/d^2$	d	$1/d^2$
6.400	.024414	6.450	.024037	6.500	.023669	6.550	.023309
1	.024406	1	.024030	1	.023661	1	.023302
2	.024399	2	.024022	2	.023654	2	.023294
3	.024391	3	.024015	3	.023647	3	.023287
4	.024384	4	.024007	4	.023640	4	.023280
5	.024376	5	.024000	5	.023632	5	.023273
6	.024368	6	.023992	6	.023625	6	.023266
7	.024361	7	.023985	7	.023618	7	.023259
8	.024353	8	.023978	8	.023610	8	.023252
9	.024346	9	.023970	9	.023603	9	.023245
6.410	.024338	6.460	.023963	6.510	.023596	6.560	.023238
1	.024330	1	.023955	1	.023589	1	.023231
2	.024323	2	.023948	2	.023581	2	.023223
3	.024315	3	.023940	3	.023574	3	.023216
4	.024308	4	.023933	4	.023567	4	.023209
5	.024300	5	.023926	5	.023560	5	.023202
6	.024292	6	.023918	6	.023553	6	.023195
7	.024285	7	.023911	7	.023545	7	.023188
8	.024277	8	.023903	8	.023538	8	.023181
9	.024270	9	.023896	9	.023531	9	.023174
6.420	.024262	6.470	.023889	6.520	.023524	6.570	.023167
1	.024255	1	.023881	1	.023516	1	.023160
2	.024247	2	.023874	2	.023509	2	.023153
3	.024240	3	.023867	3	.023502	3	.023146
4	.024232	4	.023859	4	.023495	4	.023139
5	.024224	5	.023852	5	.023488	5	.023132
6	.024217	6	.023844	6	.023480	6	.023125
7	.024209	7	.023837	7	.023473	7	.023118
8	.024202	8	.023830	8	.023466	8	.023111
9	.024194	9	.023822	9	.023459	9	.023104
6.430	.024187	6.480	.023815	6.530	.023452	6.530	.023097
1	.024179	1	.023808	1	.023444	1	.023090
2	.024172	2	.023800	2	.023437	2	.023083
3	.024164	3	.023793	3	.023430	3	.023076
4	.024157	4	.023786	4	.023423	4	.023069
5	.024149	5	.023778	5	.023416	5	.023062
6	.024142	6	.023771	6	.023409	6	.023055
7	.024134	7	.023764	7	.023401	7	.023048
8	.024127	8	.023756	8	.023394	8	.023041
9	.024119	9	.023749	9	.023387	9	.023034
6.440	.024112	6.490	.023742	6.540	.023380	6.590	.023027
1	.024104	1	.023734	1	.023373	1	.023020
2	.024097	2	.023727	2	.023366	2	.023013
3	.024089	3	.023720	3	.023359	3	.023006
4	.024082	4	.023712	4	.023351	4	.022999
5	.024074	5	.023705	5	.023344	5	.022992
6	.024067	6	.023698	6	.023337	6	.022985
7	.024059	7	.023691	7	.023330	7	.022978
8	.024052	8	.023683	8	.023323	8	.022971
9	.024044	9	.023676	9	.023316	9	.022964

d	$1/d^2$	d	$1/d^2$	d	$1/d^2$	d	$1/d^2$
6.600	.022957	6.650	.022613	6.700	.022277	6.750	.021948
1	.022950	1	.022606	1	.022270	1	.021941
2	.022943	2	.022599	2	.022263	2	.021935
3	.022936	3	.022593	3	.022257	3	.021928
4	.022929	4	.022586	4	.022250	4	.021922
5	.022922	5	.022579	5	.022243	5	.021915
6	.022915	6	.022572	6	.022237	6	.021909
7	.022908	7	.022565	7	.022230	7	.021902
8	.022901	8	.022559	8	.022224	8	.021896
9	.022894	9	.022552	9	.022217	9	.021889
6.610	.022887	6.660	.022545	6.710	.022210	6.760	.021883
1	.022881	1	.022538	1	.022204	1	.021877
2	.022874	2	.022532	2	.022197	2	.021870
3	.022867	3	.022525	3	.022190	3	.021864
4	.022860	4	.022511	4	.022184	4	.021857
5	.022853	5	.022505	5	.022177	5	.021851
6	.022846	6	.022500	6	.022171	6	.021844
7	.022839	7	.022498	7	.022164	7	.021838
8	.022832	8	.022491	8	.022157	8	.021831
9	.022825	9	.022484	9	.022151	9	.021825
6.620	.022818	6.670	.022478	6.720	.022144	6.770	.021818
1	.022811	1	.022471	1	.022138	1	.021812
2	.022805	2	.022464	2	.022131	2	.021806
3	.022798	3	.022457	3	.022125	3	.021799
4	.022791	4	.022451	4	.022118	4	.021793
5	.022784	5	.022444	5	.022111	5	.021786
6	.022777	6	.022437	6	.022105	6	.021780
7	.022770	7	.022430	7	.022098	7	.021773
8	.022763	8	.022424	8	.022092	8	021767
9	.022756	9	.022417	9	.022085	9	.021760
6.630	.022750	6.680	.022410	6.730	.022079	6.780	.021754
1	.022743	1	.022404	1	.022072	1	.021748
2	.022736	2	.022397	2	.022065	2	.021741
3	.022729	3	.022390	3	.022059	3	.021735
4	.022722	4	.022383	4	.022052	4	.021728
5	.022715	5	.022377	5	.022046	5	.021722
6	.022708	6	.022370	6	.022039	6	.021716
7	.022702	7	.022363	7	.022033	7	.021709
8	.022695	8	.022357	8	.022026	8	.021703
9	.022688	9	.022350	9	.022020	9	.021696
6.640	.022681	6.690	.022343	6.740	.022013	6.790	.021690
1	.022674	1	.022337	1	.022007	1	.021684
2	.022667	2	.022330	2	.022000	2	.021677
3	.022661	3	.022323	3	.021993	3	.021671
4	.022654	4	.022317	4	.021987	4	.021665
5	.022647	5	.022310	5	.021980	5	.021658
6	.022640	6	.022303	6	.021974	6	.021652
7	.022633	7	.022297	7	.021967	7	.021645
8	.022627	8	.022290	8	.021961	8	.021639
9	.022620	9	.022283	9	.021954	9	.021633

d	$1/d^2$	d	$1/d^2$	d	$1/d^2$	d	$1/d^2$
6.800	.021626	6.850	.021312	6.900	.021004	6.950	.020703
1	.021620	1	.021306	1	.020998	1	.020697
2	.021614	2	.021299	2	.020992	2	.020691
3	.021607	3	.021293	3	.020986	3	.020685
4	.021601	4	.021287	4	.020980	4	.020679
5	.021595	5	.021281	5	.020974	5	.020673
6	.021588	6	.021274	6	.020968	6	.020667
7	.021582	7	.021268	7	.020961	7	.020661
8	.021576	8	.021262	8	.020955	8	.020655
9	.021569	9	.021256	9	.020949	9	.020649
6.810	.021563	6.860	.021250	6.910	.020943	6.960	.020643
1	.021556	1	.021243	1	.020937	1	.020637
2	.021550	2	.021237	2	.020931	2	.020632
3	.021544	3	.021231	3	.020925	3	.020626
4	.021538	4	.021225	4	.020915	4	.020620
5	.021531	5	.021219	5	.020913	5	.020614
6	.021525	6	.021213	6	.020907	6	.020608
7	.021519	7	.021206	7	.020901	7	.020602
8	.021512	8	.021200	8	.020895	8	.020596
9	.021506	9	.021194	9	.020889	9	.020590
6.820	.021500	6.870	.021188	6.920	.020883	6.970	.020584
1	.021493	1	.021182	1	.020877	1	.020578
2	.021487	2	.021176	2	.020871	2	.020572
3	.021481	3	.021169	3	.020865	3	.020567
4	.021474	4	.021163	4	.020859	4	.020561
5	.021468	5	.021157	5	.020853	5	.020555
6	.021462	6	.021151	6	.020847	6	.020549
7	.021456	7	.021145	7	.020841	7	.020543
8	.021449	8	.021139	8	.020835	8	.020537
9	.021443	9	.021126	9	.020829	9	.020531
6.830	.021437	6.880	.021126	6.930	.020823	6.980	.020525
1	.021430	1	.021120	1	.020817	1	.020519
2	.021424	2	.021114	2	.020811	2	.020514
3	.021418	3	.021108	3	.020805	3	.020508
4	.021412	4	.021102	4	.020799	4	.020502
5	.021405	5	.021096	5	.020793	5	.020496
6	.021399	6	.021089	6	.020787	6	.020490
7	.021393	7	.021083	7	.020781	7	.020484
8	.021387	8	.021077	8	.020775	8	.020478
9	.021380	9	.021071	9	.020769	9	.020472
6.840	.021374	6.890	.021065	6.940	.020763	6.990	.020467
1	.021368	1	.021059	1	.020757	1	.020461
2	.021362	2	.021053	2	.020751	2	.020455
3	.021355	3	.021047	3	.020745	3	.020449
4	.021349	4	.021041	4	.020739	4	.020443
5	.021343	5	.021034	5	.020733	5	.020437
6	.021337	6	.021028	6	.020727	6	.020432
7	.021330	7	.021022	7	.020721	7	.020426
8	.021324	8	.021016	8	.020715	8	.020420
9	.021318	9	.021010	9	.020709	9	.020414

d	$1/d^2$	d	$1/d^2$	d	$1/d^2$	d	$1/d^2$
7.000	.020408	7.050	.020120	7.100	.019837	7.150	.019561
1	.020402	1	.020114	1	.019832	1	.019555
2	.020397	2	.020108	2	.019826	2	.019550
3	.020391	3	.020103	3	.019821	3	.019544
4	.020385	4	.020097	4	.019815	4	.019539
5	.020379	5	.020091	5	.019809	5	.019534
6	.020373	6	.020086	6	.019804	6	.019528
7	.020367	7	.020080	7	.019798	7	.019523
8	.020362	8	.020074	8	.019793	8	.019517
9	.020356	9	.020068	9	.019787	9	.019512
7.010	.020350	7.060	.020063	7.110	.019782	7.160	.019506
1	.020344	1	.020057	1	.019776	1	.019501
2	.020338	2	.020051	2	.019770	2	.019495
3	.020333	3	.020046	3	.019765	3	.019490
4	.020327	4	.020040	4	.019759	4	.019484
5	.020321	5	.020034	5	.019754	5	.019479
6	.020315	6	.020029	6	.019748	6	.019474
7	.020309	7	.020023	7	.019743	7	.019468
8	.020304	8	.020017	8	.019737	8	.019463
9	.020298	9	.020012	9	.019732	9	.019457
7.020	.020292	7.070	.020006	7.120	.019726	7.170	.019452
1	.020286	1	.020000	1	.019721	1	.019446
2	.020280	2	.019995	2	.019715	2	.019441
3	.020275	3	.019989	3	.019709	3	.019436
4	.020269	4	.019983	4	.019704	4	.019430
5	.020263	5	.019978	5	.019698	5	.019425
6	.020257	6	.019972	6	.019693	6	.019419
7	.020252	7	.019966	7	.019687	7	.019414
8	.020246	8	.019961	8	.019682	8	.019409
9	.020240	9	.019955	9	.019676	9	.019403
7.030	.020234	7.080	.019950	7.130	.019671	7.180	.019398
1	.020229	1	.019944	1	.019665	1	.019392
2	.020223	2	.019938	2	.019660	2	.019387
3	.020217	3	.019933	3	.019654	3	.019382
4	.020211	4	.019927	4	.019649	4	.019376
5	.020206	5	.019921	5	.019643	5	.019371
6	.020200	6	.019916	6	.019638	6	.019365
7	.020194	7	.019910	7	.019632	7	.019360
8	.020188	8	.019905	8	.019627	8	.019355
9	.020183	9	.019899	9	.019621	9	.019349
7.040	.020177	7.090	.019893	7.140	.019616	7.190	.019344
1	.020171	1	.019888	1	.019610	1	.019338
2	.020165	2	.019882	2	.019605	2	.019333
3	.020160	3	.019877	3	.019599	3	.019328
4	.020154	4	.019871	4	.019594	4	.019322
5	.020148	5	.019865	5	.019588	5	.019317
6	.020143	6	.019860	6	.019583	6	.019312
7	.020137	7	.019854	7	.019577	7	.019306
8	.020131	8	.019849	8	.019572	8	.019301
9	.020125	9	.019843	9	.019566	9	.019295

d	$1/d^2$	d	$1/d^2$	d	$1/d^2$	d	$1/d^2$
7.200	.019290	7.250	.019025	7.300	.018765	7.350	.018511
1	.019285	1	.019020	1	.018760	1	.018506
2	.019279	2	.019014	2	.018755	2	.018501
3	.019274	3	.019009	3	.018750	3	.018496
4	.019269	4	.019004	4	.018745	4	.018491
5	.019263	5	.018999	5	.018740	5	.018486
6	.019258	6	.018994	6	.018734	6	.018481
7	.019253	7	.018988	7	.018729	7	.018476
8	.019247	8	.018983	8	.018724	8	.018471
9	.019242	9	.018978	9	.018719	9	.018466
7.210	.019237	7.260	.018973	7.310	.018714	7.360	.018461
1	.019231	1	.018967	1	.018709	1	.018456
2	.019226	2	.018962	2	.018704	2	.018451
3	.019221	3	.018957	3	.018699	3	.018445
4	.019215	4	.018952	4	.018693	4	.018440
5	.019210	5	.018946	5	.018688	5	.018435
6	.019205	6	.018941	6	.018683	6	.018430
7	.019199	7	.018936	7	.018678	7	.018425
8	.019194	8	.018931	8	.018673	8	.018420
9	.019189	9	.018926	9	.018668	9	.018415
7.220	.019183	7.270	.018920	7.320	.018663	7.370	.018410
1	.019178	1	.018915	1	.018658	1	.018405
2	.019173	2	.018910	2	.018653	2	.018400
3	.019167	3	.018905	3	.018648	3	.018395
4	.019162	4	.018900	4	.018642	4	.018391
5	.019157	5	.018894	5	.018637	5	.018386
6	.019152	6	.018889	6	.018632	6	.018381
7	.019146	7	.018884	7	.018627	7	.018376
8	.019141	8	.018879	8	.018622	8	.018371
9	.019136	9	.018874	9	.018617	9	.018366
7.230	.019130	7.280	.018868	7.330	.018612	7.380	.018361
1	.019125	1	.018863	1	.018607	1	.018356
2	.019120	2	.018858	2	.018602	2	.018351
3	.019115	3	.018853	3	.018597	3	.018346
4	.019109	4	.018848	4	.018592	4	.018341
5	.019104	5	.018843	5	.018587	5	.018336
6	.019099	6	.018837	6	.018582	6	.018331
7	.019093	7	.018832	7	.018576	7	.018326
8	.019088	8	.018827	8	.018571	8	.018321
9	.019083	9	.018822	9	.018566	9	.018316
7.240	.019078	7.290	.018817	7.340	.018561	7.390	.018311
1	.019072	1	.018812	1	.018556	1	.018306
2	.019067	2	.018806	2	.018551	2	.018301
3	.019062	3	.018801	3	.018546	3	.018296
4	.019056	4	.018796	4	.018541	4	.018291
5	.019051	5	.018791	5	.018536	5	.018286
6	.019046	6	.018786	6	.018531	6	.018281
7	.019041	7	.018781	7	.018526	7	.018276
8	.019035	8	.018776	8	.018521	8	.018271
9	.019030	9	.018770	9	.018516	9	.018266

d	$1/d^2$	d	$1/d^2$	d	$1/d^2$	d	$1/d^2$
7.400	.018262	7.450	.018017	7.500	.017778	7.550	.017543
1	.018257	1	.018012	1	.017773	1	.017538
2	.018252	2	.018008	2	.017768	2	.017534
3	.018247	3	.018003	3	.017764	3	.017529
4	.018242	4	.017998	4	.017759	4	.017525
5	.018237	5	.017993	5	.017754	5	.017520
6	.018232	6	.017988	6	.017749	6	.017515
7	.018227	7	.017983	7	.017745	7	.017511
8	.018222	8	.017979	8	.017740	8	.017506
9	.018217	9	.017974	9	.017735	9	.017501
7.410	.018212	7.460	.017969	7.510	.017730	7.560	.017497
1	.018207	1	.017964	1	.017726	1	.017492
2	.018202	2	.017959	2	.017721	2	.017487
3	.018198	3	.017954	3	.017716	3	.017483
4	.018193	4	.017950	4	.017712	4	.017478
5	.018188	5	.017945	5	.017707	5	.017474
6	.018183	6	.017940	6	.017702	6	.017469
7	.018178	7	.017935	7	.017697	7	.017464
8	.018173	8	.017930	8	.017693	8	.017460
9	.018168	9	.017926	9	.017688	9	.017455
7.420	.018163	7.470	.017921	7.520	.017683	7.570	.017451
1	.018158	1	.017916	1	.017679	1	.017446
2	.018153	2	.017911	2	.017674	2	.017441
3	.018149	3	.017906	3	.017669	3	.017437
4	.018144	4	.017902	4	.017665	4	.017432
5	.018139	5	.017897	5	.017660	5	.017427
6	.018134	6	.017892	6	.017655	6	.017423
7	.018129	7	.017887	7	.017650	7	.017418
8	.018124	8	.017883	8	.017646	8	.017414
9	.018119	9	.017878	9	.017641	9	.017409
7.430	.018114	7.480	.017873	7.530	.017636	7.580	.017405
1	.018109	1	.017868	1	.017632	1	.017400
2	.018105	2	.017863	2	.017627	2	.017395
3	.018100	3	.017859	3	.017622	3	.017391
4	.018095	4	.017854	4	.017618	4	.017386
5	.018090	5	.017849	5	.017613	5	.017382
6	.018085	6	.017844	6	.017608	6	.017377
7	.018080	7	.017840	7	.017604	7	.017372
8	.018075	8	.017835	8	.017599	8	.017368
9	.018071	9	.017830	9	.017594	9	.017363
7.440	.018066	7.490	.017825	7.540	.017590	7.590	.017359
1	.018061	1	.017821	1	.017585	1	.017354
2	.018056	2	.017816	2	.017580	2	.017350
3	.018051	3	.017811	3	.017576	3	.017345
4	.018046	4	.017806	4	.017571	4	.017340
5	.018041	5	.017802	5	.017566	5	.017335
6	.018037	6	.017797	6	.017562	6	.017331
7	.018032	7	.017792	7	.017557	7	.017327
8	.018027	8	.017787	8	.017552	8	.017322
9	.018022	9	.017783	9	.017548	9	.017318

d	$1/d^2$	d	$1/d^2$	d	$1/d^2$	d	$1/d^2$
7.600	.017313	7.650	.017087	7.700	.016866	7.750	.016649
1	.017308	1	.017083	1	.016862	1	.016645
2	.017304	2	.017079	2	.016857	2	.016641
3	.017299	3	.017074	3	.016853	3	.016636
4	.017295	4	.017070	4	.016849	4	.016632
5	.017290	5	.017065	5	.016844	5	.016628
6	.017286	6	.017061	6	.016840	6	.016624
7	.017281	7	.017056	7	.016836	7	.016619
8	.017277	8	.017052	8	.016831	8	.016615
9	.017272	9	.017047	9	.016827	9	.016611
7.610	.017268	7.660	.017043	7.710	.016823	7.760	.016606
1	.017263	1	.017038	1	.016818	1	.016602
2	.017258	2	.017034	2	.016814	2	.016598
3	.017254	3	.017031	3	.016809	3	.016594
4	.017249	4	.017025	4	.016805	4	.016589
5	.017245	5	.017021	5	.016801	5	.016585
6	.017240	6	.017016	6	.016796	6	.016581
7	.017236	7	.017012	7	.016792	7	.016577
8	.017231	8	.017007	8	.016788	8	.016572
9	.017227	9	.017003	9	.016783	9	.016568
7.620	.017222	7.670	.016998	7.720	.016779	7.770	.016564
1	.017218	1	.016994	1	.016775	1	.016559
2	.017213	2	.016990	2	.016770	2	.016555
3	.017209	3	.016985	3	.016766	3	.016551
4	.017204	4	.016981	4	.016762	4	.016547
5	.017200	5	.016976	5	.016757	5	.016542
6	.017195	6	.016972	6	.016753	6	.016538
7	.017191	7	.016967	7	.016749	7	.016534
8	.017186	8	.016963	8	.016744	8	.016530
9	.017182	9	.016959	9	.016740	9	.016525
7.630	.017177	7.680	.016954	7.730	.016736	7.780	.016521
1	.017173	1	.016950	1	.016731	1	.016517
2	.017168	2	.016945	2	.016727	2	.016513
3	.017164	3	.016941	3	.016723	3	.016508
4	.017159	4	.016937	4	.016718	4	.016504
5	.017155	5	.016932	5	.016714	5	.016500
6	.017150	6	.016928	6	.016710	6	.016496
7	.017146	7	.016923	7	.016705	7	.016491
8	.017141	8	.016919	8	.016701	8	.016487
9	.017137	9	.016915	9	.016697	9	.016483
7.640	.017132	7.690	.016910	7.740	.016692	7.790	.016479
1	.017128	1	.016906	1	.016688	1	.016475
2	.017123	2	.016901	2	.016684	2	.016470
3	.017119	3	.016897	3	.016679	3	.016466
4	.017114	4	.016893	4	.016675	4	.016462
5	.017110	5	.016888	5	.016671	5	.016458
6	.017105	6	.016884	6	.016667	6	.016453
7	.017101	7	.016879	7	.016662	7	.016449
8	.017096	8	.016875	8	.016658	8	.016445
9	.017092	9	.016871	9	.016654	9	.016441

d	$1/d^2$	d	$1/d^2$	d	$1/d^2$	d	$1/d^2$
7.800	.016437	7.850	.016228	7.900	.016023	7.950	.015822
1	.016432	1	.016224	1	.016019	1	.015818
2	.016428	2	.016220	2	.016015	2	.015814
3	.016424	3	.016215	3	.016011	3	.015810
4	.016420	4	.016211	4	.016007	4	.015806
5	.016416	5	.016207	5	.016003	5	.015802
6	.016411	6	.016203	6	.015999	6	.015798
7	.016407	7	.016199	7	.015995	7	.015794
8	.016403	8	.016195	8	.015991	8	.015790
9	.016399	9	.016191	9	.015987	9	.015786
7.810	.016394	7.860	.016187	7.910	.015983	7.960	.015782
1	.016390	1	.016182	1	.015979	1	.015778
2	.016386	2	.016178	2	.015975	2	.015775
3	.016382	3	.016174	3	.015970	3	.015771
4	.016378	4	.016170	4	.015966	4	.015767
5	.016374	5	.016166	5	.015962	5	.015763
6	.016369	6	.016162	6	.015958	6	.015759
7	.016365	7	.016158	7	.015954	7	.015755
8	.016361	8	.016154	8	.015950	8	.015751
9	.016357	9	.016150	9	.015946	9	.015747
7.820	.016353	7.870	.016145	7.920	.015942	7.970	.015743
1	.016348	1	.016141	1	.015938	1	.015739
2	.016344	2	.016137	2	.015934	2	.015735
3	.016340	3	.016133	3	.015930	3	.015731
4	.016336	4	.016129	4	.015926	4	.015727
5	.016332	5	.016125	5	.015922	5	.015723
6	.016328	6	.016121	6	.015918	6	.015719
7	.016323	7	.016117	7	.015914	7	.015715
8	.016319	8	.016113	8	.015910	8	.015711
9	.016315	9	.016109	9	.015906	9	.015707
7.830	.016311	7.880	.016105	7.930	.015902	7.980	.015703
1	.016307	1	.016100	1	.015898	1	.015699
2	.016303	2	.016096	2	.015894	2	.015696
3	.016298	3	.016092	3	.015890	3	.015692
4	.016294	4	.016088	4	.015886	4	.015688
5	.016290	5	.016084	5	.015882	5	.015684
6	.016286	6	.016080	6	.015878	6	.015680
7	.016282	7	.016076	7	.015874	7	.015676
8	.016278	8	.016072	8	.015870	8	.015672
9	.016273	9	.016068	9	.015866	9	.015668
7.840	.016269	7.890	.016064	7.940	.015862	7.990	.015664
1	.016265	1	.016060	1	.015858	1	.015660
2	.016261	2	.016056	2	.015854	2	.015656
3	.016257	3	.016052	3	.015850	3	.015652
4	.016253	4	.016047	4	.015846	4	.015648
5	.016249	5	.016043	5	.015842	5	.015645
6	.016244	6	.016039	6	.015838	6	.015641
7	.016240	7	.016035	7	.015834	7	.015637
8	.016236	8	.016031	8	.015830	8	.015633
9	.016232	9	.016027	9	.015826	9	.015629

d	$1/d^2$	d	$1/d^2$	d	$1/d^2$	d	$1/d^2$
8.000	.015625	8.050	.015432	8.100	.015242	8.150	.015055
1	.015621	1	.015428	1	.015238	1	.015051
2	.015617	2	.015424	2	.015234	2	.015048
3	.015613	3	.015420	3	.015230	3	.015044
4	.015609	4	.015416	4	.015227	4	.015040
5	.015605	5	.015412	5	.015223	5	.015037
6	.015602	6	.015409	6	.015219	6	.015033
7	.015598	7	.015405	7	.015215	7	.015029
8	.015594	8	.015401	8	.015212	8	.015026
9	.015590	9	.015397	9	.015208	9	.015022
8.010	.015586	8.060	.015393	8.110	.015204	8.160	.015018
1	.015582	1	.015389	1	.015200	1	.015015
2	.015578	2	.015386	2	.015197	2	.015011
3	.015574	3	.015382	3	.015193	3	.015007
4	.015570	4	.015378	4	.015189	4	.015004
5	.015567	5	.015374	5	.015185	5	.015000
6	.015563	6	.015370	6	.015182	6	.014996
7	.015559	7	.015367	7	.015178	7	.014993
8	.015555	8	.015363	8	.015174	8	.014989
9	.015551	9	.015359	9	.015170	9	.014985
8.020	.015547	8.070	.015355	8.120	.015167	8.170	.014982
1	.015543	1	.015351	1	.015163	1	.014978
2	.015539	2	.015348	2	.015159	2	.014974
3	.015536	3	.015344	3	.015155	3	.014971
4	.015532	4	.015340	4	.015152	4	.014967
5	.015528	5	.015336	5	.015148	5	.014963
6	.015524	6	.015332	6	.015144	6	.014960
7	.015520	7	.015329	7	.015140	7	.014956
8	.015516	8	.015325	8	.015137	8	.014952
9	.015512	9	.015321	9	.015133	9	.014949
8.030	.015508	8.080	.015317	8.130	.015129	8.180	.014945
1	.015505	1	.015313	1	.015126	1	.014941
2	.015501	2	.015310	2	.015122	2	.014938
3	.015497	3	.015306	3	.015118	3	.014934
4	.015493	4	.015302	4	.015114	4	.014930
5	.015489	5	.015298	5	.015111	5	.014927
6	.015485	6	.015294	6	.015107	6	.014923
7	.015481	7	.015291	7	.015103	7	.014919
8	.015478	8	.015224	8	.015100	8	.014916
9	.015474	9	.015283	9	.015096	9	.014912
8.040	.015470	8.090	.015279	8.140	.015092	8.190	.014908
1	.015466	1	.015276	1	.015088	1	.014905
2	.015462	2	.015272	2	.015085	2	.014901
3	.015458	3	.015268	3	.015081	3	.014898
4	.015455	4	.015264	4	.015077	4	.014894
5	.015451	5	.015260	5	.015074	5	.014890
6	.015447	6	.015257	6	.015070	6	.014887
7	.015443	7	.015253	7	.015066	7	.014883
8	.015439	8	.015249	8	.015063	8	.014879
9	.015435	9	.015245	9	.015059	9	.014876

d	$1/d^2$	d	$1/d^2$	d	$1/d^2$	d	$1/d^2$
8.200	.014872	8.250	.014692	8.300	.014516	8.350	.014343
1	.014868	1	.014689	1	.014512	1	.014339
2	.014865	2	.014685	2	.014509	2	.014336
3	.014861	3	.014682	3	.014505	3	.014332
4	.014858	4	.014678	4	.014502	4	.014329
5	.014854	5	.014675	5	.014498	5	.014325
6	.014850	6	.014671	6	.014495	6	.014322
7	.014847	7	.014667	7	.014491	7	.014319
8	.014843	8	.014664	8	.014488	8	.014315
9	.014840	9	.014660	9	.014484	9	.014312
8.210	.014836	8.260	.014657	8.310	.014481	8.360	.014308
1	.014832	1	.014653	1	.014477	1	.014305
2	.014829	2	.014650	2	.014474	2	.014301
3	.014825	3	.014646	3	.014471	3	.014298
4	.014821	4	.014643	4	.014467	4	.014295
5	.014818	5	.014639	5	.014464	5	.014291
6	.014814	6	.014636	6	.014460	6	.014288
7	.014811	7	.014632	7	.014457	7	.014284
8	.014807	8	.014628	8	.014453	8	.014281
9	.014803	9	.014625	9	.014450	9	.014278
8.220	.014800	8.270	.014621	8.320	.014446	8.370	.014274
1	.014796	1	.014618	1	.014443	1	.014271
2	.014793	2	.014614	2	.014439	2	.014267
3	.014789	3	.014611	3	.014436	3	.014264
4	.014785	4	.014607	4	.014432	4	.014260
5	.014782	5	.014604	5	.014429	5	.014257
6	.014778	6	.014600	6	.014425	6	.014254
7	.014775	7	.014597	7	.014422	7	.014250
8	.014771	8	.014593	8	.014418	8	.014247
9	.014767	9	.014590	9	.014415	9	.014243
8.230	.014764	8.280	.014586	8.330	.014412	8.380	.014240
1	.014760	1	.014583	1	.014408	1	.014237
2	.014757	2	.014579	2	.014405	2	.014233
3	.014753	3	.014576	3	.014401	3	.014230
4	.014750	4	.014572	4	.014398	4	.014226
5	.014746	5	.014569	5	.014394	5	.014223
6	.014742	6	.014565	6	.014391	6	.014220
7	.014739	7	.014561	7	.014387	7	.014216
8	.014735	8	.014558	8	.014384	8	.014213
9	.014732	9	.014554	9	.014380	9	.014210
8.240	.014728	8.290	.014551	8.340	.014377	8.390	.014206
1	.014724	1	.014547	1	.014374	1	.014203
2	.014721	2	.014544	2	.014370	2	.014199
3	.014717	3	.014540	3	.014367	3	.014196
4	.014714	4	.014537	4	.014363	4	.014193
5	.014710	5	.014533	5	.014360	5	.014189
6	.014707	6	.014530	6	.014356	6	.014186
7	.014703	7	.014526	7	.014353	7	.014182
8	.014700	8	.014523	8	.014349	8	.014179
9	.014696	9	.014519	9	.014346	9	.014176

d	$1/d^2$	d	$1/d^2$	d	$1/d^2$	d	$1/d^2$
8.400	.014172	8.450	.014005	8.500	.013841	8.550	.013679
1	.014169	1	.014002	1	.013838	1	.013676
2	.014166	2	.013998	2	.013834	2	.013673
3	.014162	3	.013995	3	.013831	3	.013670
4	.014159	4	.013992	4	.013828	4	.013667
5	.014155	5	.013989	5	.013825	5	.013663
6	.014152	6	.013985	6	.013821	6	.013660
7	.014149	7	.013982	7	.013818	7	.013657
8	.014145	8	.013979	8	.013815	8	.013654
9	.014142	9	.013975	9	.013812	9	.013651
8.410	.014139	8.460	.013972	8.510	.013808	8.560	.013647
1	.014135	1	.013969	1	.013805	1	.013644
2	.014132	2	.013965	2	.013802	2	.013641
3	.014129	3	.013962	3	.013799	3	.013638
4	.014125	4	.013959	4	.013795	4	.013635
5	.014122	5	.013956	5	.013792	5	.013632
6	.014118	6	.013952	6	.013789	6	.013628
7	.014115	7	.013949	7	.013786	7	.013625
8	.014112	8	.013946	8	.013782	8	.013622
9	.014108	9	.013942	9	.013779	9	.013619
8.420	.014102	8.470	.013939	8.520	.013776	8.570	.013616
1	.014102	1	.013936	1	.013773	1	.013612
2	.014098	2	.013932	2	.013769	2	.013609
3	.014095	3	.013929	3	.013766	3	.013606
4	.014092	4	.013926	4	.013763	4	.013603
5	.014088	5	.013923	5	.013760	5	.013600
6	.014085	6	.013919	6	.013757	6	.013597
7	.014082	7	.013916	7	.013753	7	.013593
8	.014078	8	.013913	8	.013750	8	.013590
9	.014075	9	.013909	9	.013747	9	.013587
8.430	.014072	8.480	.013906	8.530	.013744	8.580	.013584
1	.014068	1	.013903	1	.013740	1	.013581
2	.014065	2	.013900	2	.013737	2	.013578
3	.014062	3	.013896	3	.013734	3	.013574
4	.014058	4	.013893	4	.013731	4	.013571
5	.014055	5	.013890	5	.013728	5	.013568
6	.014052	6	.013887	6	.013724	6	.013565
7	.014048	7	.013883	7	.013721	7	.013562
8	.014045	8	.013880	8	.013718	8	.013559
9	.014042	9	.013877	9	.013715	9	.013555
8.440	.014038	8.490	.013873	8.540	.013711	8.590	.013552
1	.014035	1	.013870	1	.013708	1	.013549
2	.014032	2	.013867	2	.013705	2	.013546
3	.014028	3	.013864	3	.013702	3	.013543
4	.014025	4	.013860	4	.013699	4	.013540
5	.014022	5	.013857	5	.013695	5	.013537
6	.014018	6	.013854	6	.013692	6	.013533
7	.014015	7	.013851	7	.013689	7	.013530
8	.014012	8	.013847	8	.013686	8	.013527
9	.014008	9	.013844	9	.013683	9	.013524

d	$1/d^2$	d	$1/d^2$	d	$1/d^2$	d	$1/d^2$
8.600	.013521	8.650	.013365	8.700	.013212	8.750	.013061
1	.013518	1	.013362	1	.013209	1	.013058
2	.013515	2	.013359	2	.013206	2	.013055
3	.013511	3	.013356	3	.013203	3	.013052
4	.013508	4	.013353	4	.013200	4	.013049
5	.013505	5	.013350	5	.013197	5	.013046
6	.013502	6	.013346	6	.013194	6	.013043
7	.013499	7	.013343	7	.013191	7	.013040
8	.013496	8	.013340	8	.013188	8	.013037
9	.013493	9	.013337	9	.013184	9	.013034
8.610	.013489	8.660	.013334	8.710	.013181	8.760	.013031
1	.013486	1	.013331	1	.013178	1	.013028
2	.013483	2	.013328	2	.013175	2	.013025
3	.013480	3	.013325	3	.013172	3	.013023
4	.013477	4	.013322	4	.013169	4	.013020
5	.013474	5	.013319	5	.013166	5	.013017
6	.013471	6	.013316	6	.013163	6	.013014
7	.013468	7	.013313	7	.013160	7	.013011
8	.013464	8	.013310	8	.013157	8	.013008
9	.013461	9	.013306	9	.013154	9	.013005
8.620	.013458	8.670	.013303	8.720	.013151	8.770	.013002
1	.013455	1	.013300	1	.013148	1	.012999
2	.013452	2	.013297	2	.013145	2	.012996
3	.013449	3	.013294	3	.013142	3	.012993
4	.013446	4	.013291	4	.013139	4	.012990
5	.013443	5	.013288	5	.013136	5	.012987
6	.013439	6	.013285	6	.013133	6	.012984
7	.013436	7	.013282	7	.013130	7	.012981
8	.013433	8	.013279	8	.013127	8	.012978
9	.013430	9	.013276	9	.013124	9	.012975
8.630	.013427	8.680	.013273	8.730	.013121	8.780	.012972
1	.013424	1	.013270	1	.013118	1	.012969
2	.013421	2	.013267	2	.013115	2	.012966
3	.013418	3	.013264	3	.013112	3	.012963
4	.013415	4	.013261	4	.013109	4	.012960
5	.013411	5	.013257	5	.013106	5	.012957
6	.013408	6	.013254	6	.013103	6	.012954
7	.013405	7	.013251	7	.013100	7	.012951
8	.013402	8	.013248	8	.013097	8	.012949
9	.013399	9	.013245	9	.013094	9	.012946
8.040	.013396	8.690	.013242	8.740	.013091	8.790	.012943
1	.013393	1	.013239	1	.013088	1	.012940
2	.013390	2	.013236	2	.013085	2	.012937
3	.013387	3	.013233	3	.013082	3	.012934
4	.013384	4	.013230	4	.013079	4	.012931
5	.013380	5	.013227	5	.013076	5	.012928
6	.013377	6	.013224	6	.013073	6	.012925
7	.013374	7	.013221	7	.013070	7	.012922
8	.013371	8	.013218	8	.013067	8	.012919
9	.013368	9	.013215	9	.013064	9	.012916

d	$1/d^2$	d	$1/d^2$	d	$1/d^2$	d	$1/d^2$
8.800	.012913	8.850	.012768	8.900	.012625	8.950	.012484
1	.012910	1	.012765	1	.012622	1	.012481
2	.012907	2	.012762	2	.012619	2	.012478
3	.012904	3	.012759	3	.012616	3	.012476
4	.012901	4	.012756	4	.012613	4	.012473
5	.012899	5	.012753	5	.012610	5	.012470
6	.012896	6	.012750	6	.012608	6	.012467
7	.012893	7	.012748	7	.012605	7	.012464
8	.012890	8	.012745	8	.012602	8	.012462
9	.012887	9	.012742	9	.012599	9	.012459
8.810	.012884	8.860	.012739	8.910	.012596	8.960	.012456
1	.012881	1	.012736	1	.012594	1	.012453
2	.012878	2	.012733	2	.012591	2	.012451
3	.012875	3	.012730	3	.012588	3	.012448
4	.012872	4	.012727	4	.012585	4	.012445
5	.012869	5	.012725	5	.012582	5	.012442
6	.012866	6	.012722	6	.012579	6	.012439
7	.012863	7	.012719	7	.012577	7	.012437
8	.012861	8	.012716	8	.012574	8	.012434
9	.012858	9	.012713	9	.012571	9	.012431
8.820	.012855	8.870	.012710	8.920	.012568	8.970	.012428
1	.012852	1	.012707	1	.012565	1	.012426
2	.012849	2	.012704	2	.012562	2	.012423
3	.012846	3	.012702	3	.012560	3	.012420
4	.012843	4	.012699	4	.012557	4	.012417
5	.012840	5	.012696	5	.012554	5	.012415
6	.012837	6	.012693	6	.012551	6	.012412
7	.012834	7	.012690	7	.012548	7	.012409
8	.012831	8	.012687	8	.012546	8	.012406
9	.012829	9	.012684	9	.012543	9	.012403
8.830	.012826	8.880	.012682	8.930	.012540	8.980	.012401
1	.012823	1	.012679	1	.012537	1	.012398
2	.012820	2	.012676	2	.012534	2	.012395
3	.012817	3	.012673	3	.012532	3	.012392
4	.012814	4	.012670	4	.012529	4	.012390
5	.012811	5	.012667	5	.012526	5	.012387
6	.012808	6	.012664	6	.012523	6	.012384
7	.012805	7	.012662	7	.012520	7	.012381
8	.012802	8	.012659	8	.012518	8	.012379
9	.012800	9	.012656	9	.012515	9	.012376
8.840	.012797	8.890	.012653	8.940	.012512	8.990	.012373
1	.012794	1	.012650	1	.012509	1	.012371
2	.012791	2	.012647	2	.012506	2	.012368
3	.012788	3	.012645	3	.012504	3	.012365
4	.012785	4	.012642	4	.012501	4	.012362
5	.012782	5	.012639	5	.012498	5	.012359
6	.012779	6	.012636	6	.012495	6	.012357
7	.012776	7	.012633	7	.012492	7	.012354
8	.012773	8	.012630	8	.012490	8	.012351
9	.012771	9	.012628	9	.012487	9	.012348

d	$1/d^2$	d	$1/d^2$	d	$1/d^2$	d	$1/d^2$
9.000	.012346	9.050	.012210	9.100	.012076	9.150	.011944
1	.012343	1	.012207	1	012073	1	.011942
2	.012340	2	.012204	2	.012071	2	.011939
3	.012337	3	.012202	3	.012068	3	.011936
4	.012335	4	.012199	4	.012065	4	.011934
5	.012332	5	.012196	5	.012063	5	.011931
6	.012329	6	.012193	6	.012060	6	.011929
7	.012326	7	.012191	7	.012057	7	.011926
8	.012324	8	.012188	8	.012055	8	.011923
9	.012321	9	.012185	9	.012052	9	.011921
9.010	.012318	9.060	.012183	9.110	.012049	9.160	.011918
1	.012316	1	.012180	1	.012047	1	.011916
2	.012313	2	.012177	2	.012044	2	.011913
3	.012310	3	.012175	3	.012011	3	.011910
4	.012307	4	.012172	4	.012039	4	.011908
5	.012305	5	.012169	5	.012036	5	.011905
6	.012302	6	.012167	6	.012033	6	.011903
7	.012299	7	.012164	7	.012031	7	.011900
8	.012296	8	.012161	8	.012028	8	.011897
9	.012294	9	.012159	9	.012026	9	.011895
9.020	.012291	9.070	.012156	9.120	.012023	9.170	.011892
1	.012288	1	.012153	1	.012020	1	.011890
2	.012286	2	.012150	2	.012018	2	.011887
3	.012283	3	.012148	3	.012015	3	.011884
4	.012280	4	.012145	4	.012012	4	.011882
5	.012277	5	.012142	5	.012010	5	.011879
6	.012275	6	.012140	6	.012007	6	.011877
7	.012272	7	.012137	7	.012004	7	.011874
8	.012269	8	.012134	8	.012002	8	.011871
9	.012267	9	.012132	9	.011999	9	.011869
9.030	.012264	9.080	.012129	9.130	011997	9.180	.011866
1	.012261	1	.012126	1	.011994	1	.011864
2	.012258	2	.012124	2	.011991	2	.011861
3	.012256	3	.012121	3	.011989	3	.011859
4	.012253	4	.012118	4	.011986	4	.011856
5	.012250	5	.012116	5	.011983	5	.011853
6	.012248	6	.012113	6	.011981	6	.011851
7	.012245	7	.012110	7	.011978	7	.011848
8	.012242	8	.012108	8	.011976	8	.011846
9	.012239	9	.012105	9	.011973	9	.011843
9.040	.012237	9.090	.012102	9.140	.011970	9.190	.011840
1	.012234	1	.012100	1	.011968	1	.011838
2	.012231	2	.012097	2	.011965	2	.011835
3	.012229	3	.012094	3	.011963	3	.011833
4	.012226	4	.012092	4	.011960	4	.011830
5	.012223	5	.012089	5	.011957	5	.011828
6	.012220	6	.012086	6	.011955	6	.011825
7	.012218	7	.012084	7	.011952	7	.011822
8	.012215	8	.012081	8	.011949	8	.011820
9	.012212	9	.012078	9	.011947	9	.011817

d	$1/d^2$	d	$1/d^2$	d	$1/d^2$	d	$1/d^2$
9.200	.011815	9.250	.011687	9.300	.011562	9.350	.011439
1	.011812	1	.011685	1	.011560	1	.011436
2	.011810	2	.011682	2	.011557	2	.011434
3	.011807	3	.011680	3	.011555	3	.011431
4	.011804	4	.011677	4	.011552	4	.011429
5	.011802	5	.011675	5	.011550	5	.011426
6	.011799	6	.011672	6	.011547	6	.011424
7	.011797	7	.011670	7	.011545	7	.011422
8	.011794	8	.011667	8	.011542	8	.011419
9	.011792	9	.011665	9	.011540	9	.011417
9.210	.011789	9.260	.011662	9.310	.011537	9.360	.011414
1	.011787	1	.011660	1	.011535	1	.011412
2	.011784	2	.011657	2	.011532	2	.011409
3	.011781	3	.011655	3	.011530	3	.011407
4	.011779	4	.011652	4	.011527	4	.011405
5	.011776	5	.011650	5	.011525	5	.011402
6	.011774	6	.011647	6	.011522	6	.011400
7	.011771	7	.011645	7	.011520	7	.011397
8	.011769	8	.011642	8	.011517	8	.011395
9	.011766	9	.011640	9	.011515	9	.011392
9.220	.011764	9.270	.011637	9.320	.011512	9.370	.011390
1	.011761	1	.011634	1	.011510	1	.011387
2	.011758	2	.011632	2	.011508	2	.011385
3	.011756	3	.011629	3	.011505	3	.011383
4	.011753	4	.011627	4	.011503	4	.011380
5	.011751	5	.011624	5	.011500	5	.011378
6	.011748	6	.011622	6	.011498	6	.011375
7	.011746	7	.011619	7	.011495	7	.011373
8	.011743	8	.011617	8	.011493	8	.011370
9	.011741	9	.011614	9	.011490	9	.011368
9.230	.011738	9.280	.011612	9.330	.011488	9.380	.011366
1	.011736	1	.011609	1	.011485	1	.011363
2	.011733	2	.011607	2	.011483	2	.011361
3	.011730	3	.011604	3	.011480	3	.011358
4	.011728	4	.011602	4	.011478	4	.011356
5	.011725	5	.011599	5	.011475	5	.011354
6	.011723	6	.011597	6	.011473	6	.011351
7	.011720	7	.011594	7	.011471	7	.011349
8	.011718	8	.011592	8	.011468	8	.011346
9	.011715	9	.011589	9	.011466	9	.011344
9.240	.011713	9.290	.011587	9.340	.011463	9.390	.011341
1	.011710	1	.011584	1	.011461	1	.011339
2	.011708	2	.011582	2	.011458	2	.011337
3	.011705	3	.011579	3	.011456	3	.011334
4	.011703	4	.011577	4	.011453	4	.011332
5	.011700	5	.011574	5	.011451	5	.011329
6	.011697	6	.011572	6	.011448	6	.011327
7	.011695	7	.011569	7	.011446	7	.011325
8	.011692	8	.011567	8	.011444	8	.011322
9	.011690	9	.011565	9	.011441	9	.011320

d	$1/d^2$	d	$1/d^2$	d	$1/d^2$	d	$1/d^2$
9.400	.011317	9.450	.011198	9.500	.011080	9.550	.010965
1	.011315	1	.011196	1	.011078	1	.010962
2	.011313	2	.011193	2	.011076	2	.010960
3	.011310	3	.011191	3	.011073	3	.010958
4	.011308	4	.011188	4	.011071	4	.010955
5	.011305	5	.011186	5	.011069	5	.010953
6	.011303	6	.011184	6	.011066	6	.010951
7	.011301	7	.011181	7	.011064	7	.010949
8	.011298	8	.011179	8	.011062	8	.010946
9	.011296	9	.011177	9	.011059	9	.010944
9.410	.011293	9.460	.011174	9.510	.011057	9.560	.010942
1	.011291	1	.011172	1	.011055	1	.010939
2	.011288	2	.011170	2	.011052	2	.010937
3	.011286	3	.011167	3	.011050	3	.010935
4	.011284	4	.011165	4	.011048	4	.010933
5	.011281	5	.011162	5	.011045	5	.010930
6	.011278	6	.011160	6	.011043	6	.010928
7	.011276	7	.011158	7	.011041	7	.010926
8	.011274	8	.011155	8	.011038	8	.010923
9	.011272	9	.011153	9	.011036	9	.010921
9.420	.011269	9.470	.011151	9.520	.011034	9.570	.010919
1	.011267	1	.011148	1	.011032	1	.010916
2	.011265	2	.011146	2	.011029	2	.010914
3	.011262	3	.011144	3	.011027	3	.010912
4	.011260	4	.011141	4	.011025	4	.010910
5	.011257	5	.011139	5	.011022	5	.010909
6	.011255	6	.011137	6	.011020	6	.010905
7	.011253	7	.011134	7	.011018	7	.010903
8	.011250	8	.011132	8	.011015	8	.010901
9	.011248	9	.011129	9	.011013	9	.010898
9.430	011245	9.480	.011127	9.530	.011011	9.580	.010896
1	.011243	1	.011125	1	.011008	1	.010894
2	.011241	2	.011122	2	.011006	2	.010891
3	.011238	3	.011120	3	.011004	3	.010889
4	.011236	4	.011118	4	.011001	4	.010887
5	.011234	5	.011115	5	.010999	5	.010885
6	.011231	6	.011113	6	.010997	6	.010882
7	.011230	7	.011111	7	.010995	7	.010880
8	.011226	8	.011108	8	.010992	8	.010878
9	.011224	9	.011106	9	.010990	9	.010876
9.440	.011222	9.490	.011104	9.540	.010988	9.590	.010873
1	.011219	1	.011101	1	.010985	1	.010871
2	.011217	2	.011099	2	.010983	2	.010869
3	.011215	3	.011097	3	.010981	3	.010866
4	.011212	4	.011094	4	.010978	4	.010864
5	.011210	5	.011092	5	.010976	5	.010862
6	.011207	6	.011090	6	.010974	6	.010860
7	.011205	7	.011087	7	.010972	7	.010857
8	.011203	8	.011085	8	.010969	8	.010855
9	.011200	9	.011083	9	.010967	9	.010853

d	$1/d^2$	d	$1/d^2$	d	$1/d^2$	d	$1/d^2$
9.600	.010851	9.650	.010738	9.700	.010628	9.750	.010519
1	.010848	1	.010736	1	.010626	1	.010517
2	.010846	2	.010734	2	.010624	2	.010515
3	.010844	3	.010732	3	.010622	3	.010513
4	.010842	4	.010730	4	.010619	4	.010511
5	.010839	5	.010727	5	.010617	5	.010509
6	.010837	6	.010725	6	.010615	6	.010506
7	.010835	7	.010723	7	.010613	7	.010504
8	.010833	8	.010721	8	.010611	8	.010502
9	.010830	9	.010718	9	.010608	9	.010500
9.610	.010828	9.660	.010716	9.710	.010606	9.760	.010498
1	.010827	1	.010714	1	.010604	1	.010496
2	.010824	2	.010712	2	.010602	2	.010494
3	.010821	3	.010710	3	.010600	3	.010491
4	.010819	4	.010707	4	.010598	4	.010489
5	.010817	5	.010705	5	.010595	5	.010487
6	.010815	6	.010703	6	.010593	6	.010485
7	.010812	7	.010701	7	.010591	7	.010483
8	.010810	8	.010698	8	.010589	8	.010481
9	.010808	9	.010696	9	.010587	9	.010479
9.620	.010806	9.670	.010694	9.720	.010584	9.770	.010476
1	.010803	1	.010692	1	.010582	1	.010474
2	.010801	2	.010690	2	.010580	2	.010472
3	.010799	3	.010688	3	.010578	3	.010470
4	.010797	4	.010685	4	.010576	4	.010468
5	.010794	5	.010683	5	.010574	5	.010466
6	.010792	6	.010681	6	.010571	6	.010464
7	.010790	7	.010679	7	.010569	7	.010461
8	.010788	8	.010676	8	.010567	8	.010459
9	.010785	9	.010674	9	.010565	9	.010457
9.630	.010783	9.680	.010672	9.730	.010563	9.780	.010455
1	.010781	1	.010670	1	.010560	1	.010453
2	.010779	2	.010668	2	.010558	2	.010451
3	.010776	3	.010665	3	.010556	3	.010448
4	.010774	4	.010663	4	.010554	4	.010446
5	.010772	5	.010661	5	.010552	5	.010444
6	.010770	6	.010659	6	.010550	6	.010442
7	.010768	7	.010657	7	.010548	7	.010440
8	.010765	8	.010654	8	.010545	8	.010438
9	.010763	9	.010652	9	.010543	9	.010436
9.640	.010762	9.690	.010650	9.740	.010541	9.790	.010434
1	.010759	1	.010648	1	.010539	1	.010431
2	.010756	2	.010646	2	.010537	2	.010429
3	.010754	3	.010643	3	.010534	3	.010427
4	.010752	4	.010641	4	.010532	4	.010425
5	.010750	5	.010639	5	.010530	5	.010423
6	.010747	6	.010637	6	.010528	6	.010421
7	.010745	7	.010635	7	.010526	7	.010419
8	.010743	8	.010632	8	.010524	8	.010416
9	.010741	9	.010630	9	.010522	9	.010414

d	1/d²	d	1/d²	d	1/d²	d	1/d²
9.800	.010412	9.850	.010307	9.900	.010203	9.950	.010101
1	.010410	1	.010305	1	.010201	1	.010099
2	.010408	2	.010303	2	.010199	2	.010097
3	.010406	3	.010301	3	.010197	3	.010095
4	.010404	4	.010298	4	.010195	4	.010093
5	.010402	5	.010296	5	.010193	5	.010091
6	.010400	6	.010294	6	.010191	6	.010088
7	.010397	7	.010292	7	.010189	7	.010086
8	.010395	8	.010290	8	.010186	8	.010084
9	.010393	9	.010288	9	.010184	9	.010082
9.810	.010391	9.860	.010286	9.910	.010182	9.960	.010080
1	.010389	1	.010284	1	.010180	1	.010078
2	.010387	2	.010282	2	.010178	2	.010076
3	.010385	3	.010280	3	.010176	3	.010074
4	.010383	4	.010278	4	.010174	4	.010072
5	.010381	5	.010276	5	.010172	5	.010070
6	.010378	6	.010273	6	.010170	6	.010068
7	.010376	7	.010271	7	.010168	7	.010066
8	.010374	8	.010269	8	.010166	8	.010064
9	.010372	9	.010267	9	.010164	9	.010062
9.820	.010370	9.870	.010265	9.920	.010162	9.970	.010060
1	.010368	1	.010263	1	.010160	1	.010058
2	.010366	2	.010261	2	.010158	2	.010056
3	.010364	3	.010259	3	.010156	3	.010054
4	.010362	4	.010257	4	.010154	4	.010052
5	.010359	5	.010255	5	.010152	5	.010050
6	.010357	6	.010253	6	.010150	6	.010048
7	.010355	7	.010251	7	.010148	7	.010046
8	.010353	8	.010248	8	.010146	8	.010044
9	.010351	9	.010246	9	.010144	9	.010042
9.830	.010349	9.880	.010244	9.930	.010141	9.980	.010040
1	.010347	1	.010242	1	.010139	1	.010038
2	.010345	2	.010240	2	.010137	2	.010036
3	.010342	3	.010238	3	.010135	3	.010034
4	.010340	4	.010236	4	.010133	4	.010032
5	.010338	5	.010234	5	.010131	5	.010030
6	.010336	6	.010232	6	.010129	6	.010028
7	.010334	7	.010230	7	.010127	7	.010026
8	.010332	8	.010228	8	.010125	8	.010024
9	.010330	9	.010226	9	.010123	9	.010022
9.840	.010328	9.890	.010224	9.940	.010121	9.990	.010020
1	.010326	1	.010222	1	.010119	1	.010018
2	.010324	2	.010220	2	.010117	2	.010016
3	.010322	3	.010217	3	.010115	3	.010014
4	.010320	4	.010215	4	.010113	4	.010012
5	.010317	5	.010213	5	.010111	5	.010010
6	.010315	6	.010211	6	.010109	6	.010008
7	.010313	7	.010209	7	.010107	7	.010006
8	.010311	8	.010207	8	.010105	8	.010004
9	.010309	9	.010205	9	.010103	9	.010002
						10.000	.010000

Appendix 3
Extrapolation function $\frac{1}{2}\left(\frac{\cos^2\theta}{\sin\theta} + \frac{\cos^2\theta}{\theta}\right)$

Table of $\dfrac{1}{2}\left(\dfrac{\cos^2\theta}{\sin\theta} + \dfrac{\cos^2\theta}{\theta}\right)$ as a function of θ in the useful range,

$\theta = 10$ to $89°$

$\theta°$	.0	.1	.2	.3	.4	.5	.6	.7	.8	.9
10	5.572	5.513	5.456	5.400	5.345	5.291	5.237	5.185	5.134	5.084
1	5.034	4.986	4.939	4.892	4.846	4.800	4.756	4.712	4.669	4.627
2	4.585	4.544	4.504	4.464	4.425	4.386	4.348	4.311	4.274	4.238
3	4.202	4.167	4.133	4.098	4.065	4.032	3.999	3.967	3.935	3.903
4	3.872	3.842	3.812	3.782	3.753	3.724	3.695	3.667	3.639	3.612
5	3.584	3.558	3.531	3.505	3.479	3.454	3.429	3.404	3.379	3.355
6	3.331	3.307	3.284	3.260	3.237	3.215	3.192	3.170	3.148	3.127
7	3.105	3.084	3.063	3.042	3.022	3.001	2.981	2.962	2.942	2.922
8	2.903	2.884	2.865	2.847	2.828	2.810	2.792	2.774	2.756	2.738
9	2.721	2.704	2.687	2.670	2.653	2.636	2.620	2.604	2.588	2.572
20	2.556	2.540	2.525	2.509	2.494	2.479	2.464	2.449	2.434	2.420
1	2.405	2.391	2.376	2.362	2.348	2.335	2.321	2.307	2.294	2.280
2	2.267	2.254	2.241	2.228	2 215	2.202	2.189	2.177	2.164	2.152
3	2.140	2.128	2.116	2.104	2.092	2 080	2.068	2.056	2.045	2.034
4	2.022	2.011	2.000	1.989	1.978	1.967	1.956	1.945	1.934	1.924
5	1.913	1.903	1.892	1.882	1.872	1.861	1.851	1.841	1.831	1.821
6	1.812	1.802	1.792	1.782	1.773	1.763	1.754	1.745	1.735	1.726
7	1.717	1.708	1.699	1.690	1.681	1.672	1.663	1.654	1.645	1.637
8	1.628	1.619	1.611	1.602	1.594	1.586	1.577	1.569	1.561	1.553
9	1.545	1.537	1.529	1.521	1.513	1.505	1.497	1.489	1.482	1.474
30	1.466	1.459	1.451	1.444	1.436	1.429	1.421	1.414	1.407	1.400
1	1.392	1.385	1.378	1.371	1.364	1.357	1.350	1.343	1.336	1.329
2	1.323	1.316	1.309	1.302	1.296	1.289	1.282	1.276	1.269	1.263
3	1.256	1.250	1.244	1.237	1.231	1.225	1.218	1.212	1.206	1.200
4	1.194	1.188	1.182	1.176	1.170	1.164	1.158	1.152	1.146	1.140
5	1.134	1.128	1.123	1.117	1.111	1.106	1.100	1.094	1.088	1.083
6	1.078	1.072	1.067	1.061	1.056	1.050	1.045	1.040	1.034	1.029
7	1.024	1.019	1.013	1.008	1.003	0.998	0.993	0.988	0.982	0.977
8	0.972	0.967	0.962	0.958	0.953	0.948	0.943	0.938	0.933	0.928
9	0.924	0.919	0.914	0.909	0.905	0.900	0.895	0.891	0.886	0.881
40	0.877	0.872	0.868	0.863	0.859	0.854	0.850	0.845	0.841	0.837
1	0.832	0.828	0.823	0.819	0.815	0.810	0.806	0.802	0.798	0.794
2	0.789	0.785	0.781	0.777	0.773	0.769	0.765	0.761	0.757	0.753
3	0.749	0.745	0.741	0.737	0.733	0.729	0.725	0.721	0.717	0.713
4	0.709	0.706	0.702	0.698	0.694	0.690	0.687	0.683	0.679	0.676
5	0.672	0.668	0.665	0.661	0.657	0.654	0.650	0.647	0.643	0.640
6	0.636	0.632	0.629	0.625	0.622	0.619	0.615	0.612	0.608	0.605
7	0.602	0.598	0.595	0.591	0.588	0.585	0.582	0.578	0.575	0.572
8	0.569	0.565	0.562	0.559	0.556	0.553	0.549	0.546	0.543	0.540
9	0.537	0.534	0.531	0.528	0.525	0.522	0.518	0.515	0.512	0.509

$\theta°$	.0	.1	.2	.3	.4	.5	.6	.7	.8	.9
50	0.506	0.504	0.501	0.498	0.495	0.492	0.489	0.486	0.483	0.480
1	0.477	0.474	0.472	0.469	0.466	0.463	0.460	0.458	0.455	0.452
2	0.449	0.447	0.444	0.441	0.439	0.436	0.433	0.430	0.428	0.425
3	0.423	0.420	0.417	0.415	0.412	0.410	0.407	0.404	0.402	0.399
4	0.397	0.394	0.392	0.389	0.387	0.384	0.382	0.379	0.377	0.375
5	0.372	0.370	0.367	0.365	0.363	0.360	0.358	0.356	0.353	0.351
6	0.349	0.346	0.344	0.342	0.339	0.337	0.335	0.333	0.330	0.328
7	0.326	0.324	0.322	0.319	0.317	0.315	0.313	0.311	0.309	0.306
8	0.304	0.302	0.300	0.298	0.296	0.294	0.292	0.290	0.288	0.286
9	0.284	0.282	0.280	0.278	0.276	0.274	0.272	0.270	0.268	0.266
60	0.264	0.262	0.260	0.258	0.256	0.254	0.252	0.250	0.249	0.247
1	0.245	0.243	0.241	0.239	0.237	0.236	0.234	0.232	0.230	0.229
2	0.227	0.225	0.223	0.221	0.220	0.218	0.216	0.215	0.213	0.211
3	0.209	0.208	0.206	0.204	0.203	0.201	0.199	0.198	0.196	0.195
4	0.193	0.191	0.190	0.188	0.187	0.185	0.184	0.182	0.180	0.179
5	0.177	0.176	0.174	0.173	0.171	0.170	0.168	0.167	0.165	0.164
6	0.162	0.161	0.160	0.158	0.157	0.155	0.154	0.152	0.151	0.150
7	0.148	0.147	0.146	0.144	0.143	0.141	0.140	0.139	0.138	0.136
8	0.135	0.134	0.132	0.131	0.130	0.128	0.127	0.126	0.125	0.123
9	0.122	0.121	0.120	0.119	0.117	0.116	0.115	0.114	0.112	0.111
70	0.110	0.109	0.108	0.107	0.106	0.104	0.103	0.102	0.101	0.100
1	0.099	0.098	0.097	0.096	0.095	0.094	0.092	0.091	0.090	0.089
2	0.088	0.087	0.086	0.085	0.084	0.083	0.082	0.081	0.080	0.079
3	0.078	0.077	0.076	0.075	0.075	0.074	0.073	0.072	0.071	0.070
4	0.069	0.068	0.067	0.066	0.065	0.065	0.064	0.063	0.062	0.061
5	0.060	0.059	0.059	0.058	0.057	0.056	0.055	0.055	0.054	0.053
6	0.052	0.052	0.051	0.050	0.049	0.048	0.048	0.047	0.046	0.045
7	0.045	0.044	0.043	0.043	0.042	0.041	0.041	0.040	0.039	0.039
8	0.038	0.037	0.037	0.036	0.035	0.035	0.034	0.034	0.033	0.032
9	0.032	0.031	0.031	0.030	0.029	0.029	0.028	0.028	0.027	0.027
80	0.026	0.026	0.025	0.025	0.024	0.023	0.023	0.023	0.022	0.022
1	0.021	0.021	0.020	0.020	0.019	0.019	0.018	0.018	0.017	0.017
2	0.017	0.016	0.016	0.015	0.015	0.015	0.014	0.014	0.013	0.013
3	0.013	0.012	0.012	0.012	0.011	0.011	0.010	0.010	0.010	0.010
4	0.009	0.009	0.009	0.008	0.008	0.008	0.007	0.007	0.007	0.007
5	0.006	0.006	0.006	0.006	0.005	0.005	0.005	0.005	0.005	0.004
6	0.004	0.004	0.004	0.003	0.003	0.003	0.003	0.003	0.003	0.002
7	0.002	0.002	0.002	0.002	0.002	0.002	0.001	0.001	0.001	0.001
8	0.001	0.001	0.001	0.001	0.001	0.001	0.001	0.000	0.000	0.000

Appendix 4
Extrapolation function $\cos^2 \theta$

Table of cos² θ as a function of θ in the useful range, θ = 45 to 90°
(Note that this table can be used to look up sin² θ in the range θ = 45 to 0°)

θ	.0	.1	.2	.3	.4	.5	.6	.7	.8	.9
45°	.5000	.4983	.4965	.4948	.4930	.4913	.4895	.4878	.4860	.4843
6	.4826	.4808	.4791	.4773	.4756	.4738	.4721	.4703	.4686	.4669
7	.4651	.4634	.4616	.4599	.4582	.4564	.4547	.4529	.4512	.4495
8	.4477	.4460	.4443	.4425	.4408	.4391	.4373	.4356	.4339	.4321
9	.4304	.4287	.4270	.4252	.4235	.4218	.4201	.4183	.4166	.4149
50	.4132	.4115	.4097	.4080	.4063	.4046	.4029	.4012	.3995	.3978
1	.3960	.3943	.3926	.3909	.3892	.3875	.3858	.3841	.3824	.3807
2	.3790	.3773	.3757	.3740	.3723	.3706	.3689	.3672	.3655	.3639
3	.3622	.3605	.3588	.3572	.3555	.3538	.3521	.3505	.3488	.3472
4	.3455	.3438	.3422	.3405	.3389	.3372	.3356	.3339	.3323	.3306
5	.3290	.3274	.3257	.3241	.3224	.3208	.3192	.3176	.3159	.3143
6	.3127	.3111	.3095	.3079	.3062	.3046	.3030	.3014	.2998	.2982
7	.2966	.2950	.2934	.2919	.2903	.2887	.2871	.2855	.2840	.2824
8	.2808	.2792	.2777	.2761	.2746	.2730	.2715	.2699	.2684	.2668
9	.2653	.2637	.2622	.2607	.2591	.2576	.2561	.2545	.2530	.2515
60	.2500	.2485	.2470	.2455	.2440	.2425	.2410	.2395	.2380	.2365
1	.2350	.2336	.2321	.2306	.2291	.2277	.2262	.2248	.2233	.2219
2	.2204	.2190	.2175	.2161	.2146	.2132	.2118	.2104	.2089	.2075
3	.2061	.2047	.2033	.2019	.2005	.1991	.1977	.1963	.1949	.1935
4	.1922	.1908	.1894	.1881	.1867	.1853	.1840	.1826	.1813	.1799
5	.1786	.1773	.1759	.1746	.1733	.1720	.1707	.1693	.1680	.1667
6	.1654	.1641	.1628	.1616	.1603	.1590	.1577	.1565	.1552	.1539
7	.1527	.1514	.1502	.1489	.1477	.1464	.1452	.1440	.1428	.1415
8	.1403	.1391	.1379	.1367	.1355	.1343	.1331	.1320	.1308	.1296
9	.1284	.1273	.1261	.1249	.1238	.1226	.1215	.1204	.1192	.1181
70	.1170	.1159	.1147	.1136	.1125	.1114	.1103	.1092	.1082	.1071
1	.1060	.1049	.1039	.1028	.1017	.1007	.0996	.0986	.0976	.0965
2	.0955	.0945	.0934	.0924	.0914	.0904	.0894	.0884	.0874	.0865
3	.0855	.0845	.0835	.0826	.0816	.0807	.0797	.0788	.0778	.0769
4	.0760	.0751	.0741	.0732	.0723	.0714	.0705	.0696	.0687	.0679
5	.0670	.0661	.0653	.0644	.0635	.0627	.0618	.0610	.0602	.0593
6	.0585	.0577	.0569	.0561	.0553	.0545	.0537	.0529	.0521	.0514
7	.0506	.0498	.0491	.0483	.0476	.0468	.0461	.0454	.0447	.0439
8	.0432	.0425	.0418	.0411	.0404	.0397	.0391	.0384	.0377	.0371
9	.0364	.0358	.0351	.0345	.0338	.0332	.0326	.0320	.0314	.0308
80	.0302	.0296	.0290	.0284	.0278	.0272	.0267	.0261	.0256	.0250
1	.0245	.0239	.0234	.0229	.0224	.0218	.0213	.0208	.0203	.0199
2	.0194	.0189	.0184	.0180	.0175	.0170	.0166	.0161	.0157	.0153
3	.0149	.0144	.0140	.0136	.0132	.0128	.0124	.0120	.0117	.0113
4	.0109	.0106	.0102	.0099	.0095	.0092	.0089	.0085	.0082	.0079
5	.0076	.0073	.0070	.0067	.0064	.0062	.0059	.0056	.0054	.0051
6	.0049	.0046	.0044	.0042	.0039	.0037	.0035	.0033	.0031	.0020
7	.0027	.0020	.0024	.0022	.0021	.0019	.0018	.0016	.0015	.0013
8	.0012	.0011	.0010	.0009	.0008	.0007	.0006	.0005	.0004	.0004
9	.0003	.0002	.0002	.0001	.0001	.0001	.0000	.0000	.0000	.0000

Appendix 5
Extrapolation function $\sin^2 \theta$

Table of sin² θ as a function of θ in the useful range, θ = 45 to 90°
(Note that this table can be used to look up cos² θ in the range θ = 45 to 0°)

θ	.0	.1	.2	.3	.4	.5	.6	.7	.8	.9
45°	.5000	.5017	.5035	.5052	.5070	.5087	.5105	.5122	.5140	.5157
6	.5174	.5192	.5209	.5227	.5244	.5262	.5279	.5297	.5314	.5331
7	.5349	.5366	.5384	.5401	.5418	.5436	.5453	.5471	.5488	.5505
8	.5523	.5540	.5557	.5575	.5592	.5609	.5627	.5644	.5661	.5679
9	.5696	.5713	.5730	.5748	.5765	.5782	.5799	.5817	.5834	.5851
50	.5868	.5885	.5903	.5920	.5937	.5954	.5971	.5988	.6005	.6022
1	.6040	.6057	.6074	.6091	.6108	.6125	.6142	.6159	.6176	.6193
2	.6210	.6227	.6243	.6260	.6277	.6294	.6311	.6328	.6345	.6361
3	.6378	.6395	.6412	.6428	.6445	.6462	.6479	.6495	.6512	.6528
4	.6545	.6562	.6578	.6595	.6611	.6628	.6644	.6661	.6677	.6694
5	.6710	.6726	.6743	.6759	.6776	.6792	.6808	.6824	.6841	.6857
6	.6873	.5889	.6905	.6921	.6938	.6954	.6970	.6986	.7002	.7018
7	.7034	.7050	.7066	.7081	.7097	.7113	.7129	.7145	.7160	.7176
8	.7192	.7208	.7223	.7239	.7254	.7270	.7285	.7301	.7316	.7332
9	.7347	.7363	.7378	.7393	.7409	.7424	.7439	.7455	.7470	.7485
60	.7500	.7515	.7530	.7545	.7560	.7575	.7590	.7605	.7620	.7635
1	.7650	.7664	.7679	.7694	.7709	.7723	.7738	.7752	.7767	.7781
2	.7796	.7810	.7825	.7839	.7854	.7868	.7882	.7896	.7911	.7925
3	.7939	.7953	.7967	.7981	.7995	.8009	.8023	.8037	.8051	.8065
4	.8078	.8092	.8106	.8119	.8133	.8147	.8160	.8174	.8187	.8201
5	.8214	.8227	.8241	.8254	.8267	.8280	.8293	.8307	.8320	.8333
6	.8346	.8359	.8372	.8384	.8397	.8410	.8423	.8435	.8448	.8461
7	.8473	.8486	.8498	.8511	.8523	.8536	.8548	.8560	.8572	.8585
8	.8597	.8609	.8621	.8633	.8645	.8657	.8669	.8680	.8692	.8704
9	.8716	.8727	.8739	.8751	.8762	.8774	.8785	.8796	.8808	.8819
70	.8830	.8841	.8853	.8864	.8875	.8886	.8897	.8908	.8918	.8929
1	.8940	.8951	.8961	.8972	.8983	.8993	.9004	.9014	.9024	.9035
2	.9045	.9055	.9066	.9076	.9086	.9096	.9106	.9116	.9126	.9135
3	.9145	.9155	.9165	.9174	.9184	.9193	.9203	.9212	.9222	.9231
4	.9240	.9249	.9259	.9268	.9277	.9286	.9295	.9304	.9313	.9321
5	.9330	.9339	.9347	.9356	.9365	.9373	.9382	.9390	.9398	.9407
6	.9415	.9423	.9431	.9439	.9447	.9455	.9463	.9471	.9479	.9486
7	.9494	.9502	.9509	.9517	.9524	.9532	.9539	.9546	.9553	.9561
8	.9568	.9575	.9582	.9589	.9596	.9603	.9609	.9616	.9623	.9629
9	.9636	.9642	.9649	.9655	.9662	.9668	.9674	.9680	.9686	.9692
80	.9698	.9704	.9710	.9716	.9722	.9728	.9733	.9739	.9744	.9750
1	.9755	.9761	.9766	.9771	.9776	.9782	.9787	.9792	.9797	.9801
2	.9806	.9811	.9816	.9820	.9825	.9830	.9834	.9839	.9843	.9847
3	.9851	.9856	.9860	.9864	.9868	.9872	.9876	.9880	.9883	.9887
4	.9891	.9894	.9898	.9901	.9905	.9908	.9911	.9915	.9918	.9921
5	.9924	.9927	.9930	.9933	.9936	.9938	.9941	.9944	.9946	.9949
6	.9951	.9954	.9956	.9958	.9961	.9963	.9965	.9967	.9969	.9971
7	.9973	.9974	.9976	.9978	.9979	.9981	.9982	.9984	.9985	.9987
8	.9988	.9989	.9990	.9991	.9992	.9993	.9994	.9995	.9996	.9996
9	.9997	.9998	.9998	.9999	.9999	.9999	1.0000	1.0000	1.0000	1.0000

Index

329